Waste Management Policies and Practices in BRICS Nations

Waste Management Policies and Practices in BRICS Nations

Edited By

Pardeep Singh, Yulia Milshina, Kangming Tian, Anwesha Borthakur, Pramit Verma and Ajay Kumar

CRC Press
Taylor & Francis Group
Boca Raton London New York

CRC Press is an imprint of the Taylor & Francis Group, an **informa** business

First edition published 2021
by CRC Press
6000 Broken Sound Parkway NW, Suite 300, Boca Raton, FL 33487-2742

and by CRC Press
2 Park Square, Milton Park, Abingdon, Oxon, OX14 4RN

ISBN: 9780367440992 (hbk)
ISBN: 9781003007579 (ebk)
ISBN: 9781032040776 (pbk)

Typeset in Times
by Deanta Global Publishing Services, Chennai, India

Contents

Preface

BRICS nations are the emerging economy of the world. They comprise 41% of the world's population and 27% of the world's land area. Increasing population, urbanization, industrialization and uses of chemical fertilizer and pesticide in agriculture for enhanced productivity of food, especially in India and China, to support large populations harm the natural environment. The rise in the living standard of the human population has increased the environmental pollution manyfold, resulting in the huge generation of biodegradable and non-biodegradable waste simultaneously, which has contaminated our natural resources such as soil, water and air. It has led to undesirable effects on the environment and human health.

The rapid increase in urbanization, population and industrialization causes adverse effects on the environment by producing various types of waste. Waste management is a serious problem not only in the BRICS countries but on a global scale. The sustainable management of waste is a challenge for society. There is an urgent need to address these waste management-related issues. Because the waste generated from diverse sources causes harmful effects on soil, water and air, the need of the hour is to work on the waste-related management issues, especially in rapidly urbanizing and densely populated BRICS nations.

Globally, various technologies are adopted for the treatment of industrial wastewater, toxic waste and the conversion of waste to energy. However, a strong policy framework for its proper management is lacking. Moreover, limited space, more stringent global waste disposal regulations and public consciousness have made many of the present policies outdated. Therefore, there is an urgent need to develop sustainable management policies for industries and municipalities for waste management.

The book is divided into four themes to present a holistic perspective from BRICS nations. The first theme deals with municipal solid waste management and sustainability practices. It includes case studies from Brazil, China, South Africa, Russia and India. One of the chapters deals with the economic analysis of glass waste recycling. The second theme is focussed on treatment technologies and the management of waste, particularly innovative technologies. One of the chapters gives a comparative account of leachate treatment in BRICS nations. The third theme is on policy and laws for waste management. It deals with governance, a policy framework for waste management and minimization. The fourth theme is focused on hazardous waste management. It includes case studies from Brazil, China and Russia, on the impact of hazardous waste on the environment and its management.

The book provides a comprehensive account of current and innovative waste management techniques, treatment technology, policies and laws, impacts and waste minimization techniques. Its focus on BRICS nations also reflects the global development trajectories. It would prove to be an invaluable companion for research and the student community.

Editors

Dr Pardeep Singh

Dr Pardeep Singh is presently working as Assistant Professor (Department of Environmental Science, PGDAV College University of Delhi, New Delhi, India). He obtained his master's degree from the Department of Environmental Science, Banaras Hindu University, Varanasi, India. He obtained his doctorate degree from the Indian Institute of Technology (Banaras Hindu University) Varanasi, India. The area of his doctoral research is the degradation of organic pollutants through various indigenous isolated microbes and by using various types of photocatalyst. He has published more than 45 papers in international journals in the field of waste management.

Dr Yulia Milshina (Russia)

Dr Yulia Milshina is currently working as a Research Expert at the Institute for Statistical Studies and Economics of Knowledge, National Research University Higher School of Economics, Moscow, Russia. Her field of expertise is in economic development focussing on developmental and public economics. She has published research articles and book chapters in leading research journals. She is presently engaged in research related to public-private partnerships, government-corporate partnerships, potential for collaboration and technology applications in changing environments. She is also working in the field of application of information and technology frameworks in environmental management.

Dr Kangming Tian (China)

Dr Kangming Tian is currently working as a Scientist at Tianjin University of Science and Technology, Tianjin, China. He obtained his PhD degree in 'Fermentation engineering' from Jiangnan University, Wuxi, China, in 2013. His doctoral thesis work is focused on the biosynthesis of monomers for biodegradable materials production from crude glycerol. He has constructed genetically engineered strains for high-purity D-lactic acid and high-purity L-lactic acid fermentation for high-quality polylactic acid (PLA) production; the technology of the same has been transferred to the industry. He has published more than 20 research papers in the relevant field. His present research focus is on establishing and improving the new technology of metabolic engineering breeding for industrial microbial strains represented by the molecular switch. He is also currently working on the establishment and the application of new technology for the large-scale production of efficient biocatalysts, and the development of new technologies for the production of functional oligosaccharides.

Dr Anwesha Borthakur (Belgium)

Dr Anwesha Borthakur is a Marie Skłodowska Curie Postdoctoral Fellow at the Katholieke Universiteit Leuven (KU Leuven) in Belgium. She completed her PhD on electronic waste management from Jawaharlal Nehru University (JNU), New Delhi, India. Presently, she is working on the e-waste policy approaches of India and South Africa as a part of her postdoctoral fellowship. She is an author of more than 35 peer-reviewed publications including research papers in high impact factored journals. Her research interests primarily involve waste or discard studies, science and technology policy and sustainability studies. She has been working on e-waste in India for nearly a decade now. She has travelled widely to take part and present her work in a number of recognized conferences/workshops/summer and winter schools in different parts of the world.

Pramit Verma

Pramit Verma is currently working as Senior Research Fellow at the Institute of Environment and Sustainable Development, Banaras Hindu University, Varanasi. He completed his MPhil at the same institute in 2014. He is currently working in the field of urban ecology. He has also worked in the field of urban climate disaster risk reduction and management. He has published six research articles and a technical report while participating in various national and international conferences, workshops and seminars. He is also working on the socioeconomic dynamics governing urban metabolism and urban land use in medium-sized cities of India.

Dr Ajay Kumar (Israel)
Dr Ajay Kumar is currently working as a visiting scientist in the Agriculture Research Organization, Ministry of Agriculture and Rural Development Volcani Centre, Rishon, Lezziyon, Israel. Dr Kumar completed his doctoral research from the Department of Botany, Institute of Science, Banaras Hindu University, and Varanasi, India. In his 9 years of research, he has published more than 60 research articles/book chapters in international and national journals of repute. He has a wide area of research, especially in plant–microbe interactions, endophytes related to medicinal plants, microbial inoculants and cyanobacteria–pesticides interactions.

List of Contributors

Ojo Jeremiah Akinribide
University of Johannesburg, Doornfontein Campus
Johannesburg
South Africa

Ingrid R.F.S. Alves
Military Institute of Engineering (IME)
Defense Engineering Program
Rio de Janeiro, RJ, Brazil

Olusola Olaitan Ayeleru
University of Johannesburg, Doornfontein Campus
Johannesburg, South Africa

Isabelli Dias Bassin
Federal University of Rio de Janeiro
School of Chemistry
Department of Biochemical Engineering
Rio de Janeiro, RJ, Brazil

João Paulo Bassin
Federal University of Rio de Janeiro
Chemical Engineering Program – COPPE
Civil Engineering Program – COPPE
Rio de Janeiro, RJ, Brazil

André Cavalcante da Silva Batalhão
Center for Environmental and Sustainability Research (CENSE)
NOVA University Lisbon
Caparica, Portugal

Adriana Cristina Ferreira Caldana
School of Economics
Business Administration and Accounting at Ribeirão Preto
University of São Paulo
Ribeirão Preto, Brazil

Armando Borges de Castilhos Junior
Departamento de Engenharia Sanitária e Ambiental (ENS), Centro Tecnológico (CTC), Campus Universidade Federal de Santa Catarina (UFSC)
Trindade, Universidade Federal de Santa Catarina
Florianópolis, Santa Catarina
Brazil

Sisanda Dlova
University of Johannesburg, Doornfontein Campus
Johannesburg
South Africa

Luíza Santana Franca
Federal University of Rio de Janeiro
Chemical Engineering Program – COPPE
Rio de Janeiro, RJ, Brazil

Kurian Joseph
Centre for Environmental Studies Department of Civil Engineering
Anna University, CEG Campus
Chennai, India

O.P. Karthikeyan
University of Houston
Houston, Texas

Williams Kehinde Kupolati
Tshwane University of Technology
Pretoria
South Africa

Qian Li
Wuxi Pure Future Environmental Biotechnology Co., Ltd
School of Environment and Civil Engineering
Wuxi, Jiangsu Province
People's Republic of China

Marcelo Seleme Matias
Universidade Federal De Santa Catarina (UFSC) Departamento de Engenharia Sanitária e Ambiental (ENS), Centro Tecnológico (CTC), Campus Trindade, Universidade Federal de Santa Catarina
Florianópolis, SC – Caixa
Brazil

Yulia Milshina
National Research University
Higher School of Economics
Moscow, Russia

Monika
Institute of Environment and Sustainable Development
Banaras Hindu University
Varanasi, India

Freeman Ntuli
Botswana International University of Science and Technology
Palapye, Botswana

Oluwasayo F. Olorundare
University of Johannesburg, Doornfontein Campus
Johannesburg
South Africa

Felix Ndubisi Okonta
University of Johannesburg, Auckland Park Campus
Johannesburg
South Africa

Peter Apata Olubambi
University of Johannesburg, Doornfontein Campus
Johannesburg
South Africa

K. Palanivelu
Centre for Climate Change and Disaster Management
Department of Civil Engineering
Anna University, CEG Campus
Chennai, India

Pankaj Pathak
SRM University Andhra Pradesh
India

Daria Pavlova
National Research University
Higher School of Economics
Moscow, Russia

Isabella Pimentel Pincelli
Departamento de Engenharia Sanitária e Ambiental (ENS), Centro Tecnológico (CTC), Campus Universidade Federal de Santa Catarina (UFSC)
Trindade, Universidade Federal de Santa Catarina
Florianópolis, Santa Catarina
Brazil

Liliana Proskuryakova
National Research University
Higher School of Economics
Moscow, Russia

A. Ramachandran
Centre for Climate Change and Disaster Management
Department of Civil Engineering
Anna University, CEG Campus
Chennai, India

C. Ramprasad
School of Civil Engineering
SASTRA Deemed to be University
Tamil Nadu, India

Rangabhashiyam S
Department of Biotechnology
School of Chemical and Biotechnology
SASTRA Deemed University
Tamilnadu, India

Erimar Pontes Santiago
Federal University of Amazonas
Civil Engineering Department
Manaus, AM, Brazil

S. Sri Shalini
Centre for Climate Change and Disaster Management
Department of Civil Engineering
Anna University, CEG Campus
Chennai, India

Prashant Sharma
Institute of Environment and Sustainable Development
Banaras Hindu University
Varanasi, India

Archana Singh
Institute of Environment and Sustainable Development
Banaras Hindu University
Varanasi, India

Gopal Shankar Singh
Institute of Environment and Sustainable Development
Banaras Hindu University
Varanasi, India

Denilson Teixeira
Federal University of Goiás
Goiânia, Brazil

Zhiyang Xu
Wuxi Pure Future Environmental Biotechnology Co., Ltd
School of Environment and Civil Engineering
People's Republic of China

Abhinav Yadav
Institute of Environment and Sustainable Development
Banaras Hindu University
Varanasi, India

Binghua Yan
Hunan Agricultural University
Changsha, China

Yunlong Zhou
Wuxi Pure Future Environmental Biotechnology Co., Ltd
School of Environment and Civil Engineering
People's Republic of China

Part I

MSW

1 Waste Management in Brazil
Challenges for Municipal Solid Waste

André C. S. Batalhão, Denilson Teixeira and Adriana Cristina Ferreira Caldana

CONTENTS

1.1 INTRODUCTION AND OBJECTIVES

Municipal solid waste management is a global challenge, especially in developing countries. In the BRICS countries (Brazil, Russia, India, China, and South Africa) this challenge is associated with increasing urbanization, constant population growth, increased per capita consumption and the exponential generation of municipal solid waste. In these countries, the proper management of municipal solid waste is a serious concern due to budgetary inability as a result of the high costs associated with its management, as well as the lack of technical knowledge about the main preventive and corrective factors, affecting the different stages of waste management.

The main aim of this chapter is to develop an analysis around solid waste management, conducting a critical assessment of a set of challenges identified in the literature. For this, an inductive-exploratory study was carried out, consisting of specific literature research and qualitative analysis of published documents. This analysis focused on concepts, approaches and case studies dealing with solid waste management. This was explored through a constructive perspective discussing what could be some of the new frontiers of the theme. This chapter is organized as follows: After a brief introduction to the research overview and context, the first section presents a general background, including the challenges for Brazilian municipalities. In the following section, there is a discussion on the importance of using indicators for solid waste management at the municipal level. In the last section, the final remarks are presented, as well as the contribution to the readers and the scientific community, including the practical and theoretical implications of the subject.

1.2 GENERAL BACKGROUND OF WASTE MANAGEMENT IN BRAZIL

Solid waste management is one of the most urgent and complex socio-economic and environmental challenges of our time, incessantly increased by population growth, lifestyle, consumption, technological change and government policy. This topic has emerged as one of the most challenging service sectors in the 21st century for municipal authorities (Coelho et al., 2016; Zaman, 2013).

In Brazil, the lack of effective enforcement of legal provisions and sufficient financial resources has perpetuated a considerable and growing shortfall in key points of the law: Maximizing the utilization and recovery of discarded materials and eradicating improper disposal practices, still present in all federation units and with negative health impacts on millions of Brazilians citizens (ABRELPE, 2017). In this context, some important issues raised: The legislation relevant to the theme and the current scenario in the country regarding urban solid waste.

Considered an important regulatory framework for solid waste in the country, Law number 12.305 of 2 August 2010 established the National Solid Waste Policy (NSWP). This normative structure is responsible for a number of advances in solid waste management in the country, for example, this policy: institutes guidelines, plans, and targets for all units of the federation; explores several terms such as final environmentally appropriate destination and integrated management; advocates for the adoption of decentralized systems, within an integrated plan, that identifies the problems, indicating solutions and technological alternatives, and establishes the deadlines; determines the shared responsibility of civil society, private enterprise and government for possible solutions to the solid waste problem; states that the waste management should be performed properly, and that the municipalities should include urban and rural areas in the planning of it; establishes general guidelines applicable to all types of solid waste except radioactive waste; defines for each Brazilian region a goal related to the reduction of recyclables and organic materials sent to landfill; determines the closure of improper disposal sites for solid waste within four years and sets a period for the completion of the Integrated Solid Waste Management Plans (Brazil, 2010).

A second necessary point for understanding the context of solid waste management in Brazil is to know the scenario of solid waste generation and disposal. The amount of solid waste collected in the country in 2017 was 71.6 million tons, recording a collection coverage rate of 91.2% for the country, which shows that 6.9 million tons of waste was not subject to consequently had an improper destination (ABRELPE, 2017). Regarding the final disposal of the municipal solid waste (MSW) collected, there was no progress compared to the previous year's scenario, maintaining virtually the same proportion which follows for appropriate and inappropriate locations, with about 42.3 million tons of MSW, or 59.1% of that collected, disposed of in landfills. The remainder, which corresponds to 40.9% of the collected waste, was dumped in inappropriate places by 3,352 Brazilian municipalities, totalling an additional 29 million tons of waste in inappropriate destinations, which do not have the necessary systems and measures to protect the environment against damage and degradation, with direct damage to the health of millions of people (ABRELPE, 2017). Regarding selective collection, data show that 98% of people see recycling as important for the future of the country and 94% agree that the correct way to dispose of waste is by separating materials that can be recycled. On the other hand, this perception is not reflected in behaviour: 75% revealed they did not separate their waste at home, and one of the possible reasons that lead to this is the lack of information, as 66% of respondents claimed to know little or nothing about selective collection (ABRELPE, 2017).

In Brazil, therefore, the solid waste management policy of 2010, based on previous legal history, is considered as the regulatory framework on the issue. Its widely discussed conceptual bases take into account the principles of responsible solid waste management and reduction (UNEP, 2011). The municipalities are responsible for the preparation, implementation and consolidation of the solid waste management plan (SWMP). Besides the characterization and diagnosis of the current situation of waste generation and disposal in the municipality, the SWMP aims to implement the necessary instruments for the consolidation of the solid waste management process in the municipality. However, only 12 states (44%) and 2,325 cities (42%) have completed their respective plans (Brazil, 2018).

Although there have been improvements in MSW management in Brazil compared to the early years of this century, they have been insufficient to achieve the goals established by the NSWP for the purposes of implementing integrated MSW management (Alfaia et al., 2017). The scenario presented defines the main challenges of solid waste management in the country, highlighting key issues such as the proper disposal of waste. However, a diversity of research topics has contributed to the understanding and incorporation of instruments for the consolidation of solid waste management.

Some examples of recent research, among many others, indicate the different ways of addressing the challenges of solid waste management in the country for different territorial limits: Large cities, rural communities, slums ('favelas'), watersheds. The diversity of studies permits a clearer understanding of how territories are managed and who the actors are in the management process. This allows the potentialities and challenges of the territory to be observed, and the necessary paths for integrated solid waste management to be proposed. For example, Coelho et al. (2016) investigated sustainable waste management solutions for the city of Rio de Janeiro in Brazil, based on a life cycle approach. The results indicate the urgency in implementing new strategies for a more environmentally friendly and sustainable municipal solid waste management system. Investment in source separated waste collection and materials recovery is the most environmentally friendly MSW strategy for Rio de Janeiro. In addition, scenarios with an emphasis on materials recovery presented higher environmental benefits than the alternatives focused on energy generation.

Another study discusses determinants of consumer intentions and behaviour in e-waste recycling in the major metropolitan areas of Brazil, where the consumption of electronic devices and appliances has significantly increased in the past decade. The majority of respondents have positive intentions towards recycling electronic appliances. In contrast, only a minority of respondents actually adopt adequate recycling practices connected to e-waste, a behaviour which is socially skewed among the higher income echelons of Brazilian society (Echegaray and Hansstein, 2017).

João Pessoa is one of the pioneering Brazilian cities to implement door-to-door selective collection programmes to study the effect of policy decisions in the last decade on selective collection. Ibáñez-Forésa et al. (2018) analysed the evolution of the municipal solid waste management (MSWM) systems of the municipalities. The results highlighted the importance of spreading and improving awareness-raising campaigns, and the standardization of registering information on the MSWM system is a basic issue in measuring the system and facilitating the undertaking of future analyses.

Lima and Paulo (2018) studied solid waste management in Quilombola communities. A Quilombola is an Afro-Brazilian resident of quilombo settlements first established by escaped slaves in Brazil before abolition in 1888. The results showed that the major risk to which residents are vulnerable is the appearance of vectors that can cause various types of diseases. The most suitable scenario in the studied communities would be the combination of household composting and source separation of dry waste.

Another example of a study of urban solid waste management in Brazilian communities was performed by Azevedo et al. (2019). Through a longitudinal case study conducted in the largest slum of Brazil (Rocinha), these research findings indicated that to tackle environmental problems related to SWM in such areas, it is necessary to solve basic social issues related to education, security and infrastructure, with the integration and support of government, local community and industry.

This other example aimed to analyse how information and communication technologies (ICT) and the internet, in particular, have been applied to support MSWM in the municipalities of the Alto Tietê Basin. The results reveal limited use of ICT in most municipalities and its misapplication for disseminating important information to the population about adequate procedures of separation and storage of waste for selective collection. This may hinder the effectiveness of this public service and increase the health risks for collectors who work in the recycling of these materials (Klein et al., 2018).

Despite policy improvements and considerable progress in research, they were insufficient to achieve the goals established by the NPSW to implement integrated municipal solid waste management (Alfaia et al., 2017).

We must emphasize that among different approaches and perspectives for the study and analysis of integrated solid waste management, the construction of indicators is a fundamental point in monitoring and directing the development of the NPSW implementation process. Solid waste management is an important strategy for sustainability, and implies overcoming a series of challenges. Universities, governments and businesses need to work in an integrated manner, creating appropriate technologies and public policies to meet these challenges.

1.3 CHALLENGES FOR MUNICIPAL SOLID WASTE IN BRAZILIAN MUNICIPALITIES

Compliance with the guidelines of the Brazilian solid waste legislation requires public governments to increase their knowledge in the municipal solid waste management sector. To consolidate solid waste management and reduce its challenges, it is essential to share information with the general public with transparency and regularity, communicating the topics and issues discussed, encouraging participation. Fratta et al. (2019) pointed out that one of the main challenges was little stakeholder engagement, where there was little interest in answering the questionnaires sent out in the municipalities. This may decrease the quality of public decisions, as relevant information may be left out of the data generation process, as stressed by Domingues et al. (2018). Guerrero et al. (2013) identified challenges for cities in 22 developing countries on 4 continents and concluded that municipal initiatives should be coordinated with stakeholders, research and education institutions and national governments to improve the existing precarious waste disposal situation.

In this case, lack of engagement leads to another challenge: Lack of data. The difficulty of building and supporting a municipal-level database is a persistent limitation that makes it impossible to develop indicators. Silva et al. (2019) argued that this difficulty reveals the reality of many cities in Brazil, caused by the lack of investments in infrastructure, technical conditions and human resources for processing information. Lack of engagement diminishes or cancels the participation of different layers of society, while also decreasing the usability of locally generated data and information systems.

Cifrian et al. (2015) and Wilson et al. (2015) have argued that municipal waste information systems cannot be restricted to data collection and indicator generation solely on routine waste management, but should be used to support different decision-making processes and overcome challenges related to integrated management. Agol et al. (2014) stressed that there are methodological challenges and opportunities in applying these indicators to assess long-term impacts that require systematic collection, monitoring and evaluation.

Fratta et al. (2019) concluded that it is necessary to build a shared responsibility with social ramifications, making decentralization of management a decisive political factor in the decision-making of municipal waste managers, contributing directly to the raising of society's awareness and the implementation of various governmental actions. Another challenge pointed out in this study was the emerging need for landfill eradication in municipalities, requiring new treatment models with sustainable alternatives and improvements in key issues of local environmental management, contributing to the reduction of waste costs.

As noted earlier, although there have been improvements in waste management in Brazil over the past 20 years, they have not been sufficient to achieve the goals set by national policy to implement integrated management. This highlights the need for direct modifications to the waste sector structure, increasing collection and treatment coverage, stimulating civil society participation and replacing the dumps with proper treatment and final disposal of solid waste. According to Alfaia et al. (2017), in order to achieve effective and efficient solid waste management in Brazil, the challenges are anticipated as follows: (i) change of culture and population habits regarding conscious consumption and solid waste generation; (ii) implementation of the reduction and sorting of solid waste from cities in the generation sources; (iii) closing of all open dumps and (iv) landfill reduction of organic municipal waste.

Public managers should consider different strategies and options for solid waste management, taking into account the direct and indirect challenges of each alternative, generating a reliable decision-making process in the municipal context and contributing to the analysis of public policies, processes and sectoral assessment (Li et al., 2015). The current global concern is to minimize the environmental impacts associated with solid waste generation, disposal and treatment. In Brazil, this is a worrying and urgent task, as efforts by national authorities to discuss and resolve this problem have recently begun, as highlighted by Alfaia et al. (2017).

The international trend is for solid waste generation to continue to increase as household consumption and population growth increase, making it a major challenge for urban areas, especially in regions with large population concentrations (e.g. metropolitan areas). The most populous regions of Brazil are the northeast and southeast regions, concentrating 69% of the country's inhabitants (IBGE, 2019). Consequently, these are the regions that generate the most solid waste, together producing 74% of all waste generated in Brazil (ABRELPE, 2016). Figure 1.1 presents aggregated information on solid waste generation, disposal and treatment for all regions of Brazil.

Another relevant factor for solid waste management is recycling. According to Toso and Alem (2014), what differentiates Brazil from other countries is the way solid waste is separated for recycling. In Brazil, waste is divided into subgroups of materials (recyclable and non-recyclable), while other countries classify it by type (paper, plastic, etc.). Final disposal is restricted to waste that cannot be treated or recovered, classified as discarded (Brasil, 2010). In Brazil, the selective collection is carried out through Voluntary Delivery Points, by actions of public authorities and with the informal sector. Since the adoption of the National Solid Waste Policy, the social inclusion of the informal sector through associations and cooperatives has been increasingly encouraged throughout the country. According to Ezeah et al. (2013), Brazil is a reference in social inclusion for collectors of recyclable materials through cooperatives and associations. However, competition with other waste collectors, health and safety at work and discrimination among the population are still inherent challenges in this activity (Nemadire et al., 2017).

In this sense, Ferri et al. (2015) argued that the integration between waste collectors and the global management system is not effective, often due to the low level of municipal government capacity to carry out this integrated activity. The formalization of disorganized individuals in dumps can contribute to the efficiency of waste collection and disposal, in addition to strengthening cooperatives. The formalization of workers can help combat exploitation by child and slave labour and prevent occupational health risks (Velis, 2017). Authorities and stakeholders should be aware of the global issues affecting waste management, providing systematic and regular information to the population to promote awareness and social inclusion in the municipality's recycling processes, adding value to local management.

Related to key challenges, the improper collection and disposal of solid waste at the municipal level can lead to negative outcomes that endanger environmental quality and human health, directly affecting economic issues, as highlighted by Taghipour et al. (2016). Much research published in the scientific literature has been concerned with quantifying the effect of waste flow on the environment (Ferronato and Torretta, 2019), using indicators as one of the main solid waste management tools for municipalities (Deus et al., 2019; Greene and Tonjes, 2014; Ferreira et al., 2017; Freitas and Magrini, 2017).

In this context, indicators allow public managers to monitor public services provided to the population (Wilson et al., 2012), including socio-economic impacts (Deus et al., 2017), waste collection and transportation costs (Sanjeevi and Shahabudeen, 2015) and the capacity of landfills (Ghanbari et al., 2012). These metrics should be technically oriented, with a high impact on society, including decision-makers and end-users. As discussed by El Said and Aghezzaf (2018), indicator monitoring can facilitate the identification of challenges and the development of best practices, revealing new opportunities for improvement to promote key changes in sectoral public policies.

1.4 THE IMPORTANCE OF INDICATORS FOR MUNICIPAL SOLID WASTE MANAGEMENT

In developing an efficient and integrated municipal-level waste management system, it is essential to assess the existing performance of available information that the national data system has. In the absence of a well-established and correctly standardized benchmarking process

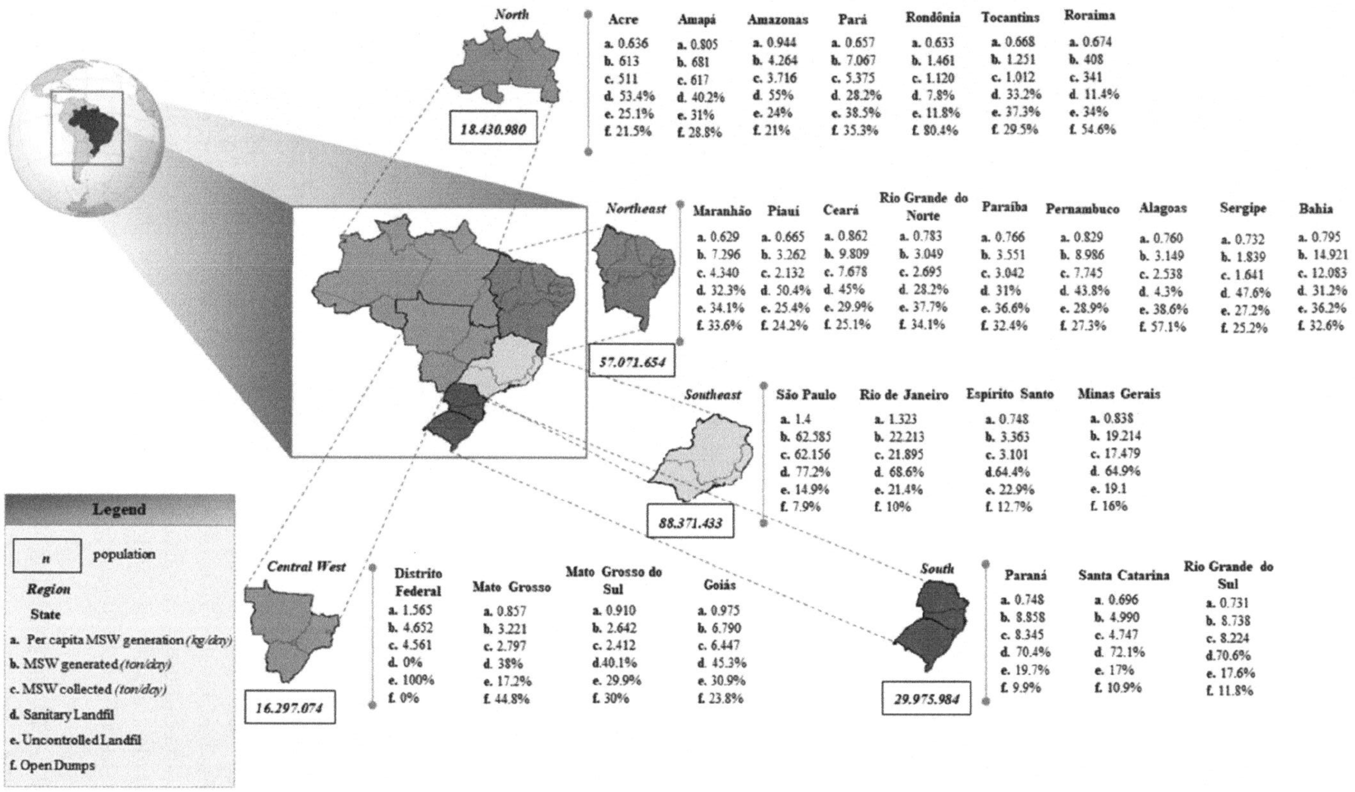

FIGURE 1.1 Information on municipal solid waste by region in Brazil. (Based on Nascimento et al., 2019, using data from IBGE, 2019 and ABRELPE, 2016.)

(AlHumid et al., 2019a), smaller municipalities in particular will function without knowing their level of service (Zotos et al., 2009; Haider et al., 2016). Consequently, solid waste managers will hardly know what their low-performance processes are, and therefore will not build proactive actions to improve them, making operational improvement difficult. On the other hand, positive results also need efforts to maintain a good level and management, reinforcing their potential in the local territory and their goals. According to AlHumid et al. (2019b), this can have a strong impact on the main components for municipal solid waste management systems, which are: Human resources, physical assets, financial capacity, citizen service, operations, technical resources and environmental concern.

To achieve municipal solid waste management objectives, it is important to assess the degree of performance and compliance of each component by assessing the entire system based on a set of specific indicators. Greene and Tonjes (2014) supported the idea that the indicators can contribute to the elaboration of a metric, aiming to guide the decision-makers, allowing a comparative analysis between the performance levels of the waste management systems. They also cited that the use of these indicators could contribute to the creation of new policies for the efficient development of sustainable solid waste management.

There are different initiatives in the international literature where indicators have been used as an auxiliary tool in the solid waste management process. The study by Zaman (2014) identified and classified the main indicators for 'zero waste' management systems, classifying them into seven different domains: Socio-cultural, economic, management, geo-administrative, environmental, political and organizational. Based on UN-Habitat solid waste management guidelines, Wilson et al. (2015) presented a set of indicators for integrated sustainable waste management for cities, allowing cities to be compared through performance benchmarking. This study also presented a mixed approach, with a set of quantitative indicators and composite qualitative indicators. In the same line, El Said and Aghezzaf (2018) suggested a list of measurable indicators covering the basic elements of solid waste management based on a qualitative assessment of quantitative indicators.

Gamberini et al. (2013) presented common engineering indices, and their values were analysed for Italian communities, focusing on the development of an accessible database, with information on demand and the cost profiles necessary for waste management. Based on quantitative indicators, Cucchiella et al. (2014) used multiobjective mathematical programming to evaluate the effectiveness of municipal waste management, focusing mainly on the landfill rate. In this same approach, Yildirim et al. (2018) highlighted that many indicator studies, specifically based on multicriteria decision-making, were mainly used to select suitable sites for the construction of landfills, proving to be limited. The study by Cifrian et al. (2013) proposed a set of individual indicators to track the progress of urban solid waste in the Cantabria region, Spain, using the carbon footprint method, assessing waste management strategies and technologies from the perspective of the impacts on global climate change.

Brazil is a developing country with a continental territory, where institutionalizing, implementing and standardizing solid waste management is still a major challenge. Approximately 62% of Brazilian municipalities do not have the financial resources to finance proper waste disposal. The southern and southeastern regions of Brazil are the most developed areas of the country, and are also the regions where most solid waste is disposed of correctly, accumulating 88% and 51% respectively. The northeast region, one of the most populous, has the worst rate of adequate disposal: 11% (ABRELPE, 2016). Thus, it is clear the lack of instruments and incentives in the Brazilian territory, revealing an important gap.

In the Brazilian context, different initiatives identified in the most recent literature can be highlighted. Silva et al. (2019) suggested a methodological proposal to select a relevant set of indicators to analyse the level of sustainability of solid waste management in medium and large cities in southern Brazil. The authors gathered 18 different methods identified in the literature with different solid waste management indicators, totalling 396 indicators. After applying the predefined selection criteria, 49 indicators were selected and applied in a case study.

The study by Fratta et al. (2019) aimed to address the challenges of solid waste management in three municipalities in the southeast region of Brazil. To assess the level of sustainability of these municipalities, the authors collected primary and secondary data, generated an integrated diagnosis of the results and finally presented future perspectives for the management of the municipalities studied. Barros and Silveira (2019) developed a solid waste management assessment in the metropolitan region of Belo Horizonte, Minas Gerais State, to support the decision-making processes of public managers. Sixteen indicators were selected to support the evaluation, selected from national and international documents, focusing on the applicability to the context of the municipalities studied. Both studies, directly or indirectly, explored the qualitative approach to deepen or validate the research. This denotes the need for the application of mixed approach tools to support municipal administrations in decision-making regarding the setting of goals and instruments for managing, monitoring and planning the delivery and execution of municipal public services. The complementarity of qualitative analyses will be further discussed in the next section.

The diversity of use and application of specific waste indicators is wide and transversal, relating to existing themes in the municipal space, such as water quality and availability, air quality and economic and social inclusion, among others. According to Cervantes et al. (2018), for good solid waste management in municipalities, a group of indicators must have minimum attributes so that they can be used in any context, namely: (i) utility for different geographic levels, providing and integrating information between levels; (ii) flexibility, being applicable and adaptable to different local contexts; (iii) a mixed approach, including quantitative and qualitative indicators; (iv) methodological consistency, facilitating reliable and replicable results; and (v) transversality, with the ability to evaluate all elements of solid waste management in municipal systems.

Considering all attributes previously presented by the authors, we should also prioritize the applicability of a municipal indicator, which should be measurable with available data and understandable to the public and the municipal government. The indicators used in the management process must be politically relevant and understandable to their users.

1.5 USING SOLID WASTE INDICATORS AND INFORMATION FOR BRAZILIAN MUNICIPALITIES MANAGEMENT

Most Brazilian municipalities are unable to assess whether the operating conditions of local waste management are adequate to current standards, making it difficult to fully evaluate the activities that encompass the provision of this service. The United Nations Environment Program (UNEP, 2004) has shown that indicators can be used as a tool to facilitate access to technical and scientific information for different user and stakeholder groups.

One of the main functions of the indicators is to aggregate a large amount of data, facilitating the communication of results and conclusions. From this perspective, the creation of decision-making indicator systems through the analysis and diagnosis of solid waste management can be a coherent way to pursue environmental sustainability.

An integrated diagnosis should bring together quantitative and qualitative indicators, so that it is possible to advance to sustainable levels between sustainability actions and practices, and effective municipal solid waste management. In this case, a mixed approach may favour the deepening of technical knowledge of local problems and demands in multiple ways, identifying key indicators to measure their impact on human health and environmental quality. However, it is necessary to constantly improve evaluation processes so that current conditions can generate positive impacts in the future, as well as having positive effects on local and regional planning through proper solid waste management. Thus, it is necessary to share information transparently, building a public agenda for operational execution and monitoring the goals of the issues discussed. We can highlight two platforms of indicators and information for waste available for Brazil: the National Sanitation Information System (SNIS) and the Brazilian Institute of Geography and Statistics (IBGE).

The Brazilian Ministry of Regional Development, through the National Sanitation Information System (SNIS), provides data on solid waste, water and sewage and rainwater from the service providers and public management organizations of municipalities. SNIS is the largest and most important sanitation sector information platform in Brazil, supported by a robust database of administrative, operational, managerial, institutional, economic-financial, accounting and quality information on service delivery. This platform periodically publishes indicators and information on these three themes, enabling the cross-checking of data on these themes. The main objectives of SNIS are: (i) the planning and execution of public policies; (ii) resource application guidance; (iii) evaluation of the sanitation sector; (iv) performance evaluation of services; (v) improvement of management; guidance on regulatory and enforcement activities; and (vi) social control (Brazil, 2019).

On this platform, the theme 'Solid Waste' has a database with information from the city halls and other municipal public organizations collected since 2002. It has enabled the construction of a historical series, where data can be accessed by each municipality and can be grouped by the unit, federation, metropolitan regions and macroregions of the country, among other possible groupings. In general, the information stored has a quantitative and qualitative approach and is related to the quantity and type of solid waste collected, the number of professionals and non-professionals (autonomous collectors) who work with waste, selective collect, the final disposal of waste and the waste collection service level. In this structure, the basis of the data is the municipal body itself, which is responsible for solid waste management. Figure 1.2 briefly demonstrates the information flow of different elements of municipal solid waste management, based on the SNIS platform.

The construction of the indicators proposed by the Brazilian Institute of Geography and Statistics (IBGE) for Brazil was inspired by the international movement led by the United Nations Center for Sustainable Development, bringing together national governments, academic institutions, non-governmental organizations, United Nations system organizations and experts from around the world. Since 1992, this movement has directed a work programme consisting of various studies and

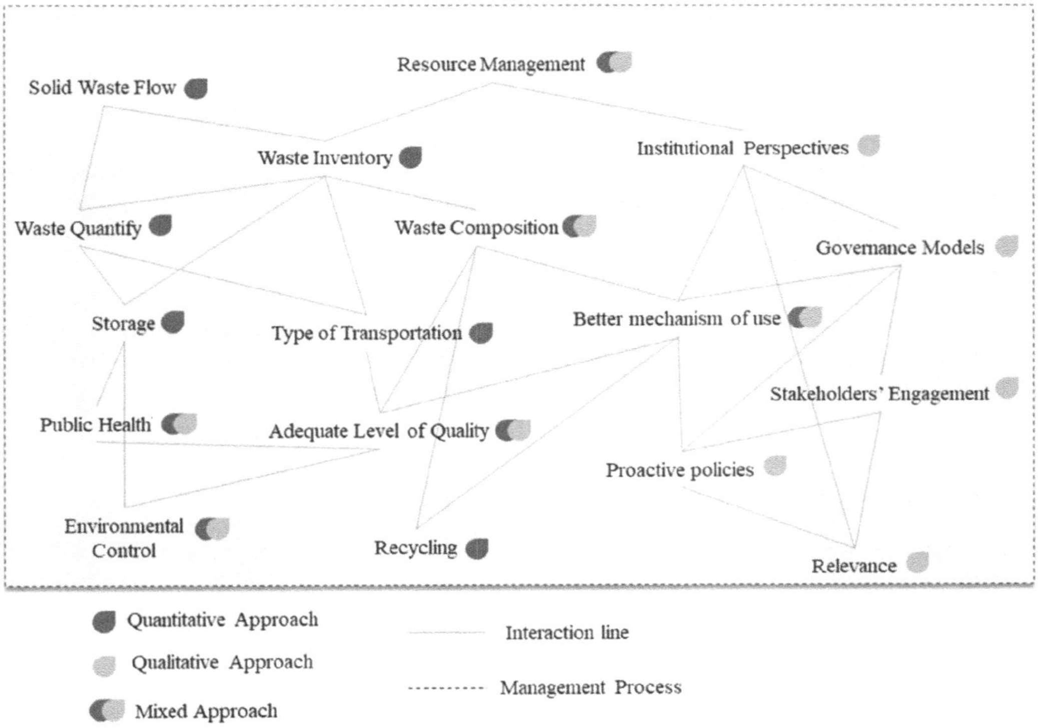

FIGURE 1.2 Flow and approach of SNIS database information for municipal solid waste management.

exchanges of information, to implement the provisions of chapters 8 and 40 of Agenda 21, which deal with the relationship between the environment, sustainable development and information for decision-making (IBGE, 2015).

The IBGE project used as a reference the recommendations of the 'Indicators of sustainable development: framework and methodologies', known as the Blue Book, adapting its content to Brazilian particularities. In the Brazilian case, in addition to the challenge of building indicators capable of characterizing and subsidizing the process of sustainable development at the national level, there is a need to express the heterogeneity of the country's regions. In this proposal, the indicators associated with the theme 'solid waste' were worked in the context of sustainable development. This enabled a broader analysis, favouring the identification of strategic information for waste management. Among the 63 indicators available, 5 indicators were identified that are directly and indirectly connected to solid waste management for Brazilian municipalities: (i) access to domestic waste collection service; (ii) recycling; (iii) radioactive waste; (iv) diseases related to inadequate environmental sanitation; and finally (v) environmental legislation.

This platform developed a matrix of relationships between the 63 indicators, demonstrating the connections between the different indicators, enabling an integrated analysis. Figure 1.3 shows the connections between the 5 waste indicators, cited in the previous paragraph, and 34 other indicators analysed under the sustainability spectrum, highlighting the multidimensional and transversal impact of the theme. According to IBGE (2015), there is no doubt that it is possible to find relationship links linking almost all the indicators presented.

In the proposal presented by IBGE (2015), many of the connections are referenced in specific consolidated studies, and others are theoretical indications, derived only from a qualitative appreciation. Using this information we can favour a systemic and broad reflection, as well as suggest quantitative and qualitative analyses or elaboration of public policies that integrate the different aspects and dimensions of sustainable development.

Information from quantitative indicators should be associated with the qualitative aspects of the analysis, where the interpretation of these indicators gives a deeper and more reliable meaning. Therefore, quantitative data must be combined with a qualitative analysis, establishing criteria for achieving goals, generating a mixed approach. According to Álvarez-Arenas and Mirón (2006), establishing links between indicators and defined goals can help add meaning to the quantitative results of the indicators analysed.

Qualitative analysis can help identify deficiencies in waste collection and storage, as indicated by Ali et al. (2019). These deficiencies include the accumulation of garbage at collection points, overflowing of garbage containers, open-air garbage burning, public and environmental health problems and lack of personal protective equipment for health personnel.

Regarding the initiatives of the Brazilian municipalities, most of the authors studied denounce the lack of clarity and objectivity in the use of information for the management process, the low level of participation of public managers and analyses focused only on service performance. This can limit not only solid waste management, but also other issues that are embedded in the municipal context.

1.6 FINAL REMARKS AND FURTHER REFLECTIONS

This chapter has presented the main challenges for municipal solid waste management in Brazil, focusing on manageable indicators as an urban management tool. It also demonstrated that the improper disposal of waste, and the constant increase in waste generation associated with population growth, can increase environmental impacts on local territories. This can cause significant changes in solid waste composition and characteristics, increasing its hazardousness. We found that most municipalities in Brazil dispose of their solid waste improperly, revealing that much needs to be done for local governments to provide public services under their responsibility, and to effectively fulfil their social and public function.

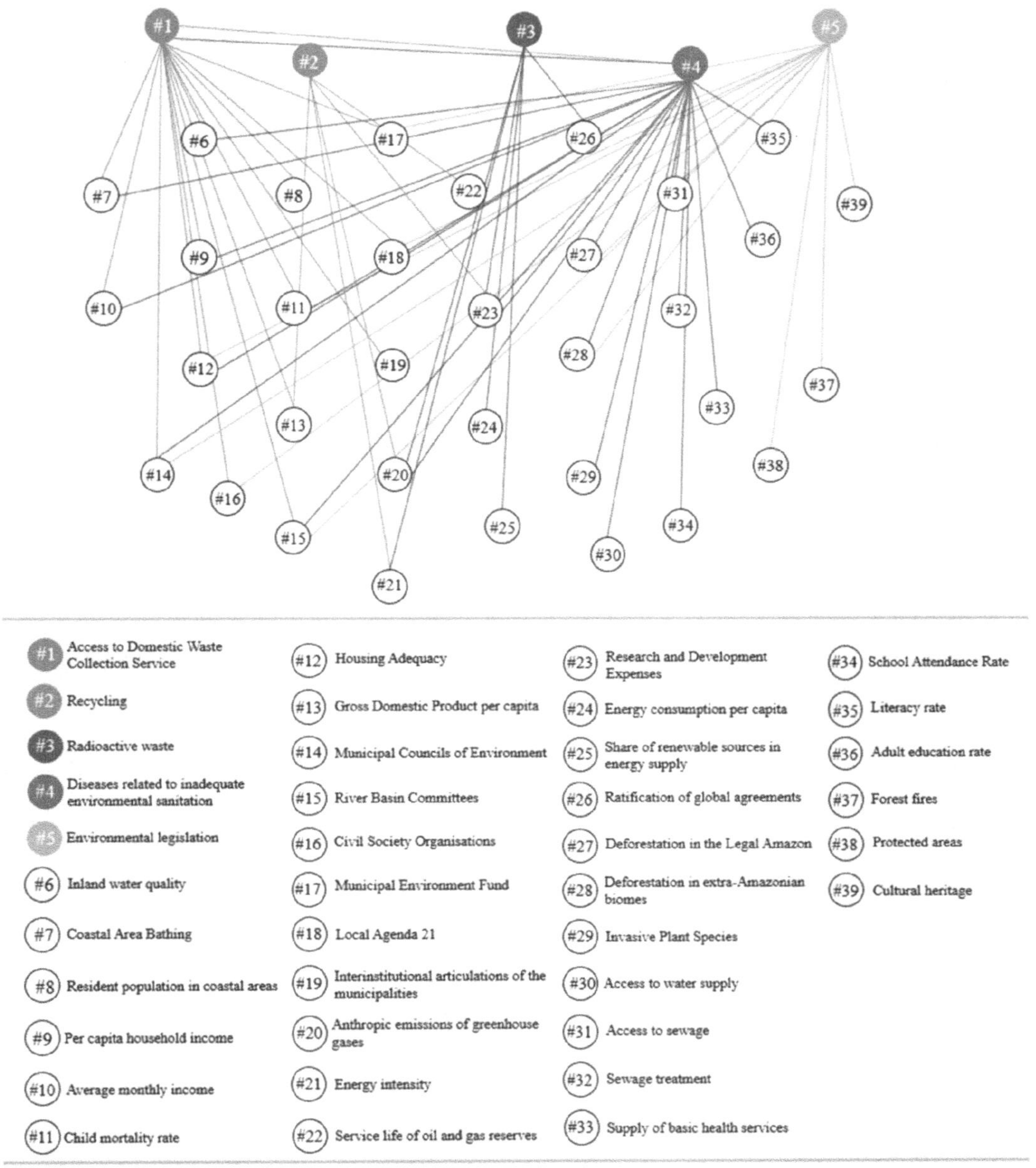

FIGURE 1.3 Relationship matrix of solid waste indicators. (Elaborated with information from IBGE, 2015.)

Good municipal solid waste management needs to address the challenges in multiple and complementary ways so that administrative and technical difficulties can be overcome to consolidate an integrated public service delivery system. A strategic action would be to formalize partnerships with autonomous collectors of recyclable material, outsourcing part of the activities of the municipal public administration. This partnership consists of carrying out the activities of collecting, sorting, processing and marketing recyclable materials, involving the environmental education of workers and other agents of society. This can decrease the flow and accumulation of solid waste with improper disposal, and potentially generate economic benefits for waste collector families. However, Brazilian municipal managers still have difficulty integrating environmental and social programmes into urban sanitation systems. Most of the selective collection of post-consumer solid waste is informal, carried out by autonomous waste collectors who work informally on city streets and dumps.

Highlighting the SNIS and IBGE platforms, the presentation of the solid waste indicators and information profile for the management of municipalities makes it possible to advance in the construction and improvement of action planning and decision-making tools, contributing to the monitoring of invested resources and targeting of public policies. It is also considered important that government agencies producing these statistical data integrate their indicators and information, promoting the consolidation of a robust information system with transparent procedures and techniques.

Solid waste management must respond to new local and global paradigms using scientific development as the basis for change. In addition, we need to define what we want to measure by tailoring the use of indicators and information to specific local situations. Available indicators and information should be used to assess what is relevant in addressing new pitfalls of false information and unauthorized agendas by society. Therefore, it is important to develop municipal information and databases collaboratively and openly, highlighting the central role of citizens facing the challenge of generating a social impact on public decisions. Challenges for solid waste management must be tackled by the general society, including volunteer individuals, non-governmental organizations, entrepreneurs and researchers with personalized and objective approaches.

REFERENCES

ABRELPE—Associação Brasileira de Empresas de Limpeza Pública e Resíduos Especiais. 2016. *Panorama dos resíduos sólidos no Brasil—2015.* http://www.abrelpe.org.br/Panorama/panorama2015.pdf (accessed October 10, 2019).

ABRELPE—Associação Brasileira de Empresas de Limpeza Pública e Resíduos Especiais. 2017. *Panorama dos resíduos sólidos no Brasil—2017.* http://www.abrelpe.org.br/Panorama/panorama2017.pdf (accessed October 10, 2019).

Agol, D., Latawiec, A.E., and Strassburg, B.B.N. 2014. Evaluating impacts of development and conservation projects using sustainability indicators: opportunities and challenges. *Environmental Impact Assessment Review* 48: 1–9. https://doi.org/10.1016/j.eiar.2014.04.001

Alfaia, R.G.S.M., Costa, A.M., and Campos, J.C. 2017. Municipal solid waste in Brazil: a review. *Waste Management & Research* 35, no. 12: 1195–1209. https://doi.org/10.1177/0734242X17735375

AlHumid, H.A, Haider, H., AlSaleem, S.S., Alinizzi, M., Shafiquzamman, M., and Sadiq, R. 2019a. Performance assessment model for municipal solid waste management systems: conceptualization, development, and implementation. *Environment* 16, no. 2: 1–25. https://doi.org/10.3390/environments6020019

AlHumid, H.A., Haider, H., AlSaleem, S.S., Shafiquzamman, M., and Sadiq, R. 2019b. Performance indicators for municipal solid waste management systems in Saudi Arabia: selection and ranking using fuzzy AHP and PROMETHEE II. *Arabian Journal of Geosciences* 12: 491. https://doi.org/10.1007/s12517-019-4645-0

Ali, M., Geng, Y., Robins, D., Cooper, D., Roberts, W., and Vogtländer, J. 2019. Improvement of waste management practices in a fast expanding sub-megacity in Pakistan, on the basis of qualitative and quantitative indicators. *Waste Management* 85: 253–263. https://doi.org/10.1016/j.wasman.2018.12.030

Álvarez-Arenas, M., and Mirón, I. 2006. A flexible framework for regional sustainable development indicators using system thinking criteria (INSURE). *Revista Internacional de Sostenibilidad Tecnología y Humanismo* 1: 41–59. https://core.ac.uk/download/pdf/39061100.pdf

Azevedo, B.D., Scavarda, L.F., and Caiado, R.G.G. 2019. Urban solid waste management in developing countries from the sustainable supply chain management perspective: a case study of Brazil's largest slum. *Journal of Cleaner Production* 233: 1377–1386. https://doi.org/10.1016/j.jclepro.2019.06.162

Barros, R.T.V., and Silveira, A.V.F. 2019. Use of sustainability indicators for the assessment of urban solid waste management in Belo Horizonte's Metropolitan Region, MG, Brazil. *Engenharia Sanitária e Ambiental* 24, no. 2: 411–423. http://dx.doi.org/10.1590/s1413-41522019177499

Brazil. 2010. *Lei No 12.305, de 2 de agosto de 2010.* Institui a Política Nacional de Resíduos Sólidos; altera a Lei no 9.605, de 12 de fevereiro de 1998; e dá outras providências. Diário Oficial da União, Brasília, August 3, 2010.

Brazil. 2018. *Municipalities that have solid waste plan.* Ministry of Environment: Brasília. https://www.mma.gov.br/informma/item/15166-54-dos-munic%C3%ADpios-t%C3%AAm-plano-de-res%C3%ADduos.html (accessed October 10, 2019).

Brazil. 2019. *National Sanitation Information System (SNIS).* Ministry of Regional Development. National Secretariat of Sanitation: Brasília, Brazil. http://app4.cidades.gov.br/ (accessed October 10, 2019).

Cervantes, D.E.T., Martínez, A.L., Hernández, M.C., and Cortázar, A.L.G. 2018. Using indicators as a tool to evaluate municipal solid waste management: a critical review. *Waste Management* 80: 51–63. https://doi.org/10.1016/j.wasman.2018.08.046

Cifrian, E., Andres, A., and Viguri, J.R. 2013. Estimating monitoring indicators and the carbon footprint of municipal solid waste management in the region of Cantabria, Northern Spain. *Waste & Biomass Valorization* 4: 271–285. https://doi.org/10.1007/s12649-012-9150-6

Cifrian, E., Andres, A., and Virguri, J.R. 2015. Developing a regional environmental information system based on macro-level waste indicators. *Ecological Indicators* 53: 258–270. https://doi.org/10.1016/j.ecolind.2015.02.010

Coelho, L.M.G., and Lange, L.C. 2016. Applying life cycle assessment to support envi-ronmentally sustainable waste management strategies in Brazil. *Resources, Conservation & Recycling* 128: 438–450. https://doi.org/10.1016/j.resconrec.2016.09.026

Cucchiella, F., D'Adamo, I., and Gastaldi, M. 2014. Strategic municipal solid waste management: a quantitative model for Italian regions. *Energy Conversion and Management* 77: 709–720. https://doi.org/10.1016/j.enconman.2013.10.024

Deus, R.M., Battistelle, R.A.G., and Silva, G.H.R. 2017. Scenario evaluation for the management of household solid waste in small Brazilian municipalities. *Clean Technologies and Environmental Policy* 19: 205–214. https://doi.org/10.1007/s10098-016-1205-0

Deus, R.M., Bezerra, B.S., and Battistelle, R.A.G. 2019. Solid waste indicators and their implications for management practice. *International Journal of Environmental Science and Technology* 16: 1129–1144. https://doi.org/10.1007/s13762-018-2163-3

Domingues, A.R., Lozano, R., and Ramos, T.B. 2018. Stakeholder-driven initiatives using sustainability indicators. In *Routledge Handbook of Sustainability Indicators*, Bell, S., Morse, S., Eds., 379–391. London: Routledge.

Echegaray, F., and Hansstein, F.V. 2017. Assessing the intention-behavior gap in electronic waste recycling: the case of Brazil. *Journal of Cleaner Production* 142: 180–190. https://doi.org/10.1016/j.jclepro.2016.05.064

El Said, S., and Aghezzaf, E.H. 2018. A progress indicator-based assessment guide for integrated municipal solid-waste management systems. *Journal of Material Cycles and Waste Management* 20, no.2: 850–863. https://doi.org/10.1007/s10163-017-0647-8

Ezeah, C., Fazakerley, J.A, and Roberts, C.L. 2013. Emerging trends in informal sector recycling in developing and transitional countries. *Waste Management* 33: 2509–2519. https://doi.org/10.1016/j.wasman.2013.06.020

Ferreira, F., Avelino, C., Bentes, I., Matos, C., and Teixeira, C.A. 2017. Assessment strategies for municipal selective waste collection schemes. *Waste Management* 59: 3–13. https://doi.org/10.1016/j.wasman.2016.10.044

Ferri, G.L., Chaves, G.L.D., and Ribeiro, G.M. 2015. Reverse logistics network for municipal solid waste management: the inclusion of waste collectors as a Brazilian legal requirement. *Waste Management* 40: 173–191. https://doi.org/10.1016/j.wasman.2015.02.036

Ferronato, N., and Torreta, V. 2019. Waste mismanagement in developing countries: a review of global issues. *International Journal of Environmental Research and Public Health* 16: 1060. https://doi.org/10.3390/ijerph16061060

Fratta, K.D.S.A, Toneli, J.T.C.L., and Antonio, G.C. 2019. Diagnosis of the management of solid urban waste of the municipalities of ABC Paulista of Brasil through the application of sustainability indicators. *Waste Management* 85: 11–17. https://doi.org/10.1016/j.wasman.2018.12.001

Freitas, L., and Magrini, A. 2017. Waste management in industrial construction: investigating contributions from industrial ecology. *Sustainability* 9: 1251. https://doi.org/10.3390/su9071251

Gamberini, R., Del Buono, D., Lolli, F., and Rimini, B. 2013. Municipal solid waste management: identification and analysis of engineering indexes representing demand and costs generated in virtuous Italian communities. *Waste Management* 33: 2532–2540. https://doi.org/10.1016/j.wasman.2013.06.003

Ghanbari, F., Sharee F.A., Monavari, M., and Zaredar, N. 2012. A new method for environmental site assessment of urban solid waste landfills. *Environmental Monitoring and Assessment* 184, no. 3: 1221–1230. https://doi.org/10.1007/s10661-011-2034-6

Greene, K.L., and Tonjes, D.J. 2014. Quantitative assessments of municipal waste management systems: using different indicators to compare and rank programs in New York State. *Waste Management* 34: 825–836. https://doi.org/10.1016/j.wasman.2013.12.020

Guerrero, L.A., Maas, G., and Hogland, W. 2013. Solid waste management challenges for cities in developing countries. *Waste Management* 33, no. 1: 220–232. https://doi.org/10.1016/j.wasman.2012.09.008

Haider, H., Sadiq, R., and Tesfamariam, S. 2016. Inter-utility performance benchmarking model for small-to-medium-sized water utilities: aggregated performance indices. *Journal of Water Resources Planning and Management* 142, no. 1: 04015039. https://doi.org/10.1061/(ASCE)WR.1943-5452.0000552

Ibáñez-Forés, V., Nóbrega, C.C., Bovea, M.D., Silva, C.M., and Virgolino, J.L.F. 2018. Influence of implementing selective collection on municipal waste management systems in developing countries: a Brazilian case study. *Resources, Conservation and Recycling* 134: 100–111. https://doi.org/10.1016/j.resconrec.2017.12.027

IBGE—Brazilian Institute of Geography and Statistics. 2015. *Indicadores de Desenvolvimento sustentável: Brasil 2015*. Rio de Janeiro: IBGE, 2015. ISBN: 978-85-240-4347-5.

IBGE—Instituto Brasileiro de Geografia e Estatística. 2019. *Projection of the population of Brazil and of the Federation Units 2019*. https://www.ibge.gov.br/apps/populacao/projecao/index.html?utm_source=portal&utm_medium=popclock&utm_campaign=novo_popclock (accessed October 13, 2019).

Klein, F.B., Gonçalves-Dias, S.L.F., and Jayourbe, M. 2018. Gestão de resíduos sólidos urbanos nos municípios da Bacia Hidrográfica do Alto Tietê: uma análise sobre o uso de TIC no acesso à informação governamental urbe. *Brazilian Journal of Urban Management* 10, no. 1: 140–153 http://dx.doi.org/10.1590/2175-3369.010.001.ao10

Li, H., Nitivattananon, V., and Li, P. 2015. Developing a sustainability assessment model to analyze China's municipal solid waste management enhancement strategy. *Sustainability* 7: 1116–1141. https://doi.org/10.3390/su7021116

Lima, P.D.M., and Paulo, P.L. 2018. Solid-waste management in the rural area of Brazil: a case study in Quilombola communities. *Journal of Material Cycles and Waste Management* 20: 1–11. https://doi.org/10.1007/s10163-018-0722-9

Nascimento, V.F., Sobral, A.C., Fehr, M., Yesiller, N., Andrade, P.R., and Ometto, J.P.H.B. 2019. Municipal solid waste disposal in Brazil: improvements and challenges. *International Journal of Environment and Waste Management* 23, no. 3: 300–318. https://doi.org/10.1504/IJEWM.2019.099007

Nemadire, S., Mapurazi, S., and Nyamadzawo, G. 2017. Formalising informal solid waste recycling at the Pomona dumpsite in Harare, Zimbabwe. *Natural Resources Forum* 41: 167–178. https://doi.org/10.1111/1477-8947.12130

Sanjeevi, V., and Shahabudeen, P. 2015. Development of performance indicators for municipal solid waste management (PIMS): a review. *Waste Management & Research* 33, no. 12: 1052–1065. https://doi.org/10.1177/0734242X15607428

Silva, L., Prietto, P.D.M., and Korf, E.P. 2019. Sustainability indicators for urban solid waste management in large and medium-sized worldwide cities. *Journal of Cleaner Production* 237: 117802. https://doi.org/10.1016/j.jclepro.2019.117802

Taghipour, H., Amjad, Z., Aslani, H., Armanfar, F., and Dehghanzadeh, R. 2016. Characterizing and quantifying solid waste of rural communities. *Journal of Material Cycles and Waste Management* 18, no. 4: 790–797. https://doi.org/10.1007/s10163-015-0365-z

Toso, E.A.V., and Alem, D. 2014. Effective location models for sorting recyclables in public management. *European Journal of Operational Research* 234: 839–860. https://doi.org/10.1016/j.ejor.2013.10.035

UNEP—United Nations Environment Programme. 2004. *The vital waste graphics*. http://www.vitalgraphics.net (accessed October 10, 2019).

UNEP—United Nations Environmental Programme. 2011. *Towards a green economy: pathways to sustainable development and poverty eradication*. Nairobi: UNEP.

Velis, C. 2017. Waste collectors in global south: informal recycling sector in a circular economy era. *Waste Management & Research* 35: 329–331. https://doi.org/10.1177/0734242X17702024

Wilson, D.C., Rodic, L., Crowing, M.J., Velis, C.A., Whiteman, A.D., Scheinberg, A., Vilches, R., Masterson, D., Stretz, J., and Oelz, B. 2015. 'Wasteaware' benchmark indicators for integrated sustainable waste management in cities. *Waste Management* 35: 329–342. https://doi.org/10.1016/j.wasman.2014.10.006

Wilson, D.C., Rodic, L., Scheinberg, A., Velis, C.A., and Alabaster, G. 2012. Comparative analysis of solid waste management in 20 cities. *Waste Management & Research* 30, no. 3: 237–254. https://doi.org/10.1177/0734242X12437569

Yildirim, V., Memisoglu, T., Bediroglu, S., and Colak, H.E. 2018. Municipal solid waste landfill site selection using multi-criteria decision making and GIS: case study of Bursa province. *Journal of Environmental Engineering and Landscape Management* 26, no. 2: 107–119. https://doi.org/10.3846/16486897.2017.1364646

Zaman, A.U. 2013. Life cycle assessment of pyrolysis-gasification as an emerging municipal solid waste treatment technology. *International Journal of Environmental Science and Technology* 10: 1029–1038. https://doi.org/10.1007/s13762-013-0230-3

Zaman, A.U. 2014. Identification of key assessment indicators of the zero waste management systems. *Ecological Indicators* 36: 682–693. https://doi.org/10.1016/j.ecolind.2013.09.024

Zotos, G., Karagiannidis, A., Zampetoglou, S., Malamakis, A., Antonopoulos, I.S., Kontogianni, S., and Tchobanoglous, G. 2009. Developing a holistic strategy for integrated waste management within municipal planning: challenges, policies, solutions and perspectives for Hellenic municipalities in the zero-waste, low-cost direction. *Waste Management* 29: 1686–1692. https://doi.org/10.1016/j.wasman.2008.11.016

2 Waste Management Policies and Practices in Brazil

Marcelo Seleme Matias, Isabella Pimentel Pincelli and Armando Borges de Castilhos Júnior

CONTENTS

2.1 INTRODUCTION

Brazil has experienced some important changes in municipal solid waste management. Oriented by Brazilian National Solid Waste Policy (PNRS), which came into force in 2010, municipalities had to organize and plan their actions for managing solid waste. The PNRS assigned responsibilities to solid waste generators and public authorities (Serra, 2015). Historically, solid waste disposal on dumpsites was a major issue in Brazil, and the PNRS has set restrictive targets for the proper disposal of solid waste until 2014. However, the participation of different actors in the different stages of waste management makes the issue not so simple to manage. Solid waste management in Brazil is a complex issue, especially considering that there are 5,570 municipalities with distinct social, environmental and economic characteristics (IBGE, 2018). In addition, there is also the social aspect represented by the recyclable waste collectors' work within the collection, classification and commercialization of recyclable waste (Fidelis et al., 2020).

Brazilian municipalities are autonomous in defining their solid waste management strategies, but some authors suggest that they are not able to manage their waste by themselves (Marino et al., 2018) and the national results in solid waste generation reduction, recycling rate increase and appropriate final disposal are not very clear (Cetrulo et al., 2018). Thus, this chapter presents an overview of the main characteristics of solid waste management in Brazil, addressing its legal structure, some performance data on generation, collection and disposal, as well as a discussion about the informal recycling sector, which is strongly present in the country.

2.2 BRAZILIAN SOLID WASTE MANAGEMENT OVERVIEW

The management of municipal solid waste in Brazil is analysed according to its legal and normative structure, where the solid waste management guidelines are presented, and according to the performance results in solid waste management in the country. Thus, the waste management sector is described, regarding its normative instruments, and, then, data on solid waste management in Brazil is presented, such as the solid waste generation per capita, the existence of collection services and other performance indicators.

2.2.1 Policies and Legal Instruments in Brazil

In Brazil, sanitation actions and framework were not considered in the urbanization process, but rather they happened as a response to the sanitation issues that arose with urbanization (Neto and Moreira, 2010). In the 1980s, sites for solid waste disposal were more scarce and public concerns about dumpsite's environmental impacts arose, and it was when solid waste management (Neto and Moreira, 2010) and legislation became a focus in Brazil. The Brazilian National Policy on Solid Waste (PNRS) – Law 12,305 from 2010 – is the standing legal framework for solid waste management in the country. Prior to the approval of the PNRS, the solid waste regulations were dispersed in several laws, decrees, ordinances and resolutions (Neto and Moreira, 2010).

In 1998 the Brazilian Law 9,605 was ratified, which regulated environmental crimes in the country (Lei 9,065, 1998). This law established penalties for the inappropriate disposal of solid waste (Lei 9,065, 1998), such as dumpsites. In the following year, the National Policy on Environmental Education, Law 9,795 from 1999, was approved, and thus, environmental conservation and sustainability were included as permanent components of the national education (Lei 9,795, 1999).

In 2007 the national guidelines for basic sanitation were established by Law 11,445 (Lei 11,445, 2007). Solid waste management was regarded by this legislation as a public basic sanitation service for the collection, transportation, transference, treatment and final destination of domestic solid waste and urban cleaning waste (Lei 11,445, 2007). Then, the PNRS was established in 2010 to regulate MSW management. The PNRS implemented principles, objectives and instruments, as well as guidelines for solid waste management in the country, including hazardous waste, imposing responsibilities on generators and public authorities (Fidelis et al., 2020). To manage solid waste in Brazil, the policy established solid waste plans as instruments for solid waste planning and governance. The PNRS requires the elaboration of integrated MSW management plans that include system planning, and management actions and targets (National Policy on Solid Waste, 2010).

The policy objectives include non-generation, reduction, reuse, recycling and treatment, as well as environmentally appropriate final disposal of residuals (National Policy on Solid Waste, 2010). Therefore, recycling should be implemented and only non-recyclable MSW should be landfilled. Campos (2014) highlighted that a relevant aspect of PNRS is the obligation of the municipalities to dispose of only the residual treatment tailings in landfills. This policy implies the expansion and creation of material recovery facilities to obtain the least amount of residual not susceptible to solid waste treatment (Campos, 2014). For recycling to become effective throughout the country, the extended producer responsibility was established. And reverse logistics systems were the instrument endorsed to implement the extended producer responsibility (National Policy on Solid Waste, 2010). All the supply chains of agrochemicals, batteries, tires, lubricating oils, fluorescent lamps, electronic products and their components are obliged to assume responsibility for the entire life cycle of their products by implementing systems of reverse logistics through sectoral agreements (Campos, 2014; Guarnieri et al., 2020).

Also, a sectoral agreement of reverse logistics for packaging in general was settled (Guarnieri et al., 2020). Decree 7.404 from 2010 created a committee for implementing reverse logistics systems (Ministry of the Environment, 2020). Regarding the relevance of the informal recyclable waste collectors' work, Brazilian legislation has developed towards the inclusion and formalization of these workers over the years. The Brazilian Occupation Classification officially recognized the profession of recyclable waste collectors in 2002 (Zolnikov et al., 2018). The recyclable waste collectors organized in associations and cooperatives were regarded as solid waste service providers within Law 11,445 (2007). The PNRS prioritizes the inclusion of recyclable waste collectors in the system of collecting, processing and commercializing recyclable solid waste management (Fidelis et al., 2020). Indeed, Law 11,445 (2007) and Law 12,305 (2010) introduced clauses that exempt municipalities from performing a bidding competition for hiring associations or cooperatives of recyclable waste collectors. Also, Decree 5,940 from 2006 determined that that recyclable solid waste generated at federal institutions must be sent to recyclable waste collectors' organizations (Pincelli et al., 2019).

And Decree 7,405, from 2010, instituted instruments for recyclable waste collectors' inclusion, such as the Pro-Catador programme and the committee for social and economic inclusion of recyclable waste collectors (Ministry of the Environment, 2020). Therefore, in Brazil the inclusion of recyclable waste collectors is a legal requirement in reverse logistics (Ferri et al., 2015). Recently, in 2017, the National Environment Council (CONAMA) established the resolution CONAMA 481 to regulate the recovery of organic waste through composting (Ministry of the Environment, 2020). As for MSW landfilling, resolution CONAMA 404 from 2008 rules landfill licensing (Ministry of the Environment, 2020). There are some CONAMA resolutions for regulating special solid wastes.

CONAMA 23 from 1996 is about hazardous solid waste definitions and treatments (Ministry of the Environment, 2020). CONAMA 313 from 2002 provides information for the national inventory of industrial solid waste (Ministry of the Environment, 2020). Health care waste treatment and disposal are ruled by CONAMA 358 from 2005 (Ministry of the Environment, 2020). Construction waste management are regulated by CONAMA 307 from 2002 and by CONAMA 448 from 2012 (Ministry of the Environment, 2020).

2.2.1.1 Integrated Solid Waste Management Plans

The strategic planning of waste management is a central task of the solid waste governance process, when goals are decided and strategies are developed (UNEP, 2015). For the planning to be accepted and owned by the community, a participatory and inclusive approach to the planning process is essential (UNEP, 2015). Solid waste management is a continuous process of planning, implementation, monitoring, evaluation and updating (UNEP, 2015). And the solid waste management plan is the documentation of the solid waste planning. In Brazil, to support solid waste management, solid waste plans are instruments of the PNRS. The PNRS established the elaboration of national, state and municipal plans on solid waste management, as well as plans on solid waste for specific activities (National Policy on Solid Waste, 2010).

The Brazilian national plan on solid waste was elaborated in 2012, but it is still a preliminary version (Brazil, 2012). Thus, Brazil lacks a consolidated national plan that could effectively orient its states and municipalities. And the elaboration of a municipal integrated plan on solid waste management is a condition for municipalities to access resources or financing from the nation for establishing projects in solid waste management (National Policy on Solid Waste, 2010).

According to the PNRS, municipal integrated MSW plans must include 19 mandatory items, including the diagnosis of solid waste management; management actions and targets for the reduction, reuse and recycling of waste; and the plan review periodicity (IBGE, 2017; National Policy on Solid Waste, 2010). IBGE's survey on basic municipal information from 2017 investigated whether the municipalities had an integrated solid waste plan (IBGE, 2017). Just over half of the municipalities (54.8%) had this plan (IBGE, 2017). The existence of this plan tends to increase with the increase of the population range of the municipalities (IBGE, 2017). Also, the Brazilian regions where the most municipalities had an integrated solid waste plan were in the Southern region – where 78.9% of municipalities had this plan – followed by the Midwest and the Southeast regions (IBGE, 2017). Planning solid waste management is essential for its proper implementation and its sustainability. In Brazil a consolidated national plan on solid waste is missing. Even so, half of

Brazilian municipalities had a solid waste plan. Still, it is necessary to strengthen strategic planning for all Brazilian municipalities, states and nation to orient solid waste management towards sustainability.

2.2.2 Solid Waste Management in Brazil

In this topic we present some performance indicators for the sector in Brazil, based on government data. The Brazilian National Sanitation Information System (SNIS) was used as a database. The SNIS is a publicly database platform created and maintained by the Brazilian Federal Government. The system contains operational, managerial and financial information on solid waste management in Brazilian municipalities. The collected data were processed and statistical analyses were performed. Besides that, as the information declared by the municipalities is not verified, sample selection criteria must be adopted to remove inconsistent data.

2.2.2.1 The Brazilian Per Capita MSW Generation Rate

Optimistic results were identified on this indicator, analysing the last few years. The Brazilian per capita household solid waste generation rate is presented graphically in Figure 2.1. An increase is observed from 2005 (0.670 kg/person·day) to 2012 (0.795 kg/person·day). From 2012, there was a drop in the growth rate, showing a decrease in the per capita household solid waste generation. The per capita household solid waste generation of each year was obtained by calculating the ratio of the sum of the collected household solid waste mass of a sample of municipalities (i) in a year (n) to the sum of the population served by the collection service of the same selected municipalities (i) of that year (n).

A piecewise linear regression was performed to identify if the decrease of per capita household solid waste generation rate noted from 2012 was statistically significant and to indicate a possible breakpoint of the evolution trend of a data series. The regression was calculated by the software StataMP 13 and showed a statistically significant breakpoint in 2011.72. The breakpoint happened with 95% confidence between 2010.637 (Aug 2010) and 2012.798 (Oct 2012), exactly in the subsequent period in which the PNRS came into force. Figure 2.2 shows graphically the result of the regression obtained. In it is shown the confidence interval of the breakpoint and the parameter c.

Although all parameters have shown p values less than 0.05, parameter c showed a slightly higher result (0.014). This parameter represents the slope of the regression after the breakpoint (β1a+β1b). Therefore, the behaviour of the regression after the breaking point was evaluated within the confidence interval obtained. The value of c, ranging from −0.0120795 to −0.0018155 (Table 2.1), as well as its graphical behaviour (Figure 2.2), confirms, with 95% confidence, a decrease in the per capita household solid waste generation rate after the breakpoint. These results show that the per capita generation of HSW in Brazil started to decline after the PNRS. It is not possible to affirm that the rupture exclusively occurred due to the PNRS, because macroeconomic aspects may have

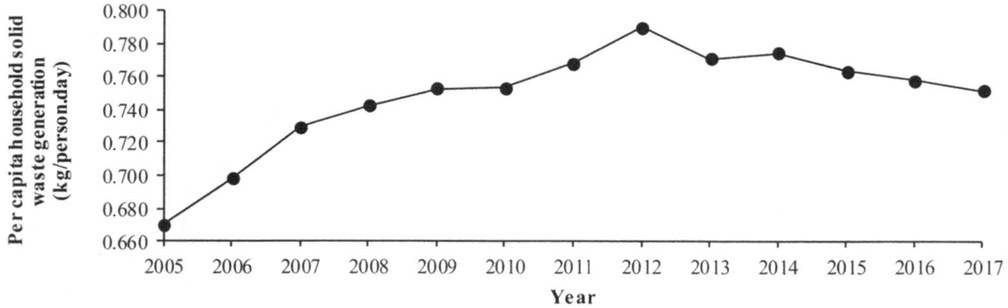

FIGURE 2.1 Per capita household solid waste generation (kg/person·day) in Brazil between 2005 and 2017.

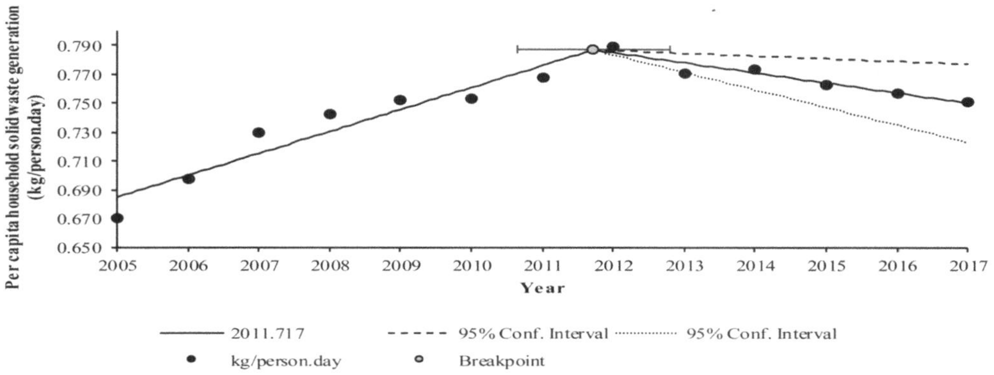

FIGURE 2.2 Segmented linear regression of per capita generation (kg/person·day) for k = 2011.717.

TABLE 2.1
Results Obtained by Segmented Linear Regression Generated Using the StataMP 13 Software for k = 2011.717

var2	Equation 1	Coef.	Std. Err	t	P>|t|	[95% Conf. Interval]	
k	k	2011.717	0.4776668	4211.55	0.000	2010.637	2012.798
a	$\beta 1a$	0.0152259	0.0017935	8.49	0.000	0.0111687	0.0192831
b	$\beta 0$	−29.84306	3.601375	−8.29	0.000	−37.98994	−21.69619
c	$\beta 1a+\beta 1b$	−0.0069475	0.0022686	−3.06	0.014	−0.0120795	−0.0018155

contributed to changes in production and consumption levels. The Brazilian gross domestic product, monitored by the Brazilian Institute of Geography and Statistics (IBGE, 2018), showed stable growth in this period, which suggests a specific study is needed to understand if a certain economic sector may have influenced this behaviour.

Once the per capita HSW generation was obtained it was possible to analyse the total mass amount of household solid waste generated in Brazil, in absolute numbers, over the last years. The per capita household solid waste generation rate has been multiplied by the total Brazilian urban population of each year (World Bank, 2018). The evolution of the total mass of household solid waste (tonnes/day) generated in Brazil is shown in Figure 2.3.

From 2012 it is noted that the generation stabilizes over the next years, despite the decline in the per capita generation. That happened due to Brazilian population growth in those same years (Table 2.2). In 2012 it was estimated that 134.48 thousand tons per day of household solid waste was generated in Brazil in urban areas, and in 2017 this number increased to 135.51 thousand tons per day (Table 2.2).

Brazil's contribution to global solid waste generation is significant, which underscores the importance of increasing management mechanisms to reduce waste generation in the coming years. Although the results showed improvement in some indexes, it is important to highlight that waste management in Brazil still lacks structural and structuring actions.

2.2.2.2 Solid Waste Generation in Brazil and Its Regional Differences

When evaluating the generation of household solid waste among the different Brazilian regions, it is possible to identify the performance of each one of them in recent years and if any region directly influenced the national standard on per capita household solid waste generation rate. Brazil has five geographical regions (Figure 2.4): The North, the Northeast, the Midwest, the Southeast and the

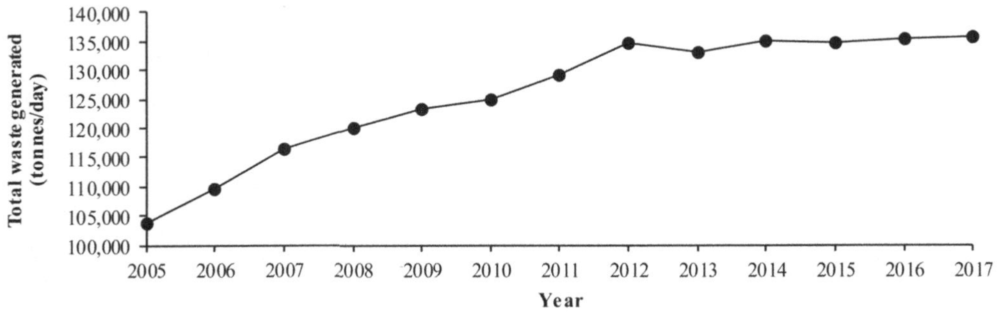

FIGURE 2.3 Daily total household solid waste generation in Brazil.

South. According to data from the last census survey carried out in the country, the Southeast region concentrates the largest population (42.13%), followed by the Northeast (27.83%), South (14.36%), North (8.32%) and Midwest (7.37%).

Figure 2.5 presents the results obtained for regional per capita household solid waste generation rates. In 2017 the regions with the highest per capita generation in Brazil were the Midwest and the Southeast. The North, Southeast and Midwest regions present a similar trend to the national standard on per capita MSW rate: A rupture break around the year 2012.

The evolution of per capita household solid waste generation in each Brazilian region is shown graphically in Figure 2.6. The figure shows the results for the years 2005, 2010, 2015 and 2017. There is a great difference between the first two years presented and a reduction of less intensity between the last two years. To assess whether this behaviour was statistically significant and whether any region directly influenced the national standard on per capita household solid waste generation rate, the existence of a breaking point in the trends of each region was also assessed.

A piecewise linear regression showed that only in the Southeast region was there a statistically significant breakpoint near to 2012 (2011,405), the period overlapping the rupture point found for the national scenario, which confirms that the Southeast Brazilian region contributed to the national trend (Table 2.3). This conclusion is consistent, since the Southeast region concentrates approximately 42% of the total Brazilian population and 46% of the Brazilian urban population (IBGE, 2010).

2.2.2.3 Mixed and Sorted Collection

MSW management is one of the major challenges for all countries in achieving sustainability. MSW management activities include its generation, separation, storage, collection, transfer, transport, processing, recovery and disposal (Zelenović Vasiljević et al., 2012; National Policy on Solid Waste, 2010). And inefficiency in these processes compromises the potential benefits of proper MSW management. Mixed collection in Brazil is when MSW is not classified and is sent to disposal (Cunha and Caixeta Filho, 2002) in landfills, controlled dumpsites or open-air dumpsites (SNIS, 2019). Therefore, mixed collection does not allow material to return to the productive chains. The PNRS determined that only non-recyclable MSW should be disposed of in landfills (National Policy on Solid Waste, 2010); thus, mixed collection should be performed for the residuals. MSW mixed collection in Brazil is undertaken by public agencies, private agencies and other agents such as philanthropic entities and large solid waste generators (SNIS, 2019).

In 2017, 98.8% of the Brazilian urban population was provided with MSW collection services (SNIS, 2019). If the rural areas are also considered, the extent of MSW collection decreases, as 57.1% of the rural population were not provided with MSW collection in 2017 (SNIS, 2019). The average amount of MSW collected in Brazil in 2017 was 0.95 kg/habitant/day, and it varied between 0.1 and 2.77 kg/habitant/day throughout the country (SNIS, 2019). The adequacy of MSW management in small municipalities remains a significant issue in Brazil because of their infrastructure

TABLE 2.2
Urban Population and Total Mass of Household Solid Waste Generated per Day in Brazil

Year	2005	2006	2007	2008	2009	2010	2011	2012	2013	2014	2015	2016	2017
Urban population (World Bank, 2018) 1,000,000 person	154.83	157.15	159.41	161.62	163.80	165.97	168.13	170.28	172.39	174.47	176.48	178.44	180.35
Per capita household solid waste generation kg/person·day	0.670	0.698	0.730	0.742	0.753	0.753	0.768	0.790	0.771	0.774	0.763	0.757	0.751
Total waste generation 1,000 tonnes/day	103.80	109.69	116.30	119.97	123.26	124.97	129.16	134.49	132.91	135.03	134.73	135.16	135.51

FIGURE 2.4 Brazilian regions.

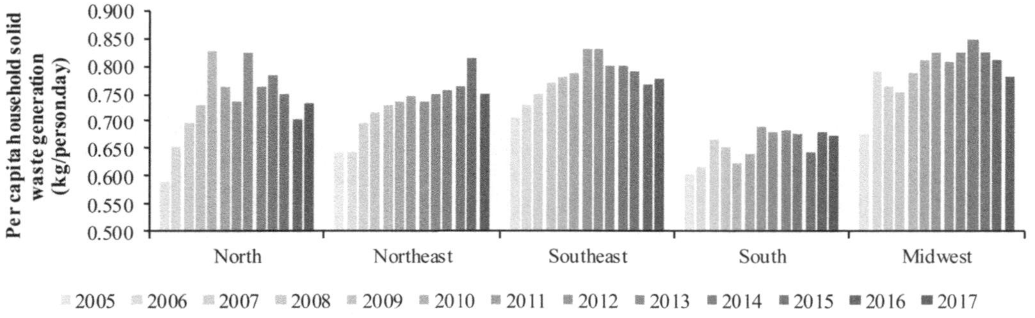

FIGURE 2.5 Evolution of per capita household solid waste generation in the Brazilian regions.

and economic limitations (Deus et al., 2020), and most of SNIS's missing information is on small municipalities (SNIS, 2019). On the other hand, selective collection is the collection of recyclable solid waste previously classified according to its composition which allows solid waste recycling and valorization.

Selective collection in Brazil is undertaken by public agencies, by private agencies and by associations or cooperatives of recyclable waste collectors performing formal selective collection services for municipalities (SNIS, 2019). In 2017, 1,256 of the Brazilian municipalities had selective collection, which represents 22.5% of the municipalities (SNIS, 2019); 2,264 of Brazilian municipalities were not served by selective collection in 2017, and 2,050 municipalities were missing this information (SNIS, 2019). Studies by the Institute of Applied Economic Research (IPEA) show that 32% of the Brazilian MSW was recyclable materials (IPEA, 2012), and they could be selectively collected to be recycled. However, the vast majority of MSW collected in Brazil is through mixed collection. Recycling is the preferred option for managing MSW, after reducing MSW generation at its source (Campos, 2014). However, most Brazilian municipalities are not provided with formal selective collection systems. Therefore, in Brazil it is necessary to promote the expansion and improvement of selective collection systems by increasing and universalizing selective waste collection services (Paes et al., 2020).

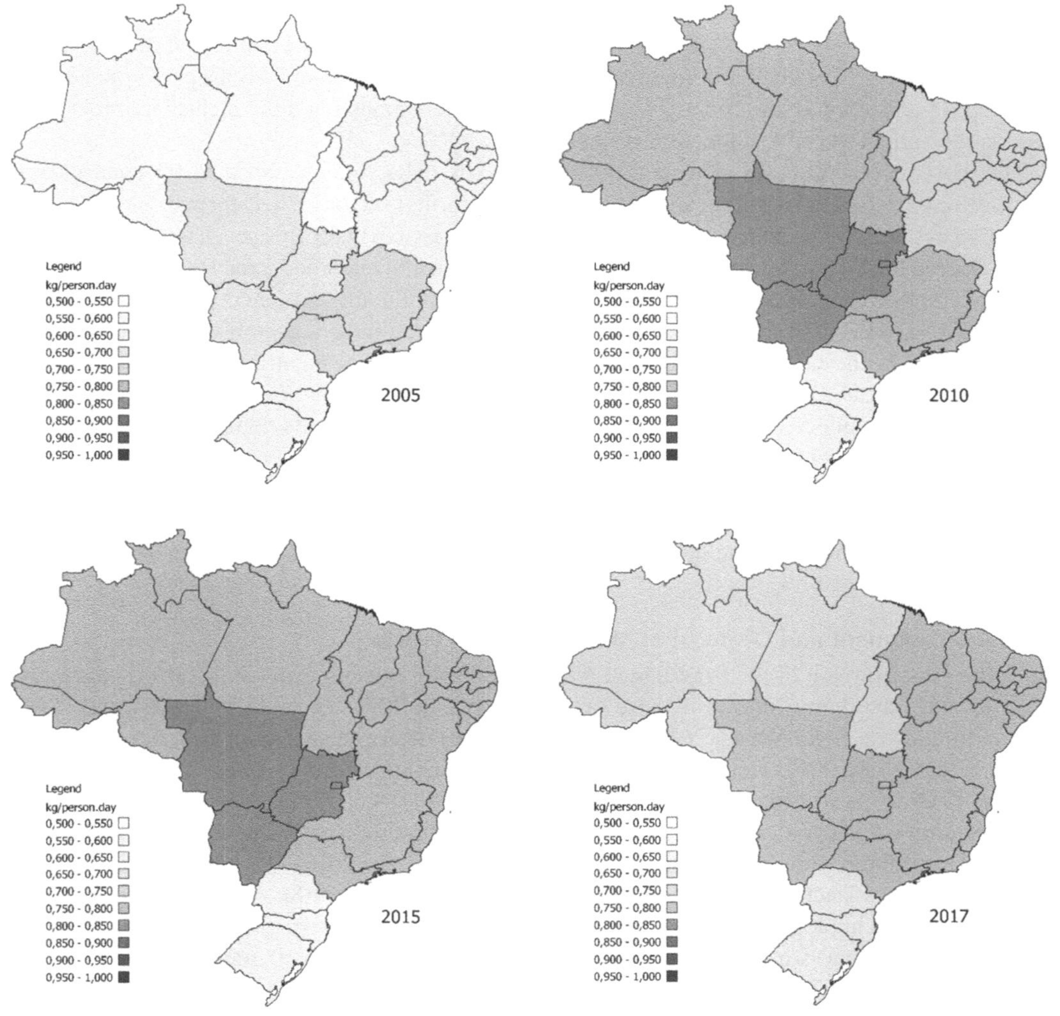

FIGURE 2.6 Evolution of per capita household solid waste generation in each Brazilian region.

Between 2005 and 2017 selective collection in Brazil collected from 0.003 to 0.009 kg of recyclable solid waste/inhabitant·day. This low collection rate is mainly due to the lack of sorted collection in many Brazilian municipalities and where the collection exists but is not available to the entire population. The relation of recyclable solid waste discharged for sorted collection to the amount of recyclable household waste generated shows that less than 10% of recyclable household

TABLE 2.3
Segmented Linear Regression Results of the per Capita Household Waste Generation in Brazilian Southeast Region

var2	Equation 1	Coef.	Std. Err	t	P>|t|	[95% Conf. Interval]	
k	k	2011.405	0.3605655	5578.47	0.000	2010.589	2012.22
a	β1a	0.0183929	0.0017696	10.39	0.000	0.0143897	0.022396
b	β0	−36.16828	3.55342	−10.18	0.000	−44.20668	−28.12989
c	β1a+β1b	−0.0108857	0.0022384	−4.86	0.001	−0.0159494	−0.005822

solid waste is made available for sorted collection, indicating a low efficacy of solid waste sorting at the household level. Private agencies performed 46% of the selective collection and the public enterprises, 17% (SNIS, 2019). Formal recyclable waste collectors' organizations performed 36% of the selective collection in the country (SNIS, 2019), which demonstrated the inclusion of recyclable waste collectors in selective collection systems.

The most successful cases of formal recyclable waste collectors are those where local governments are committed to including and organizing the informal sector in their formal selective waste collection systems, by remunerating informal sector workers fairly for this service (Gutberlet et al., 2015). Source MSW separation, by keeping materials clean and classified, enables MSW to become a resource, while the same mixed material is a potential pollutant and could be contaminated by other MSW (UNEP, 2015). Moreover, MSW mixed collection bins may emit odour pollution (Shi et al., 2020). On the other hand, after MSW classification, the concentration of odour pollution is reduced (Shi et al., 2020). Brazilian policies already determine that only non-recyclable MSW should be mixed collected and sent to disposal. Economic instruments could incentive the more sustainable management of MSW collection practices through market incentives and disincentives (UNEP, 2015). For example, a pay-as-you-throw charging system for mixed MSW would reward people for segregating their MSW on source, taxes on landfilling would discourage opting for these methods, and fiscal benefits would encourage investment in selective and recycling systems (UNEP, 2015).

2.2.2.4 Treatment and Disposal of Municipal Solid Waste

Since the establishment of the Brazilian regulations on solid waste management, efforts have been made to close the dumps and to dispose of non-recyclable MSW in landfills (Costa et al., 2018; National Policy on Solid Waste, 2010). The PNRS (2010) determined that all Brazilian dumpsites had to be closed by 2014. However, the use of landfills, controlled dumpsites* and dumpsites for disposal in Brazil is still registered.

Depending on MSW disposal, it has the potential to directly and indirectly impact the atmosphere, hydrosphere, lithosphere and the biosphere (Butt et al., 2016). The disposal of MSW in landfills generates leachate, a highly polluting liquid (Costa et al., 2018). Although there is a need to use efficient technologies for treating leachate, in Brazil the monitoring of sanitary landfills and leachate treatment performance is deficient (Costa et al., 2018). Since 2010, more than 60% of HSW has been sent to landfills in Brazil. Landfill disposal has increased over the years due to the reduction of dumpsites and controlled dumpsite disposal. Figure 2.7 and Table 2.4 show the percentage

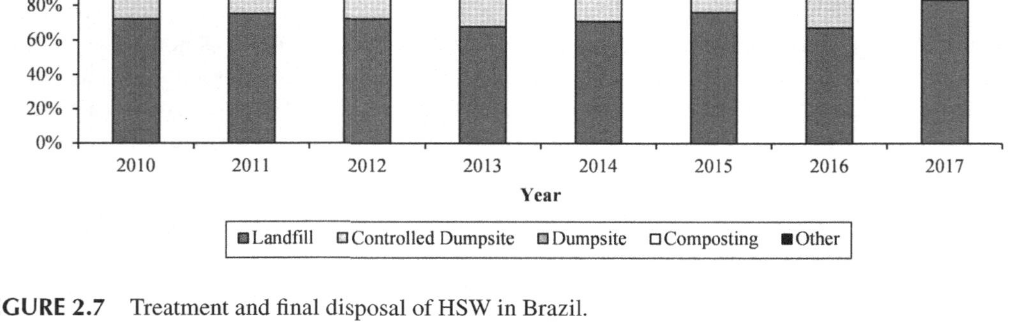

FIGURE 2.7 Treatment and final disposal of HSW in Brazil.

* Facility intended for the disposal of municipal solid waste, in which some or several types and/or objective control modalities are periodically exercised, either over the waste mass or its effluents. Thus, it is assumed that the controlled landfill is characterized by an intermediate stage between the landfill and the dumpsite (place where municipal solid waste, of all origins and nature, is simply thrown, without any kind or control over the waste and/or its effluents).

TABLE 2.4
Percentage of HSW Treatment and Final Disposal

Year	2010	2011	2012	2013	2014	2015	2016	2017
Landfill	72.07%	75.04%	72.07%	67.69%	70.65%	75.89%	66.97%	83.14%
Controlled dumpsite	23.39%	17.19%	20.41%	23.53%	16.26%	15.90%	17.69%	10.42%
Dumpsite	4.02%	7.26%	6.42%	7.88%	11.50%	6.84%	13.68%	4.68%
Composting	0.30%	0.21%	0.73%	0.22%	0.92%	0.24%	0.35%	0.30%
Other	0.23%	0.29%	0.37%	0.67%	0.68%	1.13%	1.31%	1.47%

of HSW sent to landfills, controlled dumpsites, dumpsites, composting sites and other forms of disposal or treatment.

In Brazil, the incineration of HSW is not practiced and composting has very low representativeness. Between 2010 and 2017 the amount of HSW disposed in landfills increased from 72.07% to 83.14% (Table 2.4). In the same period, HSW disposal in controlled dumpsites fell from 23.39% to 10.42%.

The overall decrease of environmentally inadequate disposal, such as dumpsites and controlled dumpsites, is possibly due to the PNRS. The year of 2014 was the deadline to ensure environmentally adequate HSW disposal in Brazil.

The amount of HSW sent for sorting in relation to the total amount of HSW generated by the municipalities analysed reached values between 1.45% and 7.56% (Figure 2.8).

2.2.2.5 Health Care Solid Waste and Construction and Demolition Solid Waste

The following treatment and final disposal units for health care solid waste and construction and demolition solid waste were analysed in Brazil: Controlled landfill, landfill, dumpsite, kiln firing, microwave or autoclave treatment unit, incineration treatment unit, health care waste specific area for disposal (ditch) in landfills or dumpsites and another unit not specified.

It is observed that there is no information on the final destination of more than 50% of health care solid waste. The final disposal methodologies adopted are landfill, controlled landfill, dump, autoclave treatment and specific ditches (Figure 2.9).

There is a lack of information on treatment and final disposal for more than 70% of C&D solid waste. Landfill was the most commonly used technology for its final disposal in 2017. Specific landfills for construction and demolition solid waste disposal have not prevailed in the national scenario (Figure 2.10).

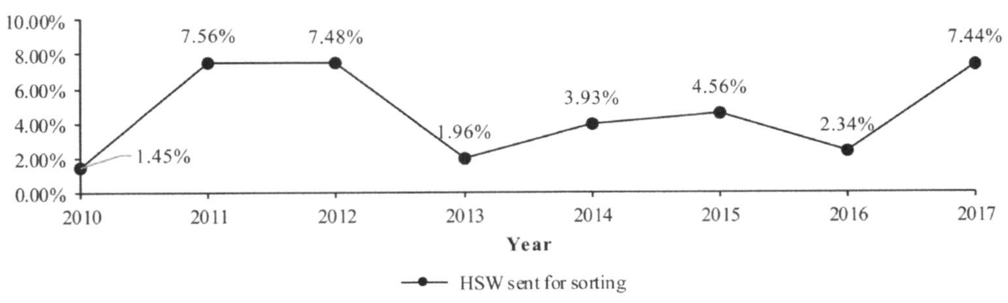

FIGURE 2.8 Household solid waste sent for sorting.

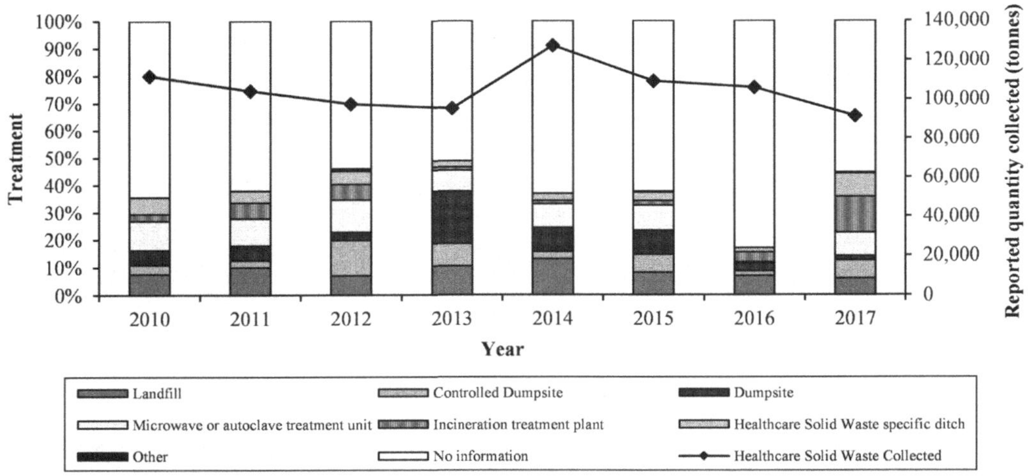

FIGURE 2.9 Health care solid waste treatment and final disposal.

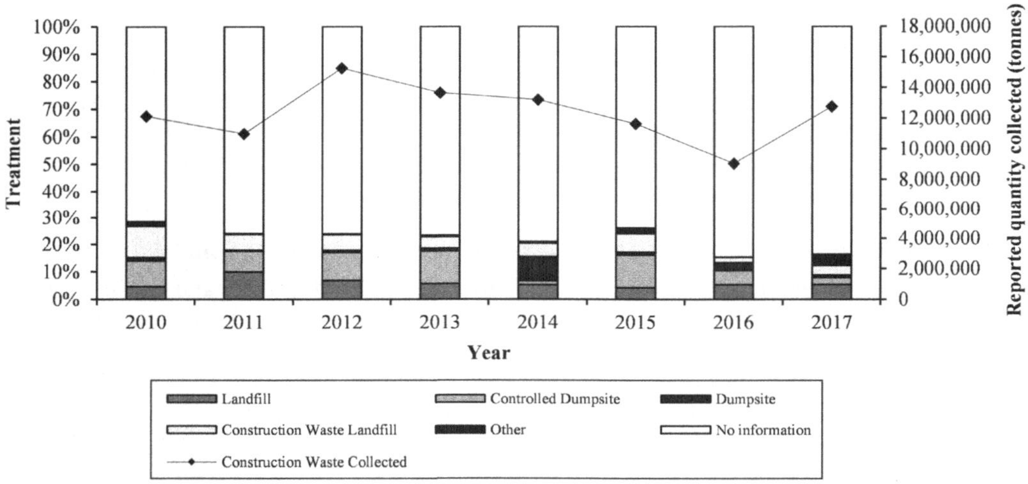

FIGURE 2.10 Treatment and final disposal of C&D solid waste.

2.3 DUMPSITES, INFORMAL WASTE MANAGEMENT AND RECYCLABLE WASTE COLLECTORS' ISSUES IN BRAZIL

MSW disposal in open-air dumpsites leads to surface water pollution through the leachate discharged from the dumps, and to groundwater pollution by the leaching of pollutants from the MSW over the years (Mohan and Joseph, 2020). Dumpsites produce air pollution, releasing carbon dioxide and methane, thus contributing as well to climate change (Mohan and Joseph, 2020). Eliminating open-air dumpsites is a priority for protecting the environment and health (UNEP, 2015). In 2000, slightly more than half of the Brazilian MSW disposal was in environmentally inadequate sites – 22.3% in controlled dumps, 30.5% in open-air dumps (IBGE, 2000). The Brazilian dumpsite Estrutural had been the largest open-air dump in Latin America for approximately 60 years (International Solid Waste Association, 2018), and the second largest open-air dump in the world (Crurvinel et al., 2019). There were about 1,200 recyclable waste collectors working at the Estrutural dump, before it was officially closed in January of 2018 (Crurvinel et al., 2019).

After new Brazilian policies and regulations on closing dumpsites, just a few of them have been effectively closed (Cruvinel el al., 2019). There were approximately 3,000 open-air dumps identified throughout Brazil in June 2017 (Zolnikov et al., 2018). It is noted that environmentally inadequate disposal in dumps decreased but it is still part of the Brazilian reality. Some of the closed dumpsites have continued to operate informally (Cruvinel el al., 2019). Informally operated dumpsites are subject to less oversight and under more hazardous conditions than during formal operation (Zolnikov et al., 2018). Recyclable waste collectors continue to work at closed dumps, being subjected to even worse conditions than before as dumpsites became officially unrecognized and unregulated (Zolnikov et al., 2018).

Moreover, closed dumps were rarely recovered (Ramos et al., 2017). Most of them were land covered and abandoned for natural vegetation growth, without an adequate environmental diagnosis and ignoring the risks to the surrounding population and environment (Ramos et al., 2017). Current improper waste management in Brazil by disposing solid waste in dumpsites causes environmental and human health impacts, through surface and groundwater contamination and air quality decay (Lima et al., 2018; Ramos et al., 2017). Moreover, recyclable waste collectors who work in this environment are exposed to various risks.

2.3.1 Informal Recyclable Waste Collection and the Recyclable Waste Collectors in Brazil

The informal recycling sector is an informal activity that involves sorting through waste to find valuable materials that can later be sold for profit (Cruvinel et al., 2019). Informal collection is undertaken by individuals or enterprises involved in the extraction of recyclable materials from MSW (Ezeah et al., 2013). The informal-sector recycling network is composed of household waste collectors, street pickers, itinerant waste buyers, MSW collection crews, dumpsite pickers and intermediate dealers (Ezeah et al., 2013). Recyclable waste collectors work on collecting, processing and commercializing the recyclable materials (Fidelis et al., 2020).

Recyclable waste collectors' informal activity can happen on the streets, informal sorting plants or even in landfills or open-air dumps (Cruvinel et al., 2019; Zolnikov et al., 2018). Recyclable waste collectors who find and sell valuable materials in dumps usually are a vulnerable population and their occupation is a means of survival (Almeida et al., 2009; Guerrero et al., 2013). Recyclable waste collectors have a low socio-economic standard of living and reside in vulnerable areas (Cruvinel et al., 2019).

Recyclable waste collectors work in unhealthy and precarious conditions; they are exposed to social, psychological, biological and environmental risks (Cruvinel et al., 2019). Each work activity exposes recyclable waste collectors to various risks. Usually, they are continuously exposed to hazardous materials, have limited access to proper protective equipment and have excessive workload (Zolnikov et al., 2018). Informal material recovery facilities have little or no institutional control; they often do not comply with legal or technical requirements or have adequate infrastructure (Campos, 2014), which contributes to an unhealthy and unsafe work environment. The exposure to contaminated water is one of the primary hazards, as it could lead to waterborne diseases (Cruvinel et al., 2019). Moreover, most recyclable waste collectors do not realize the risks they are exposed to, including serious injuries or even death (Coelho et al., 2016). Open-air dumps are the worst environment for recyclable waste collectors' activity regarding salary, health, exposure to environmental risks and psychological and social problems, as they suffer from a loss of basic human rights (Zolnikov et al., 2018).

Social exclusion increases the level of vulnerability of the recyclable waste collectors. Recyclable waste collectors suffer from social stigma, stress, depression, anxiety and dependencies on drugs and alcohol (Zolnikov et al., 2018). The informal recycling sector lies outside of state control (Ezeah et al., 2013), which makes the estimation of its performance difficult. The IPEA (2012) estimated that only 10% of the Brazilian recyclable waste collectors belonged to an association or cooperative.

The Brazilian National Movement of Waste Pickers (MNCR), which is an established national movement (Bonner, 2008) that strives for recyclable waste collectors' recognition, legitimization, self-empowerment and inclusion (Gutberlet and Baeder, 2008), estimated that there were approximately 800,000 recyclable waste collectors (MNCR, 2019). The Brazilian recycling productive chain depends on the work performance of recyclable waste collectors at the base of the system (Fidelis et al., 2020).

There has been recycling progress in Brazil, which has been based on informality and on the precarious work conditions of the recyclable waste collectors (Campos, 2014). However, informal recyclable waste collectors' sanitation and environmental services are often not recognized (IPEA, 2012). Informal recycling systems need to be integrated into formal MSW management (Ibáñez-Foréz et al., 2018; Yıldız-Geyhan et al., 2019). According to Ferronato et al. (2019), the first strategy to introduce a circular economy within developing countries is the formalization of recyclable waste collectors.

To address the informal recycling sector in Brazil, the country's policies on solid waste management have as a priority the inclusion of recyclable waste collectors into MSW management (Fidelis et al., 2020). Beyond being taken into account in solid waste management legislation, the government has established programmes for their inclusion (da Silva et al., 2019). The first programme was the Waste and Citizenship programme from 1998; later, there was the Pro-Catador programme (da Silva et al., 2019). Brazil is taking actions for recyclable waste collectors' recognition and the formalization of their activities, by supporting the creation of cooperatives and associations and by public regulation (Gutberlet, 2015).

The Brazilian recyclable waste strategies were among global best practice for integrating informal sector to waste management (Rutkowski and Rutkowski, 2015). Indeed, the PNRS has supported an improvement in the life conditions of recyclable waste collectors, but Brazilian strategies have not yet removed them from a situation of risk and vulnerability (Fidelis et al., 2020). Therefore, it is necessary in Brazil to engage and to develop recyclable waste collectors' socio-productive inclusion.

2.4 BRAZILIAN PERSPECTIVES FOR WASTE MANAGEMENT

Solid waste management in Brazil has developed in the last decades; however, the country still faces challenges for achieving sustainability in solid waste management. Brazil has not yet closed and recovered all of its dumpsites. The recycling informal sector and recyclable waste collectors' work are fundamental to Brazilian solid waste management. Thus, actions are needed to formalize and effectively include this population in solid waste management services and in society. Solid waste management regulation has improved due to the PNRS promulgation. However, municipal solid waste management is a municipal competence, and many municipalities lack the technical infrastructure and economic capital to manage their solid waste.

REFERENCES

Almeida, J.R., Elias, E.T., Magalhaes, M.A. (2009). Effect of age on the quality of life and health of waste pickers of an association in Governador Valadares, Minas Gerais, Brazil. *Revista Ciência Saúde Coletiva*, v.14, p.2169–2180. http://dx.doi.org/10.1590/S1413-81232009000600024

Bonner, C. (2008). Waste pickers without frontiers: first international and third Latin American Conference of Waste-Pickers. <https://www.wiego.org/sites/default/files/reports/files/WIEGO-WastePickers-Conf-Report-2008.pdf> Accessed on the 6th February 2020.

Brasil (2012). Plano Nacional de Resíduos Sólidos—versão preliminar. Brasília, DF. <https://www.mma.gov.br/estruturas/253/_publicacao/253_publicacao02022012041757.pdf> Accessed on the 6th February 2020.

Butt, T.E., Javadi, A.A., Nunns, M.A., Beal, C.D. (2016). Development of a conceptual framework of holistic risk assessment—landfill as a particular type of contaminated land. *Science of The Total Environment*, v.569–570, p.815–829. http://dx.doi.org/10.1016/j.scitotenv.2016.04.152

Campos, H.K.T. (2014). Recycling in Brazil: challenges and prospects. *Resources, Conservation and Recycling*, v.85, p.130–138. http://dx.doi.org/10.1016/j.resconrec.2013.10.017

Cetrulo, T.B., Marques, R.C., Cetrulo, N.M., Pinto, F.S., Moreira, R.M., Mendizábal-Cortés, A.D., Malheiros, F.T. (2018). Effectiveness of solid waste policies in developing countries: a case study in Brazil. *Journal of Cleaner Production*, v.205, p.179–187. https://doi.org/10.1016/j.jclepro.2018.09.094

Coelho, A.P.F., Beck, C.L.C., Fernandes, M.N.S., Freitas, N.Q., Prestes, F.C., Tonel, J.Z. (2016). Women waste pickers: living conditions, work, and health. *Revista Gaúcha de Enfermagem*, v.37, no.3, p.e57321. http://dx.doi.org/10.1590/1983-1447.2016.03.57321.

Costa, A.M., Alfaia, R.G.S.M., Campos, J.C. (2018). Landfill leachate treatment in Brazil—an overview. *Journal of Environmental Management*, v.232, p.110–116. https://doi.org/10.1016/j.jenvman.2018.11.006

Cruvinel, V.R.N., Zolnikov, T.R., Bashash, M., Marques, C.P.M., Scott, J.A. (2019). Waterborne diseases in waste pickers of Estrutural, Brazil, the second largest open-air dumpsite in world. *Waste Management*, v.99, p.71–78. https://doi.org/10.1016/j.wasman.2019.08.035

Cunha, V., Caixeta Filho, J.V. (2002). Gerenciamento da Coleta de Resíduos Sólidos Urbanos: Estruturação e Aplicação de Modelo Não-Linear de Programação por Metas. *Gestão & Produção*, v.9, p.143–161. http://dx.doi.org/10.1590/S0104-530X2002000200004.

Da Silva, C.L., Weins, N., Potinkara, M. (2019). Formalizing the informal? A perspective on informal waste management in the BRICS through the lens of institutional economics. *Waste Management*, v.99, p.79–89. https://doi.org/10.1016/j.wasman.2019.08.023

Deus, R.M., Mele, F.D., Bezerra, B.S., Battistelle, R.A.G. (2020). A municipal solid waste indicator for environmental impact: assessment and identification of best management practices. *Journal of Cleaner Production*, v.242, p.118433. https://doi.org/10.1016/j.jclepro.2019.118433

Ezeah, C., Fazakerley, J.A., Roberts, C.L. (2013). Emerging trends in informal sector recycling in developing and transition countries. *Waste Management*, v.33, p.2509–2519. http://dx.doi.org/10.1016/j.wasman.2013.06.020

Ferri, G.L., Chaves, G.L.D., Ribeiro, G.M. (2015). Reverse logistics network for municipal solid waste management: the inclusion of waste pickers as a Brazilian legal requirement. *Waste Management*, v.40, p.173–191. https://doi.org/10.1016/j.wasman.2015.02.036

Ferronato, N., Rada, E.C., Portillo, M.A.G., Cioca, L.I., Ragazzi, M., Torretta, V. (2019). Introduction of the circular economy within developing regions: a comparative analysis of advantages and opportunities for waste valorization. *Journal of Environmental Management*, v.230, p.366–378. https://doi.org/10.1016/j.jenvman.2018.09.095

Fidelis, R., Marco-Ferreira, A., Antunes, L.C., Komatsu, A.K. (2020) Socio-productive inclusion of scavengers in municipal solid waste management in Brazil: practices, paradigms and future prospects. *Resources, Conservation & Recycling*, v.154, no.i, p.104594. https://doi.org/10.1016/j.resconrec.2019.104594

Guarnieri, P., Cerqueira-Streit, J.A., Batista, L.C. (2020). Reverse logistics and the sectoral agreement of packaging industry in Brazil towards a transition to circular economy. *Resources, Conservation & Recycling*, v.153, p.104541. https://doi.org/10.1016/j.resconrec.2019.104541

Guerrero, L., Maas, G., Hogland, W. (2013). Solid waste management challenges for cities in developing countries. *Waste Management*, v.33, p.220–232. https://doi.org/10.1016/j.wasman.2012.09.008

Gutberlet, J. (2015). Cooperative urban mining in Brazil: collective practices in selective household waste collection and recycling. *Waste Management*, v.45, p.22–31. http://dx.doi.org/10.1016/j.wasman.2015.06.023

Gutberlet, Jutta, Baeder, Angela M. (2008). Informal recycling and occupational health in Santo André, Brazil. *International Journal of Environmental Health Research*, v.18, no.1, p.1–15. https://doi.org/10.1080/09603120701844258

Ibáñez-Forés, V., Coutinho-Nóbrega, C., Bovea, M.D., Mello-Silva, C., Lessa-Feitosa-Virgolino, J. (2018). Influence of implementing selective collection on municipal waste management systems in developing countries: a Brazilian case study. *Resources, Conservation & Recycling*, v.134, p.100–111. https://doi.org/10.1016/j.resconrec.2017.12.027

IBGE—Brazilian Institute of Geography and Statistics (2000). Pesquisa Nacional de Saneamento Básico, Brazil. <https://servicodados.ibge.gov.br/Download/Download.ashx?http=1&u=biblioteca.ibge.gov.br/visualizacao/livros/liv45.pdf> Accessed on the 6th February 2020.

IBGE—Brazilian Institute of Geography and Statistics (2010). Censo Demográfico de 2010. <https://www.ibge.gov.br/estatisticas/sociais/populacao/9662-censo-demografico-2010.html?=&t=series-historicas>.

IBGE—Brazilian Institute of Geography and Statistics (2017). Pesquisa de Informações Básicas Municipais—MUNIC. <https://biblioteca.ibge.gov.br/visualizacao/livros/liv101595.pdf> Accessed on the 6th February 2020.

IBGE—Brazilian Institute of Geography and Statistics (2018). Sistema de Constas Nacionais. <https://www.ibge.gov.br/estatisticas-novoportal/economicas/contas-nacionais/9052-sistema-de-contas-nacionais-brasil.html?=&t=resultados>.

International Solid Waste Association (2018). A visit to estructural: a major success. <http://www.iswa.org/home/news/news-detail/browse/11/article/a-visit-to-estrutural-closedumpsites-first-major-success/109/> Accessed on the 6th February 2020.

IPEA—Instituto de Pesquisa Econômica Aplicada (2012). Diagnóstico dos Resíduos Sólidos Urbanos. Brasilia, p.82. 2012. <http://repositorio.ipea.gov.br/bitstream/11058/7633/1/RP_Diagn%c3%b3stico_2013.pdf> Accessed on the 11th November 2019.

Lei 9,065 (1998). Environmental crimes law. <http://www.planalto.gov.br/ccivil_03/leis/L9605.htm> Accessed on the 6th February 2019.

Lei 9,795 (1999). National policy on environmental education. Brasília, DF. <http://www.planalto.gov.br/ccivil_03/Leis/L9795.htm> Accessed on the 6th February 2019.

Lei 11,445 (2007). Federal law on basic sanitation. Brasília, DF. <http://www.planalto.gov.br/ccivil_03/_ato2007-2010/2007/lei/l11445.htm> Accessed on the 6th February 2019.

Lima, P.M., Colvero, D.A., Gomes, A.P., Wenzel, H., Schalch, V., Cimpan, C. (2018). Environmental assessment of existing and alternative options for management of municipal solid waste in Brazil. *Waste Management*, v.78, p.857–870. https://doi.org/10.1016/j.wasman.2018.07.007

Marino, A.L., Chaves G.L.D., Junior J.L.S. (2018). Do Brazilian municipalities have the technical capacity to implement waste management at the local level? *Journal of Cleaner Production*, 188, 378–386. https://doi.org/10.1016/j.jclepro.2018.03.311

Ministry of the Environment (2020). Leis. <https://sinir.gov.br/legislacao> Accessed on the 6th February 2020.

MNCR—Movimento Nacional de Catadores(as) de Material Reciclável (2019). Quantos Catadores existem em atividade no Brasil? <http://www.mncr.org.br/sobre-o-mncr/duvidas-frequentes/quantos-catadores-existem-em-atividade-no-brasil> Accessed on the 6th February 2020.

Mohan S., Joseph C.P. (2020) Biomining: an innovative and practical solution for reclamation of open dumpsite. In: Kalamdhad A. (eds) *Recent Developments in Waste Management. Lecture Notes in Civil Engineering*, vol 57. Springer, Singapore.

National Policy on Solid Waste (2010). Lei n° 12.305 de 2010. Política Nacional de Resíduos Sólidos. Brasília, DF. <http://www.planalto.gov.br/ccivil_03/_Ato2007-2010/2010/Lei/L12305.htm> Accessed on the 11th November 2019.

Neto, P.N., Moreira, T.A. (2010). Política nacional de resíduos sólidos—reflexões a cerca do novo marco regulatório nacional. *Revista Brasileira de Ciências Ambientais*, v.15, p.10–19.

Paes, M.X., Medeiros, G.A., Mancini, S.D., Bortoleto, A.P., Oliveira, J.A.P., Kulay, L.A. (2020). Municipal solid waste management: integrated analysis of environmental and economic indicators based on life cycle assessment. *Journal of Cleaner Production*, v. 254, p.119848. https://doi.org/10.1016/j.jclepro.2019.119848

Pincelli, I.P., Meireles, S., de Castilhos Jr., A.B. (2019). Socio-productive inclusion of waste pickers on segregated solid waste collection in Brazilian universities as an instrument for sustainability promotion. In: Leal Filho, W. and Bardi, U. (eds) *Sustainability on University Campuses: learning, skills building and best practices*. Springer. https://doi.org/10.1007/978-3-030-15864-4_18

Ramos, N.F., Gomes, J.C., Castilhos Jr., A.B., Gourdon, R. (2017). Desenvolvimento de ferramenta para diagnóstico ambiental de lixões de resíduos sólidos urbanos no Brasil. *Engenharia Sanitária e Ambiental*, v.22, p.1233–1241. https://doi.org/10.1590/S1413-41522017165817

Rutkowski, J.E., Rutkowski, E.W (2015). Expanding worldwide urban solid waste recycling: the Brazilian social technology in WPs inclusion. *Waste Management & Research*, v.33, p.1084–1093. https://doi.org/10.1177/0734242X15607424

Serra, T.B. (2015). *Política de resíduos sólidos: gestão econômica, responsável e ambientalmente adequada.* São Paulo: Verbatim.

Shi, X., Zheng, G., Shao, Z., Gao, D. (2020). Effect of source-classified and mixed collection from residential household waste bins on the emission characteristics of volatile organic compounds. *Science of the Total Environment*, v.707, p.135478. https://doi.org/10.1016/j.scitotenv.2019.135478

SNIS—Sistema Nacional de Informações de Saneamento (2019). Diagnóstico do Manejo de Resíduos Sólidos Urbanos—2017. Ministério das Cidades. Brasília, p.173. 2019. <http://www.snis.gov.br/diagnostico-residuos-solidos/diagnostico-rs-2017> Accessed on the 11th November 2019.

United Nations Environment Programme—UNEP (2015). Global waste management outlook. International Solid Waste Association General Secretariat. United Nations: Tara Cannon, 2015. <https://www.unclearn.org/sites/default/files/inventory/unep23092015.pdf>. Accessed on the 11th November 2019.

World Bank (2018). The World Bank data. <https://data.worldbank.org/indicator/SP.URB.TOTL?locations=BR>.

Yıldız-Geyhan, E., Yilan, G., Altun-Çiftçioğlu,G.A., Kadırgan, M.A.N. (2019). Environmental and social life cycle sustainability assessment of different packaging waste collection systems. *Resources, Conservation & Recycling*, v.143, p.119–132. https://doi.org/10.1016/j.resconrec.2018.12.028

Zelenović Vasiljević, T., Srdjević, Z., Bajčetić, R., Vojinović Miloradov, M. (2012) GIS and the analytic hierarchy process for regional landfill site selection in transitional countries: a case study from Serbia. *Environmental Management*, v.49, p.445–458. https://doi.org/10.1007/s00267-011-9792-3

Zolnikov, T.R., Silva, R.C., Tuesta, A.A., Marques, C.P., Cruvinel, V.R.N. (2018). Ineffective waste site closures in Brazil: a systematic review on continuing health conditions and occupational hazards of waste collectors. *Waste Management*, v.80, p.26–39. https://doi.org/10.1016/j.wasman.2018.08.047

3 Economic Estimation for a Glass Waste Recycling Facility in Johannesburg, South Africa

Olusola Olaitan Ayeleru, Peter Apata Olubambi, Felix Ndubisi and Freeman Ntuli

CONTENTS

3.1 INTRODUCTION

Glass has been used for several millennia. It was first used by the Egyptians around the 15th century BC (Isa, 2008). Precisely, glass has been utilized for over 9,000 years owing to its availability and cost effectiveness. Glass has unique mechanical, chemical, thermal and optical properties, and these play a significant role when it is to be selected for an application. Studies have shown that glass is one of the materials that was synthesized at the beginning of time. The periods were specifically in the middle of 27 BC and AD 14 and later about AD 100 (Ferguson, 2006; Groninger, 2014; Roberts, 2004; U.S. Department of Energy Office of Industrial Technologies, 2002). Glass, in its original state and transparent form, is made from sand, silica, soda, ash and calcium carbonate ($CaCO_3$). Glass is an amorphous material whose structure is in-between crystalline material and liquid, and at high temperatures, glass becomes liquid. The atoms of glass are loosed and this makes it easier for them to move around freely (IEEE GlobalSpec, 2017). The atoms of glass rearrange themselves whenever the liquid is cooled down. Further, when the temperature becomes very low to a point that is referred to as fictive temperature, the atoms do not rearrange themselves (IEEE GlobalSpec, 2017). Glass can be made as strong as steel and can be processed in such a way that it will be as fragile as paper; it all depends on the consumers' requirements and specifications. The properties of glass are best determined by its composition (IEEE GlobalSpec, 2017).

Moreover, glass is produced by melting the raw materials in a furnace at a high temperature of 1500°C. The melted mixture is then cooled for solidification without crystallization (Safiuddin et al., 2010). The mixture is converted into molten form in the presence of heat and thereafter placed in a mould, and air is blown into the mould which subsequently gives shape to the container (Clean Up, 2009). Glass has been known to be hard, brittle and transparent. Glass is now part and parcel of people's lifestyles since it is widely used as a packaging material and for automobile windows, structural buildings and other applications (Care, 1999). The total quantity of glass production globally was estimated as ~115 million tons in 2007. It has been established that over 90% of glass products are sold to other sectors. In other words, glass production sectors depend on other sectors like building construction, car manufacturing and food and beverage companies for usage of their products (Wintour, 2015).

Glass takes about 1 million years to degrade at the landfill site (LS); it only takes between 8 and 12 weeks for it to be recycled. Recycling of glass is the conversion of waste glass into valuable products. When glass is recycled, the recovery of the resources is 100%, and this can be repeated several times without loss of quality or composition and energy is saved. The amount of energy saved is capable of lighting a 100-watt bulb for 4 hours (Department of Conservation: Division of Recycling, 2011; Ling et al., 2013; RCBC (Recycling Council of British Columbia), 2004; Shayan & Xu, 2004; Sustainability Victoria, 2013). A recent study has also shown that a computer can be powered for 25 minutes, a colour TV for 20 minutes or a washing machine for 10 minutes when a glass container is recycled (The Glass Recycling Company, 2014). Additionally, when a ton of glass is recycled, the extraction of about ~1.2 tons of virgin materials is avoided. When recycling takes place, CO_2 is minimized, and therefore there is a reduction in climate change. Moreover, for a ton of glass made from recyclables, about 670 kg of CO_2 is saved from entering into the atmosphere (The Glass Recycling Company, 2014). A recent study has shown that about 315 kg of CO_2 could be saved when a ton of glass is recycled (Ling et al., 2013). Glass waste recycling is a successful step to reduce the extraction of virgin materials so that sustainability can be attained. In some parts of the world, about 85% of glass waste is converted into new products. When a ton of this glass waste is recycled, emission of greenhouse gases (GHG) to the environment is reduced by 200 kilograms (kg) (Mogensen GmbH & Co. KG, 2008). Studies on the recycling of glass waste with respect to greenhouse gas emissions are scarce. This is not unconnected to the fact that the energy required in the glass waste recycling process is quite too small (Larsen et al., 2009). Additionally, a study has proven recycling as the best option compared to landfilling or incineration methods of waste management (WM) especially when it comes to the threat to the environment (Larsen et al., 2009). It also been established that the global warming factors (GWFs) for glass waste recycling into cullet have been estimated to be from –505 kg CO_2-eq. tonne^{-1} to –416 kg CO_2-eq. tonne^{-1}. This clearly shows that a significant amount of GHG emissions is saved in glass waste recycling processes (Larsen et al., 2009).

In addition, a study has shown that about ~845,000 tons of glass waste which were comprised of jars and bottles were generated in South Africa in 2014. The amount recycled was only 40%. Furthermore, about 3 million bottles were generated as waste and were sent to the landfills (GreenCape, 2016). Another study has revealed that ~1.5 million tons of glass are sent to landfills every year and when this glass is deposited at the landfills, it won't degrade for over a million years, thereby causing environmental problems (Bolden et al., 2013). Glass waste needs to be sorted from other waste streams to get quality recyclable glasses. This can be achieved at a material recovery facility (MRF) and may be beneficiated for further processing (Sustainability Victoria, 2013). For glass waste recycling to be viable and for waste glasses to be completely diverted from landfills, continuous education and campaigns on the benefits of glass waste recycling become paramount. The initiatives can begin from places like schools, market and commercial places, offices and relaxation centres (Hudson, 2012). Moreover, the establishment of drop-off centres for the collection of source-separated glass waste could also assist in getting high-quality waste glasses for recycling, and the sourced-separated waste glasses would serve as input in recycling plants (Lozier & TerMeer, 2011).

Hence, it is crucial to estimate the cost of setting up a glass waste recycling facility, and to achieve this, a parametric estimate may be employed. A parametric estimate is one of the methods of evaluating the cost of a project, and it is established based on several project-based factors (Washington State Department of Transportation, 2015). Policy makers and waste managers often centre their decisions on project estimates. Therefore, this aspect of project evaluation becomes very vital in engineering economics (Washington State Department of Transportation, 2015). To the best of our knowledge, little or no study has been carried out on an economic assessment of setting up a glass waste recycling facility in South Africa.

Therefore, this paper aims to estimate the cost of setting up a glass waste recycling facility as a baseline for the proposition for the complete diversion of glass waste from landfill sites (LSs) to recycling facilities in South Africa.

3.2 DESCRIPTION OF THE STUDY AREA

Florida, South Africa, came into being after Roodepoort was founded. Its geographical coordinates are 26° 10' 0" South, 27° 55' 0" East. The township was developed around 1888. Florida was formed through the voyage of a surveyor known as William Pritchard. The surveyor was on a mission to search for gold in the area between 1884 and 1886. He had previously mapped out Roodepoort during his search, and there he discovered a valley where there was a plethora of flowers. Thus, he was so excited by the surplus of flowers that he saw at the valley, and in view of that, he decided to name the valley 'Florida' as flowers are called in Spanish. Florida was majorly an area where the heads of mines who were generally white-coloured lived. Most of the unskilled workers were Africans and coloured races. These workers lived in areas like Roodepoort and Maraisburg which are not very far from Florida. As the township began to develop, railway workers also began to live in the area. During the pre-independence period of South Africa, the township was majorly dominated by the whites and the level of apartheid in that area was extremely high. But as soon as South Africa gained independence in 1994, the level of apartheid gradually began to be weakened and other races began to move into the area. At present, the distribution of the population of the township is about 10% Africans and 40–50% are coloured race (Ayeleru et al., 2016b; Maplandia.com, 2005–2016).

3.3 SOURCES FOR EQUIPMENT COSTS

All equipment costs were locally sourced in Johannesburg, South Africa. The main objective of this project is to make use of locally fabricated machinery which can be easily accessed and will be readily available and whose spare parts can be sourced locally. The reason for sourcing equipment locally was to exclude shipping and forwarding costs, the payment of custom duties and other importation-related costs.

3.4 PROCESSES FOR THE RECYCLING OF GLASS WASTE

Glass waste is separated into different colours. This is usually done from the point of generation or at the MRF. Coloured glass can only be recycled back to its original form. Therefore, coloured glass is separated from the clear glass or left together. Afterwards, contaminants like plastic lids and metals are removed, and glass waste is then crushed into cullet. The cullet is subjected to heat and any paper left on the glass is burnt off. The recycled cullet is subjected to a lower temperature, less than what is required for new glass processing. This is to save a considerable amount of energy. The resulting liquid that was subjected to heat is poured into moulds and a different kind of shape is formed. Sometimes, sand, soda ash and limestones are added to the cullet and a new shape is formed (WasteCare Corporation, 1997–2013). Glass waste is found in commercial and industrial

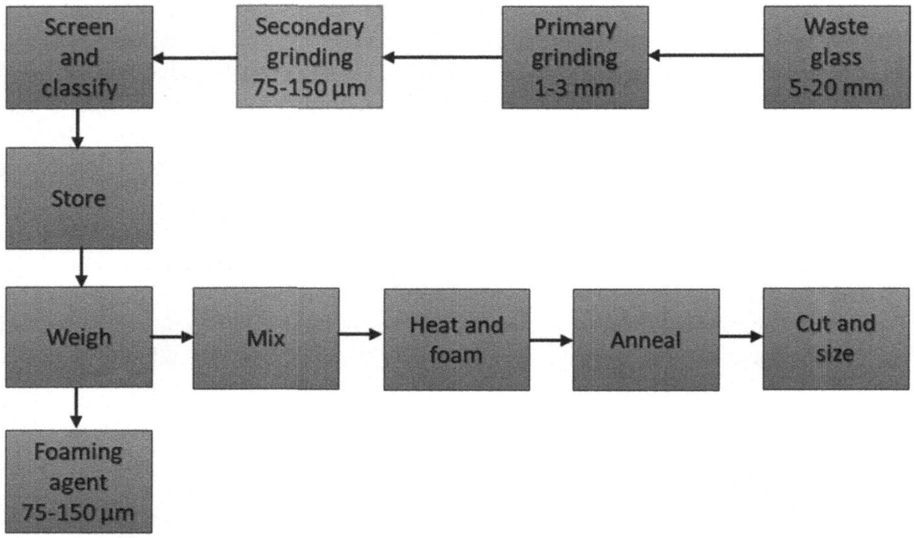

FIGURE 3.1 Flow of glass waste through the recycling process. (Redrawn from El-Haggar, 2007.)

(C&I) waste, in construction and demolition (C&D) waste and in municipal solid waste (MSW) in general (Sustainability Victoria, 2013). Figure 3.1 shows a flow chart for glass waste recycling.

3.5 METHODOLOGY

Two waste composition studies were conducted during the winter and summer seasons at the Marie Louise LS, Johannesburg, South Africa. Data were obtained from the waste characterization activities. From the results obtained, about 50 tons of glass waste are disposed of in the LS every day. The capacity utilization of the glass waste recycling facility was 95%. Hence, the economic evaluation in this study was established on 47 tons of glass wastes based on the capacity utilization. These tons of glass waste were founded on two of the services offered by the municipality. The services are round collected refuse (RCR) compacted waste and daily refuse (DR) non-compacted waste. The RCR comes from households in formal residential areas and businesses. They are regularly collected weekly. The DR originates from hotels, restaurants, fast food joints, butcher shops and street sweeping. It is collected daily to avoid unpleasant odour that may result from its decay. This can present a danger to the environment and public health. Based on the historical data of six years of annual reports of Pikitup Johannesburg (State Owned Company) (SOC) Limited (the municipality), it was established that a total ~1,114 tons of waste are received daily at the Marie Louise, and from these, the quantity of glass waste based on the waste characterization results was 49.94 tons per day (Ayeleru et al., 2018; Pikitup (Pikitup Johannesburg SOC Ltd), 2009/10, 2010/11, 2011/12, 2012/13, 2013/14, 2015/16).

3.6 RESULT

3.6.1 Glass Waste Composition by Percentage Weight

The composition of the glass waste based on the percentage weight of each of the waste streams received at the disposal facility is represented graphically as shown in Figure 3.2. Figure 3.2a and Figure 3.2b show the distribution in weight percent of glass waste during the summer and winter months for the DR. Moreover, Figure 3.2c and Figure 3.2d show the percentage weight compositions in weight during the summer and winter months for the RCR.

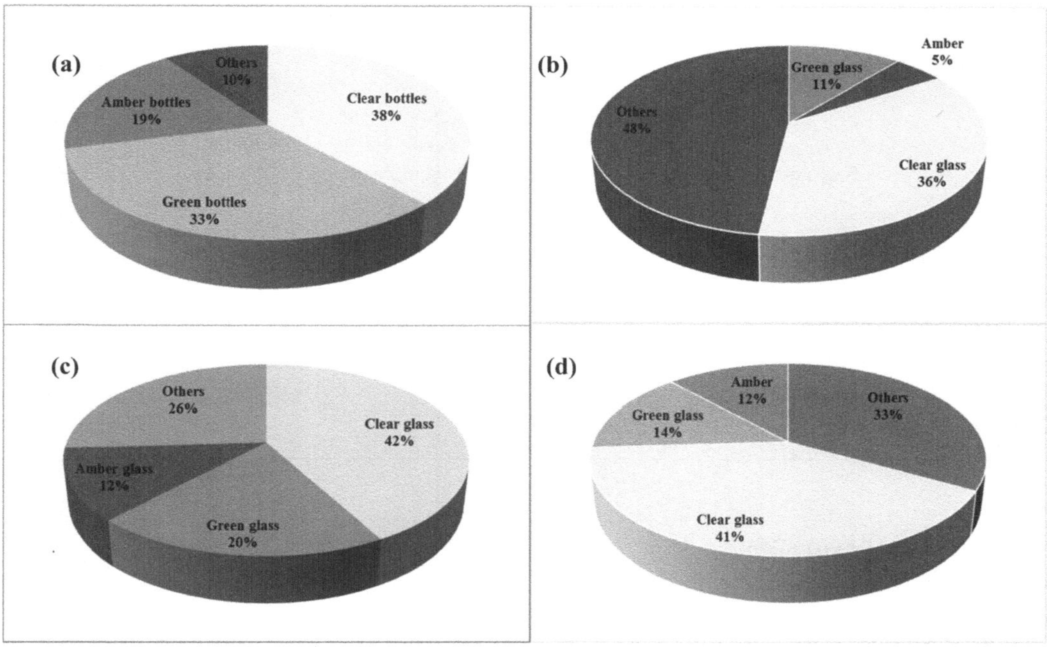

FIGURE 3.2 (a, b) Daily refuse glass waste during summer and winter; (c, d) round collected refuse glass waste during summer and winter.

3.6.2 Factors Taken into Consideration for the Cost Estimation

A few factors were considered in setting up a recycling facility. These factors include the purchased equipment cost, direct costs, indirect costs, working capital (WC), fixed capital investment (FCI) and the total capital investment (TCI). The unit/currency used for the economic analysis is in Rand (R). The current exchange rate during the economic analysis estimation was R 17.26 = USD 1.

3.6.2.1 Purchased Equipment Cost

The purchased equipment cost (PCE) is the baseline for estimating a capital investment. PCE comprises the prices of equipment, methods to adjust the capacity of equipment and methods to estimate the ancillary equipment (Peters et al., 2003). The PCE (Table 3.1) for this study was R 1,340,000.

3.6.2.2 Direct Costs (DC)

The direct cost (DC) of setting up a recycling facility involves several factors, and quite a lot of parameters are considered. These parameters include the purchased equipment cost; service facilities and yard improvement; land; installation, insulation and painting; building and electrical installation (R. K. Sinnott, 1999). The DC (Table 3.2) for this study was estimated as R 2,720,200.

3.6.2.3 Indirect Costs (IDC)

The indirect cost (IDC) of setting up a capital project also has some parameters to be considered. These parameters are the cost incurred in the construction of the facility. The parameters include engineering and supervision; construction expenses and contractor's fee; contingency and legal expenses (Peters et al., 2003). The IDC (Table 3.3) for this study was estimated to be R 843,262.

3.6.2.4 Total Capital Investment (TCI)

Total capital investment (TCI) is the sum of money required to commence the operation of a plant. Part of the operations include the purchase and installation of equipment, land, building, money

TABLE 3.1
Purchased Equipment Cost for Glass Recycling

Process	Equipment/Capacity	Quantity	Number of Personnel per Equipment	Total Number of Workers	Cost (Rand)
Glass waste recycling	Glass crushers (3 tons/hr)	2	3	6	200000
	Front-end loader	1	2	2	400000
	Weigh bridge	1	3	3	350000
	Weighing scales with capacity of 3000 kg	10	1	10	190000
	Ancillary equipment				200000
Total				**21**	**1340000**

TABLE 3.2
Total Direct Costs and Factors

	Factors Used	Rand/Year
Building	20% of PEC	268000
Land	8% of PEC	107200
Electrical installed	10% of PEC	134000
Service facilities and yard improvements	40% of PEC	536000
Installation, insulation and painting	25% of PEC	335000
Purchase equipment cost		1340000
Total		**2720200**

TABLE 3.3
Total Indirect Costs and Factors

	Factors Used	Rand/Year
Engineering and supervision	10% of DC	272020
Construction expense and contractor's fee	10% of DC	272020
Contingency	7% of DC	190414
Legal expenses	4% of DC	108808
Total		**843262**

earmarked for the purchase of raw material, the payment of wages and salaries and other overhead costs. TCI is the sum of FCI and working capital (WC) (Choy, Porter, Hui, & McKay, 2004). The WC of a capital investment comprises the amount of money earmarked for the purchase of raw materials and supplies, both finished products and semi-finished products, account receivable, payment of salaries, taxes, wages and other operating expenses (R. Sinnott & Towler, 2009). The TCI (Table 3.4) for this study was given as R 4,097,981.30. The WC was taken as 13% of TCI. The FCI was obtained as the sum of DC and IDC and this is also shown in Table 3.4.

TABLE 3.4
Total Capital Investment and Factors

	Percent	Amount (Rand)
Direct cost (including PEC)	66.4	2720200
Indirect cost	20.6	843262
Working capital	13	534519.3
Total capital investment	**100**	**4097981.3**

3.7 DISCUSSION

3.7.1 Glass Waste Compositions

Waste composition studies turn out to be vastly central due to the subsequent justifications, and these include estimation of the possibility of material reclamation, naming of the source of waste streams, acceleration of planning of waste processing equipment, evaluation of the physical, chemical and thermal properties of MSW and supervising of the conformity of the activities of landfill sites in line with national and international standards (Abd Alqader & Hamad, 2012; Ayeleru et al., 2016a, 2018; Gidarakos et al., 2006). From the results of the analysis, it was observed that more clear glass waste was consumed compared to coloured glass waste during the two seasons. The data obtained showed that about 38% clear glass was generated during the summer while the amount generated during the winter was 36% for the DR. Similarly, about 42% clear glass was generated during the summer while the amount generated during the winter was 41% for the RCR. This clearly shows that people maintained the same lifestyles and consumption patterns during the two seasons. Based on these percentages of clear glass waste, we can conclude that more raw materials will be readily available for diverse products since coloured waste glasses can only be recycled to their original form but clear waste glasses could be used for a variety of products including drinking cups, glass fibre and cullet, table tops and counters, benches, wall and floor tiles, etc. Figure 3.3 shows some of the products that can be made from recycled glasses. When glass waste is converted into cullet, it is then pounded, sorted out, washed and made ready to be combined with soda ash and sand for the preparation of new glasses.

3.7.2 Economic Estimation of Glass Waste Recycling Facility

The economic assessment of the glass waste recycling plant was based on 47 tons of glass waste (95% capacity utilization of the recycling facility). The TCI for the recycling facility was found to be R 4,097,981.30, the FCI was R 3,563,462.00, the WC was R 534,519.30, the DC was R 2,720,200, the IDC was R 843,262 and the PEC was R 1,340,000. The interest rate was assumed to be around 12% since the capital would be obtained from the banks or financial institutions. It was envisaged that about 21 potential downstream jobs would be created in this project for unemployed South African youths. The rate of unemployment has been rising steadily in the country coupled with constant population growth in cities. The rapid population growth has also been contributing exponentially to acute MSW generation. The recycling of glass waste via the proposed recycling facility would eventually lead to sustainability in the waste management sector since the recycled glasses would serve as input in the production process and the need for virgin materials would be reduced (Ayeleru et al., 2020). It has also been established from the literature that GHG is saved in glass waste recycling by way of WM. Similarly, the establishment of a recycling facility would also serve as a source of revenue for the government. Taking all these factors into consideration, it was concluded that the establishment of a glass waste recycling facility in the City of Johannesburg would contribute positively to the socio-economic development of the Republic of South Africa (RSA).

FIGURE 3.3 (a) Tables; (b) benches and (c) wall and floor tiles obtained from recycled glasses. (Images obtained with permission from Cruncher Glass Recycling Machine Manufacturers.)

3.8 CONCLUSION

An economic evaluation of a glass waste recycling plant has been successfully carried out based on the historical data and waste composition data. The results obtained showed the viability of the project and its contributions to the socio-economic development of the country. Establishing the glass waste recycling facility in Florida Township would assist in regularly feeding the recycling facility with raw materials because of its proximity to the Marie Louise LS which is one of the central factors considered when siting a firm. The introduction of a recycling facility would attract more infrastructure to the area since the company may put in place a community social responsibility (CSR) as a means of developing the area.

ACKNOWLEDGEMENTS

The authors wish to appreciate the management of **Pikitup Johannesburg (SOC) Limited** for their support during the waste composition activities. Special thanks also go to the waste pickers (reclaimers) at the site for the total support. Appreciation goes to the **Department of Chemical Engineering, University of Johannesburg**, for providing a platform and a conducive environment for research and **Cruncher Glass Recycling Machine Manufacturers** for granting us the permission to use their images in this work.

REFERENCES

Abd Alqader, A., & Hamad, J. (2012). Municipal solid waste composition determination supporting the integrated solid waste management in Gaza strip. *International Journal of Environmental Science and Development*, *3*(2), 172.

Ayeleru, O. O., Dlova, S., Akinribide, O. J., Ntuli, F., Kupolati, W. K., Marina, P. F., ... Olubambi, P. A. (2020). Challenges of plastic waste generation and management in sub-Saharan Africa: a review. *Waste Management*, *110*, 24–42.

Ayeleru, O. O., Ntuli, F., & Mbohwa, C. (2016a). *Municipal Solid Waste Composition Determination in the City of Johannesburg*. Paper presented at the Proceedings of the World Congress on Engineering and Computer Science, San Franscisco.

Ayeleru, O. O., Ntuli, F., & Mbohwa, C. (2016b). *Utilization of Organic Fraction of Municipal Solid Waste (OFMSW) as Compost: A Case Study of Florida, South Africa*. Paper presented at the Proceedings of the World Congress on Engineering and Computer Science.

Ayeleru, O. O., Okonta, F. N., & Ntuli, F. (2018). Municipal solid waste generation and characterization in the City of Johannesburg: A pathway for the implementation of zero waste. *Waste Management*, *79*, 87–97.

Bolden, J., Abu-Lebdeh, T., & Fini, E. (2013). Utilization of recycled and waste materials in various construction applications. *American Journal of Environmental Sciences*, *9*(1), 14.

Care. (1999). *Glass Recycling: An Automotive Perspective*. Retrieved from http://www.seas.columbia.edu/earth/RRC/documents/glass_recycling_an_automotive_perspection.pdf

Choy, K. K. H., Porter, J. F., Hui, C., & McKay, G. (2004). Process design and feasibility study for small scale MSW gasification. *Chemical Engineering Journal*, *105*(1), 31–41.

Clean Up. (2009). *Recycle Glass—Preserve the Environment* [Press release]. Retrieved from https://www.cleanup.org.au/PDF/au/cua_glass_recycling_factsheet_final.pdf

Department of Conservation: Division of Recycling. (2011). *Recycling: Facts and Fun*. Retrieved from http://www.calrecycle.ca.gov/Publications/Documents/BevContainer%5C2011032.pdf

El-Haggar, D. S. M. (2007). Sustainability of municipal solid waste management. In *Sustainable Industrial Design and Waste Management* (pp. 149–196). Academic Press.

Ferguson, C. (2006). *Historical Introduction to the Development of Material Science and Engineering as a Teaching Discipline*. Retrieved from the UKCME at http://www.materials.ac.uk/pub/Materials-History-Intro.pdf

Gidarakos, E., Havas, G., & Ntzamilis, P. (2006). Municipal solid waste composition determination supporting the integrated solid waste management system in the island of Crete. *Waste Management*, *26*(6), 668–679.

GreenCape. (2016). *Waste Economy—2016 Market Intelligence Report*. Retrieved from South Africa: http://greencape.co.za/assets/GreenCape-Waste-MIR-2016.pdf

Groninger. (2014). *Next Level QbD: Processing Glass Containers: Stress-Free, Contact-Less Transport a Need, Trend or Just a New Hype?* [Press release]. Retrieved from http://www.pharmaceutical-tech.com/pharmaadmin/upload/productspec_pdf/1430907441groninger_pressRelease_glass-breakage.pdf

Hudson, D. (2012). *Glass Recycling Overview*. Strategic Materials: Recycling Earth's Resources. Retrieved from http://www.ct.gov/deep/lib/deep/waste_management_and_disposal/solid_waste/transforming_matls_mgmt/summit_3/david_hudson_glass_markets_overview.pdf

IEEE GlobalSpec. (2017). *Chemical Composition of Glass*. Retrieved from Properties of Glass.

Isa, H. (2008). The need for waste management in the glass industries: a review. *Scientific Research and Essays*, *3*(7), 276–279.

Larsen, A. W., Merrild, H., & Christensen, T. H. (2009). Recycling of glass: accounting of greenhouse gases and global warming contributions. *Waste Management & Research*, *27*(8), 754–762.

Ling, T., Poon, C., & Wong, H. (2013). Management and recycling of waste glass in concrete products: current situations in Hong Kong. *Resources, Conservation and Recycling*, *70*, 25–31.

Lozier, T. R., & TerMeer, T. (2011). *Ohio Glass Recycling Study Final Report*. Retrieved from Ohio.

Maplandia.com. (2005–2016). *Florida Map—Satellite Images of Florida*. Retrieved from http://www.maplandia.com/south-africa/guateng/roodepoort/florida/

Mogensen GmbH & Co. KG. (2008). *Glass Recycling (1)*. Retrieved from Germany: http://www.mogensen.de/en/pdf/Glass_Recycling.pdf

Peters, M. S., Timmerhaus, K. D., & West, R. E. (2003). *Plant Design and Economics for Chemical Engineers* (5th ed.). New York: McGraw-Hill.

Pikitup (Pikitup Johannesburg SOC Ltd). (2009/10). *Annual Report*. Retrieved from South Africa.

Pikitup (Pikitup Johannesburg SOC Ltd). (2010/11). *Annual Report*. Retrieved from South Africa.

Pikitup (Pikitup Johannesburg SOC Ltd). (2011/12). *Annual Report*. Retrieved from South Africa.

Pikitup (Pikitup Johannesburg SOC Ltd). (2012/13). *Annual Report*. Retrieved from South Africa

Pikitup (Pikitup Johannesburg SOC Ltd). (2013/14). *Integrated Annual Report*. Retrieved from South Africa.

Pikitup (Pikitup Johannesburg SOC Ltd). (2015/16). *Business Plan*. Retrieved from South Africa.

RCBC (Recycling Council of British Columbia). (2004). *Just the Facts: Glass* [Press release]. Retrieved from http://www.recycleworks.org/pdf/kids_glass.pdf

Roberts, J. J. (2004). *Glass Waste* (Vol. 1). Great Britain: Thomas Telford.

Safiuddin, M., Jumaat, M. Z., Salam, M. A., Islam, M. S., & Hashim, R. (2010). Utilization of solid wastes in construction materials. *International Journal of the Physical Sciences*, *5*(13), 1952–1963.

Shayan, A., & Xu, A. (2004). Value-added utilisation of waste glass in concrete. *Cement and Concrete Research*, *34*(1), 81–89.

Sinnott, R., & Towler, G. (2009). *Chemical Engineering Design* (5th ed.). United Kingdom: Elsevier Ltd.
Sinnott, R. K. (1999). *Coulson & Richardson's Chemical engineering: Chemical engineering design* (3rd ed. Vol. 6). United Kingdom: Elsevier's Science & Technology.
Sustainability Victoria. (2013). *Market Summary—Recycled Glass* [Press release]
The Glass Recycling Company. (2014). *Interesting Facts about Glass & Recycling.* Retrieved from https://www.theglassrecyclingcompany.co.za/recycling-at-home-work-school/facts-about-glass/
U.S. Department of Energy Office of Industrial Technologies. (2002). *Energy and Environmental Profile of the U.S. Glass Industry.* Retrieved from Washington, DC: http://www.nrel.gov/docs/fy02osti/32135.pdf
Washington State Department of Transportation. (2015). *Cost Estimating Manual for Projects (M 3034.03).* Retrieved from http://www.wsdot.wa.gov/publications/manuals/fulltext/M3034/EstimatingGuidelines.pdf
WasteCare Corporation. (1997–2013). *Glass Recycling.* Retrieved from http://www.wastecare.com/Articles/Glass_Recycling.htm
Wintour, N. (2015). *The Glass Industry: Recent Trends and Changes in Working Conditions and Employment Relations.* Retrieved from Geneva.

4 Waste Management in India
Status and Challenges

A. Singh, G.S. Singh and P.K. Sharma

CONTENTS

4.1 INTRODUCTION

Solid waste production is increasing day by day because of rapid industrialization and urbanization, and henceforth its disposal and management are becoming a pressing issue (Gidde et al., 2008). Solid waste is produced by human actions, and these wastes are stored, handled, collected and disposed of, which can cause menaces to the surroundings and to human health (Gupta et al., 2015; Saxena et al., 2010). Solid wastes are generally categorized in very commonly known categories. For instance, solid wastes consist of industrial, domestic, commercial, institutional, agricultural and miscellaneous waste. Mostly, commercial and domestic wastes cannot be distinguished and are collectively considered as urban wastes (Syed, 2006).

Municipal solid waste is usually a blend of commercial and household garbage produced by human activities (Rajkumar et al., 2010). The nonstop haphazard disposal of municipal solid waste is increasing and is associated with urbanization, poverty, poor governance, population growth, less environmental awareness (Rachel et al., 2009) and insufficient management of environmental information. Municipal solid waste usually contains decomposable (food waste, straw, paper, yard waste and textiles), partially decomposable (disposable bibs, wood and sludge) and non-decomposable matter (plastics, rubbers, leather, metals, dust, ash from burning coal or wood) (Gupta et al., 2015). Rapid population growth and industrialization in India have resulted in mass movement from rural to urban areas, which produces thousands of tons of municipal solid waste daily. Poor collection and improper transportation are responsible for the deposition of municipal solid waste (Bundela et al., 2010). Municipal solid waste management is facing a serious phase, because of the non-availability of efficient services to dispose of and treat the huge amounts of municipal solid waste per day in metropolitan cities. Solid waste management has been deemed unsuitable by various researchers in several cities of developing countries (Mohanty et al., 2014; Das and Bhattacharya, 2013; Noorjahan et al., 2012; Jafari et al., 2010; Chatterjee, 2010; Imam et al., 2008; Chung and Carlos, 2008; Berkun et al., 2005). India has the second largest urban population in the world. Metropolitan cities (Chennai, Kolkata, Mumbai and Delhi) are home to more than 42% of India's urban individuals (Ghosh and Kansal, 2014) and generate the highest amounts of municipal solid waste due to their high habitation. In the present chapter, an effort has been made to deliver a comprehensive review of the current status and challenges of municipal solid waste in India. A literature review reveals financial issues as the most important ones limiting upgradation in solid waste management (Parthan and Milke, 2009). There is an urgent need for efficient medium-term planning at the municipal and state levels which can help to develop realistic investment projection and implementation (Hanrahan et al., 2006).

4.2 CURRENT STATUS IN INDIA

The population of India is above 1.2 billion which accounts for 17.5% of the global population and is at second position in the world (Population of India, 2017). In India it is very difficult to attain sustainability in waste management and there are big challenges in the areas of urbanization, industrialization and economic growth which have led to the production of millions of tonnes of waste (Kumar et al., 2017). The whole waste produced by the cities is around 1,351,978.27 tonnes/day (CPCB Report, 2015–16). And there are no practices, activities and proper arrangement or infrastructure for actual municipal solid waste management in the cities, and the problem is aggravated at a disturbing rate because of rapid growth in the population of the cities, as waste production is directly proportional to the country's population.

In India around 31.16% of the country's population lives in urban areas (Sudhir and Gururaja, 2012). The population migration and expansion of have led to rapid growth in waste production. In India, usually municipal solid waste is disposed of in open dumps without precautions. Hence, in Indian megacities municipal solid waste management is one of the most challenging environmental issues. The present urban municipal solid waste generation rate is about 109,598 tonnes/day and it is predicted to reach 376,639 tonnes per day by 2025 (Hoornweg and Bhada-Tada, 2012). A survey of 59 cities done by the Central Pollution Control Board (CPCB) has indicated that around 35,401 tonnes of municipal solid waste were produced per day from 59 cities in 2004–2005. Then another survey done by the Central Institute of Plastic Engineering and Technology (CIPET) found around 50,592 tonnes of municipal solid waste produced in 2010–2011 in the same 59 cities (Table 4.1). In India, the waste collection efficiency is around 70-72% in spite of the fact that 90% of the total municipal budget is for waste collection (CPCB, 2012).

4.3 WASTE CHARACTERIZATION

The solid waste characteristics vary from place to place. Different factors that impact the composition and quantity are the sources, average income level, the population, climate, social behaviour, industrial generation and waste materials market (Late and Mule, 2013). The current quantity of solid waste produced currently in Indian cities has increased from 6 million tonnes (in 1947) to 48 million tonnes (in 1997) and increased further up to 90 million tonnes (in 2009), and it is predicted to increase in 2047 to about 300 million tonnes (Gupta et al., 2015). Due to rapid economic growth and urbanization and high living standards the waste generation rate per individual in some cities may be increased. In urban lands the municipal solid waste fraction is 40–60% compostable matter and 30–50% inert materials. A lower socio-economic status usually increases relative organic waste production in municipal waste; hence rural individuals produce more organic waste than urban individuals. Similarly, the recyclables (glass, paper, metals and plastics) percentage is very low, it has been observed, because of rag pickers who collect and segregate the waste materials at the source of production and disposal points (Gupta et al., 2015).

TABLE 4.1
Municipal Solid Waste Production (Tonnes/Day) (CIPET, 2010–2011; CBCP, 2012)

Major Cities	Waste (Tonnes)
Delhi	6800
Mumbai	6500
Bangalore	3700
Chennai	4500
Kolkata	3670
Ahmedabad	2300
Bhopal	350
Bhubaneshwar	400
Chandigarh	264
Dehradun	220
Guwahati	204
Indore	720
Jammu	300
Kanpur	1600
Lucknow	1200
Patna	220
Pune	1300
Shillong	97
Srinagar	550
Varanasi	450

4.3.1 Composition of Municipal Solid Waste

Municipal waste categories are many such as rubbish, institutional waste, industrial waste, food waste, street-sweeping waste, sanitation waste and demolition and construction waste. Municipal solid waste includes decomposable organic waste (vegetable, food and fruit wastes), recyclable wastes (glass, metals, paper and plastics and so on), soiled waste (sanitary pads, blood-stained cotton and disposable syringes) and toxic materials (pesticides, medicines, used batteries and paints) (Kausal et al., 2012). Of these, plastics, papers, food waste, textiles, wood, disposable diapers, leather, bones and other organic materials are combustible materials while metal, glass and aluminium are non-combustible materials (Srivastava et al., 2014). The municipal solid waste composition at production points and collection sources was found on a wet weight basis, and it is comprised chiefly of a large portion of organic waste (40–60%), ash and fine earth (30–40%), paper (3–6%) and glass, plastic and metals (<1%). The carbon nitrogen ratio lies between 800 and 1,000 kcal/kg (Sharholy et al., 2008).

4.4 CHALLENGES OF SOLID WASTE MANAGEMENT

i) Segregation awareness necessity:

   In India a lack of garbage segregation at source has been seen and has resulted in several environmental problems, making it tough to segregate the waste at the transfer station or treatment area or in landfill. The number of urban local body employees is not suitable for the number of individuals living in that area, and the bins are also not situated appropriately. Hence, the awareness and participation of citizens needed to segregate waste at the source point, door-to-door collection and disposal at the proper place are crucial.

ii) Municipal solid waste classification:

In India due to different climatic zones, different cultures and different food habits the waste produced is of different types. Limited information is available on all cities and towns of waste produced. Hence, the policy makers depend on the little information available and are unable to provide proper solutions for the waste type generated in a particular region.

iii) Absence of proper funding level and development:

As the population increases, there is a challenge to provide a suitable setup in urban sites and landfill areas. In metropolitan cities the landfill sites are mostly running beyond their limit. And improper financial assistance aggravates the waste management problem.

iv) Rules implementation:

In urban areas the municipal solid waste management rules are not implemented properly as shown by several reports, hence it is difficult to maintain the municipal solid waste properly. There is a need for proper training and practical knowledge that would help to find blocks at implementation steps and take suitable action.

v) Financial assessment:

In the system finance plays an important role in increasing the commitment of the workforce involved in the work study. So, there is a need for such an approach by the system.

vi) Difficulties in coordination between state and centre:

Less coordination between state and central government leads to delayed report submission and also delays the action process at ground level with poor strategy.

vii) Suitable technological solutions:

In India the availability of skilled labour, understanding of new as well as best practices accessible for solid waste management and financial spurs for finding new technological solutions for proper waste management are real challenges.

viii) Waste-to-energy project failure:

India still needs improved and economically sound technologies to make waste-to-energy projects a success story.

ix) Organized sector necessity:

The lack of recycling industries and society acceptance has been ignored. And systematic organization is needed from the source of waste generation to segregation and dumping or treatment.

4.5 WASTE MANAGEMENT PRACTICES IN INDIA

4.5.1 Pyrolysis and Gasification

In these methods waste is converted into energy by the combustion of solid waste in anoxic conditions. Both gasification and pyrolysis are endothermic processes. Pyrolysis is the thermal decomposition of organic matter under pressure and temperatures between 400 and 1000°C in the absence of oxygen. Gasification is the incomplete combustion of organic matter at temperatures between 1000 and 1400°C in low amounts of oxygen (NSCA, 2002). In both processes the end products are gas (termed syngas), liquid (containing acetic acid, acetone and methanol) and char (containing carbon with inert material).

4.5.2 Incineration

This is a thermal process of waste management. In this method the combustion of unprocessed waste takes place in regulated conditions at 850°C in presence of air or oxygen (DEFRA, 2007). It is done in a closed chamber. The by-products are sulphur dioxide, particulate matter, furans, dioxins, carbon dioxide, ash, water vapour, non-combustible matter and heat. It is an exothermic method, producing heat which could be used in electricity and steam generation.

4.5.3 Composting

Composting is a method in which organic matter is decomposed aerobically under controlled conditions such as pH, humidity and temperature. The native organisms (mesophile and thermophile) convert organic material into a stabilized product termed compost (Srivastava et al., 2015). Therefore, compost is used as a soil conditioner in agricultural land (Singh and Singh, 2017).

4.5.4 Vermicomposting

This is a biological eco-friendly method in which bio-oxidation takes place to stabilize organic waste into a valuable end product known as vermicompost (Singh and Singh, 2017). Vermicomposting has combined action of micro-organisms and earthworms. Moreover, the feedstock, the microbes present in the earthworm gut, is also liable for the biochemical degradation of organic matter. Earthworms also improve the enzymatic action and change the organic waste's physical status and chemical status indirectly (Fornes et al., 2012).

4.5.5 Landfilling

Landfilling is the method in which waste is disposed of in evacuated areas. It is an essential part of any planned municipal solid waste management system. It is the final storehouse of any city's municipal solid waste after all relevant waste management options. Open dumping is the most lucid and economical sound practice applied in most developing countries. Among all the waste management practices open dumping takes care of 51% of the waste generated in Asia (World Bank, 2012).

4.5.6 Waste Application in Land

At specific locations of agricultural land, organic waste (sewage sludge) is directly applied without pre-treatment. Sewage sludge has many nutrients that improve plant growth (Singh and Agrawal, 2010). The organic part of municipal solid waste generally is abundant in phosphorus, nitrogen, micronutrients, potassium, macronutrients and organic matter, and this might support soil fertility restoration (Gonzalez et al., 2008).

4.6 CONCLUSION

This chapter reviewed the problem and issues related to solid waste generation and disposal in India. A huge amount of solid waste is generated in India; it poses a great risk to human health and the environment. The present paper emphasizes on options for waste management and to make policies needed by urban local bodies. There is a need for the realization of source segregation of waste at the production point as recyclable matter, biodegradable and inert for proper waste management. Composting plants should be installed to decrease the degradable waste to reduce the pressure on landfills. Public involvement in waste segregation, collection and management should be increased by awareness campaigns starting from schools.

REFERENCES

Berkun, M., Aras, E., Nemlioglu, S. 2005. Disposal of solid waste in Istanbul and along the black sea coast of Turkey. *Waste Management*. 25, 847–855.

Bundela, P.S., Gautam, S.P., Pandey, A.K., Awasthi, M.K., Sarsaiya, S. 2010. Municipal solid waste management in Indian cities. *International Journal of Environmental Sciences*. 1 (4), 591–605.

Central Pollution Control Board (CPCB) 2004. *Management of municipal solid waste*. Ministry of Environment and Forests (MoEF), New Delhi, India.

Central Pollution Control Board (CPCB) 2012. *Status report on municipal solid waste management*. Ministry of Environment and Forest (MoEF), New Delhi, India.

Chatterjee, R. 2010. Municipal solid waste management in Kohima city—India. *Iranian Journal of Environmental Health Science & Engineering*. 7 (2), 173–180.

Chung, S.S., Lo, C.W.H. 2008. Local waste management constraints and waste administrators in China. *Waste Management*. 28 (2), 272–281.

CPCB: Central Pollution Control Board 2015–16. http://cpcb.nic.in/uploads/hwmd/MSW_AnnualReport_2015-16.pdf. Accessed May 2017.

Das, S., Bhattacharya, B.K. 2013. Municipal solid waste characteristics and management in Kolkata, India. *International Journal of Emerging Technology & Advanced Engineering*. 3 (2), 147–152.

Department for Environment, Food and Rural Affairs 2007. Incineration of municipal solid waste, waste management technology brief, the new technologies work steam of the Defra Waste Implementation Programme.

Fornes, F., Mendoza-Hernandez, D., Garcıa-de-la-Fuente, R., Abad, M., Belda, R.M. 2012. Composting versus vermicomposting: a comparative study of organic matter evolution through straight and combined processes. *Bioresource Technology*. 118, 296–305.

Ghosh, R., Kansal, A. 2014. Urban challenges in India and the mission for a sustainable habitat. *Interdisciplina*. 2 (2), 281–304.

Gidde, M.R., Todkar, V.V., Kokate, K.K. 2008. Municipal solid waste management in emerging mega cities: a case study of Pune city. Proceedings of Indo Italian Conference on Green and Clean Environment, Pune, India (March 20–21).

Gonzalez, M., Mingorance, M., Sanchez, L., Pena, A. 2008. Pesticide adsorption on a calcareous soil modified with sewage sludge and quaternary alkyl-ammonium cationic surfactants. *Environmental Science & Pollution Research International*. 15, 8–14.

Gupta, N., Yadav, K.K., Kumar, V. 2015. A review on current status of municipal solid waste management in India. *Journal of Environmental Sciences*. 37, 206–217.

Hanrahan, D., Srivastava, S., Ramakrishna, A.S. 2006. *Improving management of municipal solid waste in India*. World Bank.

Hoornweg, D., Bhada-Tada, P. 2012. *What a waste: a global review of solid waste management*. Vol. 15. World Bank.

Imam, A., Mohammed, B., Wilson, D.C., Cheeseman, C.R. 2008. Solid waste management in Abuja, Nigeria. *Waste Management*. 28 (2), 468–472.

Jafari, A., Godini, H., Mirhousaini, S.H.J. 2010. Municipal solid waste management in KhoramAbad city and experiences. *World Academia Science Engineering and Technology*. 4, 163–168.

Kausal, R.K., Varghese, G.K., Chabukdhara, M. 2012. Municipal solid waste management in India—current state and future challenges: a review. *International Journal of Engineering, Science and Technology*. 4 (4), 1473–1489.

Kumar, S., Smith, S.R., Fowler, G., Velis, C., Kumar, S.J., Arya, S.R., Kumar, R., Cheeseman, C. 2017. Challenges & opportunities associated with waste management in India. *Royal Society Open Science*. https://doi.org/10.1098/rsos.160764.

Late, A., Mule, M.B. 2013. Composition and characterization study of solid waste from Aurangabad city. *University Journal of Environment Research and Technology*. 3 (1), 55–60.

Mohanty, C.R., Mishra, U., Beuria, P.R. 2014. Municipal solid waste management in Bhubaneswar, India—a review. *International Journal Latest Trends Engineering Technology*. 3 (3), 303–312.

Noorjahan, A., Dhanakumar, S., Mohanraj, R., Ravichandran, M. 2012. Status of heavy metals distribution in municipal solid waste in Tiruchirappalli City, India. *International Journal Applied Biological Pharmacology Technology*. 3 (3), 252–258.

NSCA. 2002. *Comparison of emissions from waste management options*. National Society for Clean Air and Environmental Protection. United Kingdom

Parthan, S., Milke, M.W. 2009. Cost analysis of municipal solid waste management in India. Proceeding of Waste MINZ, 21st Annual Conference, Christchurch, New Zealand (Oct 14–16).

Population of India 2017 Wikipedia. http://www.google.co.in/search?=india%27s+population+in+2017&oq=india+poplu&aqs=crome

Rachel, O.A., Komine, H., Yasuhara, K., Murakami, S. 2009. Municipal solid waste management in developed and developing countries—Japan and Nigeria as case studies. http://wwwgeo.civil.ibaraki.ac.jp/komine/mypapers/JGSPaper/2009/JGS2009(973)Rachel.pdf

Rajkumar, N., Subramani, T., Elango, L. 2010. Groundwater contamination due to municipal solid waste diposal—a GIS based study in Erode City. *International Journal of Environmental Sciences*. 1 (1), 39–55.

Saxena, S., Srivastava, R.K., Samaddar, A.B. 2010. Sustainable waste management issues in India. *IUP Journal of Soil and Water Sciences*. 3 (1), 72–90.

Sharholy, M., Ahmed, K., Mahmood, G., Trivedi, R.C. 2008. Municipal solid waste management in Indian cities: a review. *Waste Management*. 28, 459–476. doi:10.1016/j.wasman.2007.02.008.

Singh, A., Singh, G.S. 2017. Vermicomposting: a sustainable tool for environmental equilibria. *Environmental Quality Management*. 27, 23–40.

Singh, R.P., Agrawal, M. 2010. Effects of different sewage sludge amendment rates on growth, biomass accumulation, nutritional quality, heavy metal accumulation and yield of mungbean *Vigna radiata* L. plants. *Ecological Engineering*. 36, 969–972.

Srivastava, R., Krishna, V., Sonkar, I. 2014.Characterization and management of municipal solid waste: a case study of Varanasi city, India. *International Journal of Current Research and Academic Review*. 2, 10–16.

Srivastava, V., Ismail, S.A., Singh, P., Singh, R.P. 2015. Urban solid waste management in the developing world with emphasis on India: challenges and opportunities. *Reviews in Environmental Science and Bio/Technology*. doi:10.1007/s11157-014-9352-4.

Sudhir, H.S., Gururaja, K.V. 2012. Population crunch in India: is it urban or still rural? *Current Science*. 103 (1), 37–40.

Syed, S. 2006. Solid and liquid waste management. *Emirates Journal for Engineering Research*. 11 (2), 19–36.

World Bank. 2012. *What a waste: a global review of solid waste management*. Urban development series knowledge papers.

5 Urban Waste Management Issues in China

With Special Reference to Municipal Solid Waste

Zhiyang Xu, Yunlong Zhou, Qian Li and Haoqi Wang

CONTENTS

5.1 INTRODUCTION

China is a developing country with the largest population and the third largest land area in the world and the second largest economy in the world. In 2019, the total economic volume exceeded 99 trillion yuan. According to the *China Statistical Yearbook 2016*, 56.1% of Chinese people were city residents in 2015 (*China Statistical Yearbook 2016*, 2016). There are still many migrant workers working and living in urban areas (Tang et al., 2016). The population in urban areas accounts for more than 60% of the total population in China (Zhang et al., 2010). The rapid growth of the population and development of the economy have resulted in a sharp increase in the amount of urban waste, which brings heavy burdens to China's environmental quality and long-term sustainable development. How to deal with the increasing volume of solid waste is becoming a tremendous challenge for the Chinese government. China has tried to learn from the experiences in solid waste management from developed countries in recent years. Methods such as incinerating, composting, anaerobic digestion and landfill are applied in waste treatment. This chapter describes the characteristics, impact, classification and different treatment technologies of urban waste and problems and suggestions for the disposal of waste and relevant cases.

5.2 CHARACTERISTICS OF URBAN SOLID WASTES IN CHINA

Compared with developed countries, the contents of urban solid waste in China are more similar to other developing countries. China's urban solid wastes are mainly composed of lime soil, tiles, paper, fabrics, glass, metal, wood and bamboo, food and so on (Mosannenzadeh et al., 2017). In urban solid wastes, the content of inorganic matters in solid wastes is higher than that of organic matter, and the content of non-flammable matters is higher than that of combustible matters. The characteristics of urban solid wastes are as follows: (i) few combustible materials, (ii) few materials with high calorific value, (iii) organic materials with high water content (Zhao et al., 2014). In recent years, with the development of the economy and society and the changes in consumption, some changes have occurred in the composition of urban wastes. To be specific, organic and toxic matter has gradually increased. China is still not able to recycle some types of solid waste, which can be recycled in the US or Japan.

5.3 INFLUENCE OF SOLID WASTES ON CITIES

5.3.1 Garbage Siege

With the rapid development of the economy in modern cities, the growth rate of solid wastes has been totally unexpected. The solid wastes produced by city residents even completely surrounds the whole city, which undoubtedly seriously affects the normal and healthy life of city residents (Jiao et al., 2019; Goldstein et al., 2016). Fast-growing development brings about the serious problem of 'garbage siege', which is the focus and difficulty of many cities. Statistics show that more than 400 large and medium-sized cities in China are surrounded by waste and 150 cities have no suitable place to dump waste. The total amount of urban waste dumped over the years has been up to 7 billion tons, and the annual production is increasing at a rate of about 8.98%. About 18,300 tons of wastes are produced every day with an annual increase rate of 8% in Beijing, but Beijing only has a capacity of 10,410 tons every day. According to a survey published in 2010 by the Shanghai Academy of Social Sciences, the daily production of household wastes in Shanghai reached 20,000 tons by the end of 2010. Similarly, Guangzhou, a mega city with a population of over 10 million in the south of China, can produce up to 18,000 tons of household wastes every day.

5.3.2 Air Pollution

A large amount of domestic solid waste can be degraded over time, especially kitchen waste, and produces a very unpleasant odour, which pollutes air. In addition, some industrial waste such as coal ash is left untreated in time and then exposed to the air. Coal ash particles can diffuse into the atmosphere with the wind, which affects the quality of urban air too. Once these particles are inhaled by people, they also can harm human health.

5.3.3 Water Pollution

Failure to treat urban solid waste in time can cause urban water pollution. After the accumulation, some certain chemical reactions occur inside the waste and many toxic substances are generated under the catalysis of the external environment. These toxic substances infiltrate into the ground through the soil and then affect the groundwater source. At the same time, in rainstorms, they are brought into the river with the rainwater, causing water pollution and affecting water safety.

5.3.4 Soil Pollution

The soil pollution caused by urban solid waste is not obvious in the short term. Because cities do not need land for crop planting as in rural areas, the connection between solid waste and soil is easily ignored. However, soil is also part of the ecological environment. Harmful substances in solid waste infiltrate into the ground during the stacking process, affecting the soil structure. When people carry out greening under such soil conditions, harmful substances may spread to the plant.

5.3.5 Influence on Development

The continuous increase of municipal solid wastes has a huge impact on the pleasant living environment, which is totally inconsistent with sustainable development in constructing a socialist harmonious society. During the process of rapid urbanization and construction in China, every city faces great challenges in waste disposal.

Taking waste plastics as an example, polyethylene and polyvinyl chloride plastic films can only be completely degraded in the soil after about 300~400 years. They can damage the air permeability of the soil, reduce the water storage capacity of the soil and affect the moisture content of crops. They also can influence the nutrient absorption of crops and hinder the growth of grass roots, which causes a significant reduction in crop yields and degrades the cultivated land. In addition, heavy metal ions and toxic substances in plastic additives directly affect groundwater and plant growth through diffusion and infiltration in the soil. The incineration of plastic waste can produce harmful gases, such as hydrogen chloride, ABS, hydrogen cyanide and other harmful gases. Among them, toxic gases such as dioxin produced during the combustion of chlorides can cause malformations and death of mammals and birds and cause great damage to the ecological environment (Zhao et al., 2008; Yang et al., 2012).

In the current urbanization construction in China, each city faces a large problem of waste disposal. Taking Shanghai as an example, public opinion surveys during the Two Sessions showed that people are mostly concerned about two major issues: The environment and traffic. Garbage siege, water pollution, fog and haze and other environmental pollution problems have become the most difficult public issues to solve. It can be seen that urban solid waste is a great threat to the living environment and human health.

5.4 SOLID WASTE CLASSIFICATION

According to the different sources and hazards of solid waste, it is generally divided into three categories: Domestic solid waste, industrial solid waste and hazardous solid waste.

5.4.1 Industrial Solid Waste

Industrial solid wastes are formed in the production and operation process of industrial departments. They include the coal gangue produced in the coal mining industry, fly ash and slag from heating system boilers and power plants, steel slag, blast furnace slag, nonferrous metal metallurgical slag and red mud produced in the metallurgical industry, carbide slag, gypsum, alkaline residue from the chemical industry and tailings dam and waste rock after mining. Complex composition and large volume are the prominent characteristics of industrial waste. The many toxic substances in it can cause unimaginable harm to the human body and natural environment. With the continuous increase of engineering projects, the amount of construction waste has increased sharply and construction waste has gradually become an important part of urban solid waste. These industrial solid wastes occupy a large area and contain certain industrial chemical components, so they also must be urgently treated.

5.4.2 Domestic Waste

Domestic wastes are produced in daily life and are usually divided into rural living garbage and city life garbage. The metal, waste packaging supplies, used batteries and other living garbage are the main constituent elements of domestic wastes. The composition of domestic wastes has to do with people's daily life habits and the local climate. These domestic wastes are generated in daily life, so the amount of urban solid waste is very alarming, which adds a lot of difficulty to the treatment of urban solid waste. According to the Ministry of Construction, urban domestic waste that was collected and transported in 654 cities of China amounted to 157,000,000 tons in 2010 (Yang et al., 2012). The actual production is supposed to be 25% higher (Du et al., 2006), or, according to Zhang Lijing's conservative estimate, 30% higher. So the actual production of urban domestic waste in the 654 cities may be estimated between 196,250,000 tons and 204,100,000 tons or ~200,000,000 tons for convenience.

5.4.3 Hazardous Waste

Hazardous solid waste generally refers to the waste identified by the national hazardous waste identification indicators and identification methods as having certain hazardous characteristics (Trivedi et al., 2015). It mainly contains industrial solid wastes, urban household wastes, medical wastes, residual pesticides and so on. If not timely treated, hazardous solid wastes not only cause serious environment pollution, but also threaten people's health and safety. Hazardous solid waste presents infectious, radioactive and toxic elements, so it threatens the urban ecological environment.

5.5 MUNICIPAL SOLID WASTE DISPOSAL METHOD

Technologies for solid waste management are composting, landfilling, anaerobic digestion and incineration technology (shown in Figure 5.1).

5.5.1 Composting

In the process of composting treatment, the solid waste is successfully heated to 70°C for storage and fermentation. This process involves decomposing organic pollutants into organic nutrients by means of microorganisms in solid waste. This method is beneficial to increase the nutrients in the soil as a conditioner, compost manure and the addition of humic acid and natural pesticides for earth (Kumar, 2010). However, this treatment method is very limited and the scope of application is very small. This method is usually applied to organic solid waste which easily corrodes. If the scope of solid waste is large, it cannot be put into use. For composting, it is mandatory to maintain various stages carefully to avoid contamination (Awasthi et al., 2014). The segregation of materials at source has the highest potential to recover biodegradable waste and convert it into a resource. Further, it is necessary to determine the organic fraction for effective composting and cellulose degradation activity. Some important parameters such as waste size and proportion, temperature,

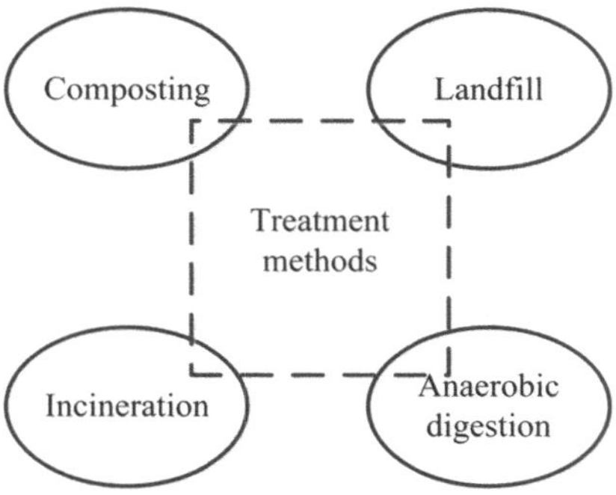

FIGURE 5.1 Various treatment technologies for solid waste.

carbon:nitrogen (C:N) ratio, aeration and moisture content are to be maintained for an effective composting process (Liu et al., 2019). The components like water, heat and air are the key to composting. The process of composting is shown in Figure 5.2. With the implementation and improvement of environmental protection policies, there is a great development space for the composting of organic solid waste. The level of composting technology and equipment in China is relatively low, which requires the introduction of foreign advanced technology, so there is a great space for cooperation. The development of composting in China is basically market-oriented, including the development of independent technology, the use of bacterial agents and the proposal of rapid composting technology. The recycling of organic solid waste is realized and the development route of organic fertilizer is promoted. These models can be used for reference by all countries in the world.

5.5.2 Incineration

Incineration is another effective way to deal with municipal solid waste. This method mainly burns solid waste through boilers and other equipment. Under high temperature, there is a chemical reaction between combustible matter and oxygen and heat is released completely. The process of incineration is shown in Figure 5.3. Carbon dioxide (CO_2), water (H_2O) and some other gases such as NOx, SOx and so on can be produced during the process. During full combustion there is oxygen in excess, no carbon monoxide (CO) is produced and the average gas temperature is 1,200°C. In the case of lack of oxygen, the reactions are characterized as incomplete combustion where the produced CO_2 reacts with C that has not been consumed and is converted to CO at higher temperatures. The overall treatment efficiency of this method is higher and the heat generated in the whole combustion process can be reused. However, the only deficiency of this method is the new pollutants produced, which have a prominent impact on the atmospheric environment.

$C + O_2 \rightarrow CO_2 + 393.77$ J (oxygen in excess)
$CxHy + (x + y/4)\ O_2 \rightarrow xCO_2 + y/2\ H_2O$
$C + CO_2 + 172.58$ J $\rightarrow 2CO$ (lack of oxygen)

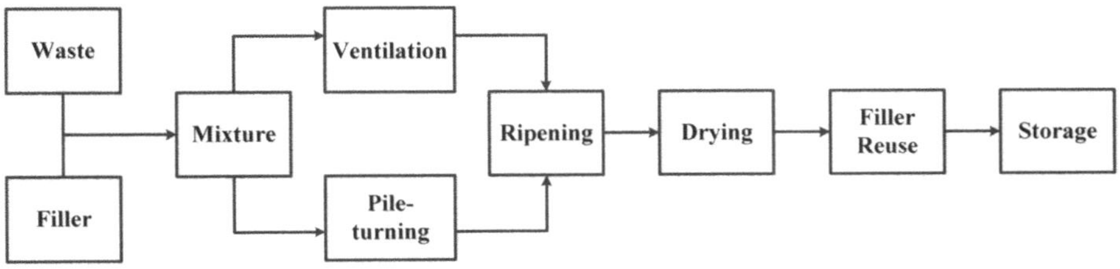

FIGURE 5.2 Schematic process flow diagram of composting.

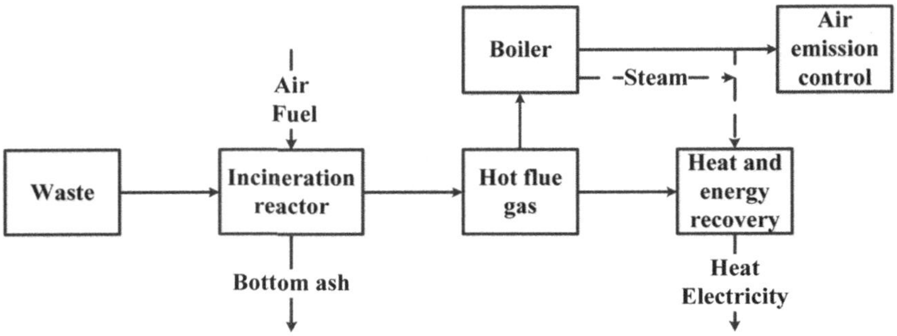

FIGURE 5.3 Schematic process flow diagram of incineration.

5.5.3 Landfill

Landfill treatment includes sanitary landfill and direct landfill. Sanitary landfill mainly refers to the treatment of solid waste in a specific range with soil cover, seepage prevention and other inverted facilities to avoid polluting the atmospheric environment and water resources. The overall cost of this method is relatively low, and considerable results can be achieved. In the process of direct landfill, solid wastes are usually put directly into the reserved space and the compaction and sealing are realized at the same time. Through the natural organic matter decomposition of solid waste, pollution of the solid wastes is reduced to the lowest degree. And the cost of this method is relatively low and operation is simple, but it can pollute urban groundwater (NBSR, 2016; Kalyani et al., 2014).

5.5.4 Anaerobic Digestion

Anaerobic digestion is a process to decompose organic waste without oxygen. The end product of anaerobic digestion contains biogas (60~70%), CO_2 (30~40%) and other impurities. This technology decreases the organic loading from agricultural waste, food waste and sludge of wastewater. However, the design of the anaerobic digester and maintaining its operating parameters are major concerns in the AD process, such as continuous or batch design, operating temperature psychrophilic, mesophilic and thermophilic, covered lagoons, complete mix and plug flow (Jain et al., 2015). There are four prime steps for anaerobic digestion which are discussed below and shown in Figure 5.4. Hydrolysis is the first step of the anaerobic digestion process. In this step, microorganisms break down the organic polymer chain into very small parts and make a simple soluble molecule with the help of enzymes. Carbohydrates, lipids and proteins are hydrolysed to sugar, fatty acids and amino acids respectively. In the process of fermentation, the fermentative microorganisms reduce the organic compounds and convert them into CO_2, H_2, acetic acid and fatty acid. In the process of acetogenesis, acetogenic microorganisms help to convert the acetic acid to acetate. The CO_2 and H_2 are directly used for the production of methane (CH_4). Methanogenesis is the final step for the anaerobic method, where methanogenic bacteria help to degrade the acetate, CO_2 and H_2 to produce the methane which is called biogas.

5.6 PROBLEMS EXISTING IN THE DISPOSAL OF MUNICIPAL SOLID WASTE

Urban waste disposal development in China has made some progress in the past 30 years. At the start of the 21st century, China could dispose of 54.2% of its total municipal solid waste and 50.4% of its total industrial solid waste (ISW). About 44% of hazardous waste was also successfully recycled (Chinese Environmental Yearbook Compiling Committee, 2011). According to *China Statistical Yearbook 2016*, the municipal solid waste harmless treatment rate has reached 94.1% and the comprehensive utilization rate of industrial solid waste was over 70% in 2015 (National

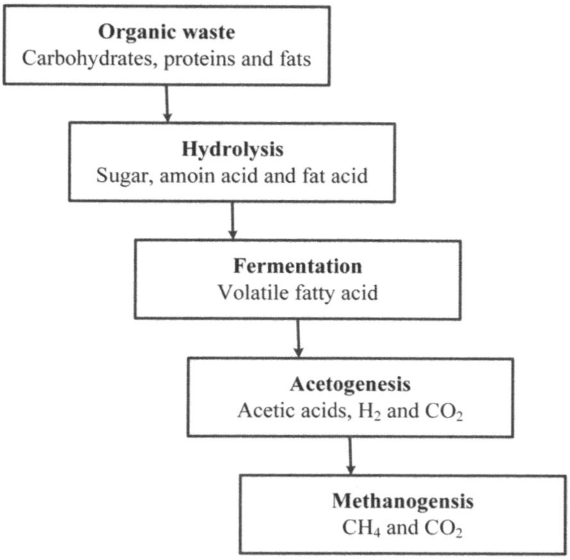

FIGURE 5.4 Mechanism of anaerobic digestion.

Bureau of Statistics of People's Republic of China, 2016). However, there are some problems in the development process of disposal.

First, laws and regulations are still not perfect. In recent decades, China has successively issued a series of laws and regulations, such as the Environmental Protection Law, Law of the People's Republic of China on the Prevention and Control of Solid Waste Pollution, Promotion Law on Cleaner Production, Promotion Law on Circular Economy and so on. However, the corresponding supporting laws and regulations are still not perfect, so local government lacks operability at the implementation level. An operable legal system matching the generation and treatment of urban waste has not yet been fully established. Although China has introduced many laws and regulations related to resource recycling and environmental pollution control, there is still a lack of corresponding implementation rules. Local governments encounter difficulties and problems in the implementation of policies, and there are no methods that can be universally applied for waste disposal and utilization. The Chinese government has also reformed the solid waste management system in recent years. From the perspective of research methodology, developed countries paid more attention to systematically investigating the application of new technology, economic performance, policy innovations and social-psychology in solid waste management. Scholars in those countries have widely used statistical and quantitative methods to assess the performance of solid waste disposal. Studies on solid waste disposal in developing countries pay more attention to direct and practical institutional influences. However, these studies typically apply a case study method which makes it hard to measure and compare performance of solid waste disposal. Although these studies can explain some phenomena and address some problems, they lack the generalizability to offer valuable suggestions to improve the performance of solid waste disposal beyond the jurisdiction under study (Ezeah et al., 2012). How can the recycling, harmlessness and high value of wastes be realized? There are still some problems such as policy refinement, management refinement, technology promotion and concept transformation.

Second, institutions and mechanisms need to be improved. For a long time, the waste management systems and mechanisms have needed to be improved. The city sanitation department acts as both 'referee' and 'athletes', which makes social forces hard to participate in the garbage collector system. At the same time, the market competition mechanism is not sound and the energy market is lacking.

Third, policy incentives and capital investment are insufficient. Affected by the macroeconomic situation and the abolition of value-added tax preferential policies in the renewable resources recycling industry, renewable resources recycling enterprises face great difficulties in their development. Glass, waste energy-saving lamps and other wastes have high recycling costs and low profits, so it is difficult to mobilize enterprises' enthusiasm by relying on the market mechanism. As a result, these low-value-added products have been pouring into domestic garbage due to the lack of reduction policy subsidies, resulting in increasing domestic garbage. At the same time, due to the high cost of waste treatment and high investment in resource utilization technology and equipment, a number of enterprises can barely survive in the early stage of government funding subsidies. Waste treatment plants cost a lot and labour costs increase gradually, which makes the cost of enterprises increase suddenly and it is difficult to ensure the normal operation of waste treatment.

Lastly, residents' real participation enthusiasm is low. For example, it is difficult to mobilize enthusiasm for garbage classification, so that people really are unwilling to participate in classification. On the one hand, the signs of garbage classification facilities in residential areas are unclear and unreasonable. On the other hand, some residents still do not have a strong awareness of garbage classification and resource recycling, resulting in unreasonable garbage classification, mixture of dry and wet garbage and inconvenience for subsequent garbage collection, transportation, classification treatment and disposal. Because some government once didn't inform the public of incineration power plant construction in time, a lot of collective rights safeguarding incidents of refuse incineration power plant were happened. In October 2009, hundreds of residents in Dashi Town, Panyu District, Guangdong Province, launched a protest against the construction of a refuse incineration power plant. At that time, about 300,000 residents of major buildings in Dashi Town expressed their opposition and firmly resisted. Subsequently, Lijiang Garden and other neighbourhoods successively launched signature campaigns against incineration. According to a survey of 12 districts within 8 km of the refuse incineration power plant in Guangdong Province, 97.1% of the residents interviewed opposed the construction of the refuse incineration power plant. Hangzhou, which enjoys the reputation of 'the most beautiful city in the world', launched a public action against the construction of Jiufeng refuse incineration power plant, which also shocked Chinese people and spread abroad. It can be seen from these cases that the environmental problems come down to the problems of resources, people, society and even politics (Jiao et al., 2019).

5.7 SUGGESTIONS (ZHOU ET AL., 2017)

5.7.1 Continue Implementing Solid Waste Disposal Policies

Maintaining and strengthening solid waste disposal legislation is essential for maintaining the current development momentum in urban solid waste management. According to the above analysis, it is obvious that a city located in a province with a resource use policy has stronger solid waste disposal capacity than others. Therefore, Chinese cities should continue implementing and reforming the resource use policy for better effectiveness. Many environmental policies have been put in place to regulate solid waste disposal; further investigations are needed to assess their effectiveness.

Government officials could ask environmental experts to participate in making resource use policy and other solid waste disposal policies to keep policy arrangements more professional and targeted. The Chinese government also should consider allowing more enterprises and communities to be involved in solid urban waste disposal to reduce official pressures and increase effectiveness through market and social powers. Besides, the government itself should take action to dispose of some specific types of urban solid waste. If certain kinds of urban solid wastes are hard to collect and recycle by voluntary or commercial actions, the government should use its administrative power to support relevant official departments and professional organizations to address them.

5.7.2 Upgrade the Industrial System

Many urban solid wastes are from industrial sectors in China. The traditional industrial system and production methods have seriously affected the environment. Controlling the development of traditional industries, especially heavy industries, is a good approach to reducing solid waste problems as well as polluting emissions. In recent years, upgrading industrial systems and developing sustainable green industries have been new trends to reduce resource and environmental burdens for long-term development.

All Chinese provincial governments and municipal governments should not only follow general environmental policies made by the central government, but also develop their own sustainable development strategies and upgrade their industrial systems according to local contexts. All the industrial upgrading actions and development of circular economies should consider local economic and environmental carrying capacity. If the local government just thinks about the urgency of upgrading the industrial system but ignores local economic/resource capacity and advantages, it can only bring new disasters for sustainable development. Only a sustainable industrial system has the potential to solve China's solid waste problems at a more fundamental level.

However, upgrading environmental equipment and industrial systems requires much money and policy support. It is hard to totally rely on local government and entrepreneurs to achieve the task since upgrading for environmental purposes makes it hard to get direct economic profit. The central government should give more preferential treatment and financial subsidies to local governments and enterprises to upgrade local industrial systems, factories and environmental equipment. Official involvement could guide the development of industrial systems in China to pay more attention to environment protection and energy saving than to economic profits. China also can combine experience from developed countries and its own characteristics to avoid unexpected problems in the industrial upgrading plan.

5.7.3 Push Forward Economic and Social Reform

On the one hand, fast social and economic growth increases the environmental burdens of solid waste disposal. On the other hand, economic and social development can impose pressures and chances on the government to pay more attention to prioritizing and solving solid waste problems.

As mentioned, economic development can offer capital and new technology for a province or a city to solve solid waste problems. However, the premise is that the province or the city should have a relatively healthy economic structure. Meanwhile, heavy industrial-based economic structure is still very common in China nowadays which increases the difficulty in implementing environmental policies.

In order to change the situation, China should abandon its focus on GDP, and refocus its development strategy on economic and social structure reform, striking a balance between economic development, social development and environmental protection. The whole economic and social structure of China should be shifted to sustainable development. Both administrative approaches and market approaches can be taken to make appropriate institutional arrangements. China should deepen the reform of economic structures and focus on environmentally friendly production methods. More detailed policies should be made and the execution of policies should be strengthened to keep the economic and social reform on the right path. Environmental offenses should be punished strictly. The government could also use multiple ways to increase the development of finance, IT and other new industries, which produce less solid waste and are environmentally friendly. Social/private capital, foreign investment and government resources should all be used to push the sustainable reform of economic and social sectors.

5.7.4 Increase Environmental Education

Environmental protection and urban solid waste disposal cannot only rely on making appropriate institutional arrangements but also require increased environmental education for students, officials, enterprise owners, farmers and anyone else in society. China needs to establish a formal environmental education system to spread environmental common sense and knowledge, and improve the environmental education level. Establishing healthy and environmentally friendly living and consumption habits should also be included in the education system. Besides, the Chinese government should establish a professional training system to offer knowledge to enterprise owners, farmers and workers about how to deal with waste and pollutants effectively and scientifically.

5.8 CASE STUDIES: WASTE SEPARATION

The growth of urban waste is a critical global issue because it poses a threat to human health and the environment if not properly dealt with. Concerns include emissions coming from waste decomposition in landfill sites and the simple fact that landfills are already full. As stated in the World Economic and Social Survey 2013, the reduction of waste production, improvement of waste collection and recycling systems are amongst the most pressing challenges of our time, and investments in strategies are needed in most cities worldwide (UN DESA, 2013). To address our resource problem the European Commission released the Waste Framework Directive (EU, 2008) and the Circular Economy package (EC, 2015). The objectives include recycling 50% of urban waste by 2020 and reusing 'waste' as a resource whenever possible, accompanied by an overall reduction of waste.

As urban residents generate large quantities of municipal wastes, their participation in the MSWMS by pre-sorting common household waste streams at source is critical to reduce the amount of contamination that could ruin batches of recycled material (Recycling Magazine, 2018). It is fundamental for improving the overall recycling performance and enabling the maximum recovery of useful resources (Varotto, 2017). Moreover, requirements for quality of recyclate are becoming more strict, and the rejection of loads due to ineffective recycling behaviour can entail large financial impacts (Jesson, 2014). Efficient recycling systems depend upon how well the public participates (Thomas, 2001). Yet, recycling rates from densely populated urban cities are under-performing in comparison with less populated areas (Miafodzyeva, 2013). Thus, local decision-makers bear a great responsibility as they determine whether the common global waste management goals will be accomplished. In the process of establishing and improving well-performing municipal solid waste management systems the understanding of fundamental social factors influencing public behaviour is commonly underestimated but of utmost importance.

5.8.1 Social Factors Influencing Household Waste Separation

5.8.1.1 Demographic Characteristics

It is obligatory to examine who has to interact with the given recycling system: The user. Recycling behaviour and related barriers vary by socio-demography; therefore the MSWMS ought to be adjusted to the characteristics of its targeted social groups. Links between household waste separation behaviour and a variety of individual socio-demographic attributes have been previously studied.

5.8.1.2 Psychological Factors

Furthermore various external and internal socio-psychological variables determine recycling behaviour. Several theoretical frameworks have been developed over the last decades by social psychologists to investigate, model and predict recycling behaviour in relation with its determinants. The following section provides an overview of the main findings and theories.

5.8.1.2.1 Perceived Convenience and Effort

One of the most important factors to increase household waste separation behaviour is convenience (Ando et al., 2005). The perception of convenience depends on the individual who is desired to participate in the sorting system. In short: The easier it is to understand and use the system, the more likely it is to be used and vice versa (perceived difficulty results in inconvenience).

Although some may not consider external factors (referred to as situational-contextual, technological or organizational) as socio-psychological, because they are related to the infrastructure and service provision, they are of significant importance in the behavioural context. It appears that they influence the internal factors (Tonglet et al., 2004). In fact, a lack of necessary infrastructure can be one of the most important barriers for households to participate in recycling activities. Even slight changes enhancing convenience can substantially impact behaviour. For source-separation it is stressed that these adjustments must happen inside the household (Bernstad et al., 2013).

The following variables bear the potential to influence convenience positively and can be modified by the local MSWMS but also the household itself:

Short distance and strategic location of collection points for recyclables
Ease of access and availability of bins
Appropriate storage space at the household
Availability of curb-side collection
High frequency of collection
Clean appearance of the recycling site
Smart visual design of collection points (colour, shape, capacity)
Perceived effort, particularly the time factor, is also linked to convenience.

5.8.1.2.2 State of Knowledge and Information

As Barr pinpoints, 'knowledge for action is a significant prerequisite for behaving in an appropriate manner and would be a significant barrier to action if levels were low' (2007). He refers to abstract and concrete knowledge, i.e. general awareness of environmental problems and knowing the specific local recycling services. The continuing change of MSWMS and population mobility pose learning barriers. In combination with varying systems in different areas and buildings, recycling appears more complex to some residents. While understanding the classification is complicated for some people in general, new material-mixes and multi-component packaging further complicate the learning of how to dispose them properly and risk the creation of ever more confusion. Confusing recycling logos on packaging and a poor understanding of waste treatment activities (people do not understand the value) exacerbate knowledge barriers (Jesson et al., 2014). This may be particularly problematic for growing numbers of 'new immigrants', as they are not initially familiar with local (recycling) regulations or potentially even the language.

5.8.1.2.3 Social Norms

Fundamental for the desired conduct of household waste separation is the intention to do so. Based on the frequently applied *theory of planned behaviour (TPB)* by Ajzen (1991), the intention to recycle can be predicted by positive attitude towards recycling, social norms representing the expectations of relevant people ('social pressure') and perceived behavioural control, which is the feeling of being able to perform the intended recycling behaviour. Social norms are determined by *local culture and context*. They control behaviour without legal enforcement; adherence is rewarded with societal approval. Applied to recycling, the influence of others (e.g. family, neighbours, friends) offers guidance and can create more positive intentions to recycle and enhance the consequent action especially in places with a high visibility of behaviour. In consonance with *Bandura's social learning theory*, proposing that in the social learning system new patterns of behaviour can be acquired through direct experience or by observing the behaviour of others, *parental behaviour* was found to impact the children's subjective norms about recycling. Thomas and Sharp (2013) suggest that the 'social pressure' arising from the knowledge that others recycle is more powerful than the one created by other people having the expectation that one should recycle. However, it appears that in well-established recycling schemes social pressure is not significantly correlated to recycling.

5.8.1.2.4 Moral Norms

Additionally internal personal norms motivate household waste separation behaviour and intention. It appears that personal or moral norms internalize social norms over time. In comparison to social norms, which are executed by expressed (dis-) approval from others, moral norms are self-enforced. Following the *norm-activation model*, altruistic behaviour is mainly driven by feelings of moral obligation to act. On this basis,

> people might develop a feeling of moral responsibility to recycle, if they are aware of the negative consequences of not recycling, if they are aware that their behaviour has a significant impact on the waste problem, if they feel capable of recycling, and if they finally experience the expectation of other people they value to participate in the recycling program (Xu et al., 2016).

Generally, residents will sort waste if they feel a personal responsibility to do so, which, when acknowledged, can be viewed as civic duty (Xu et al., 2016). Moral concerns may even lower the perceived costs of carrying out recycling. Ultimately Thomas and Sharp point out that if the expectation to recycle is included in the personal norms, the individual will identify as being a recycler.

5.8.1.2.5 Attitude and Environmental Concern

Attitude towards recycling and general environmental concern significantly influence recycling behaviour and willingness to recycle. Based on the TPB, Wan et al. (2017) explain 'experiential attitude is an individual's affective feelings toward behavior (e.g., recycling is good behavior); whereas instrumental attitude refers to an individual's evaluation of behavior's outcomes (e.g., recycling could reduce landfill burden)'. Pro-recycling attitudes are found to be the strongest motivator to recycle, influenced by 'having appropriate opportunities, facilities and knowledge' and 'by not being physically prevented to recycle (time, space, inconvenience)'.

Acknowledging the complexity of behaviour as 'a function of the organism and its environment', the attitude-behaviour-context theory was developed and states that the dependence of behaviours on attitude decreases with difficulty, time effort and costs unless the context encourages it (such as reward for or requirement of the behaviour). Applied to recycling, the external structural conditions were found to impact the attitude–behaviour relationship. Attitudinal barriers to recycling can be beliefs (no environmental/social benefits, no personal reward for efforts), common misconceptions and concerns (regarding food waste, e.g. hygiene and smell) (Jesson et al., 2014).

5.8.1.2.6 Recycling Habit

As a behaviour that is carried out on a daily basis under similar conditions, recycling has the potential to turn into habitual behaviour. Turning the established household pattern of using a single bin into the 'automatic' sorting of waste into different containers requires strong internal motivation and is affected by attitude and social norms (Li et al., 2017). Consideration of each step in the recycling process (e.g. sorting, washing, storing, disposal) is recommended to understand it. The frequency of past behaviours is important, but also a stable context – constant awareness of action is not necessary. The conceptual framework of the Waste and Resources Action Programme in the UK refers to the 'concept of recycling competence', according to which the ideal state of habitual optimum household recycling behaviour is developed by a learning cycle in which people 'move from a state of unconscious incompetence [not aware, don't care], through conscious incompetence (recognition of the need to change), to conscious competence (learning to be better) and finally to unconscious competence [embedded systematic routine]' (Jesson et al., 2014). On the other hand, these changes form an obstacle for people who have already established a habit of recycling (Jesson et al., 2014). Lack of recycling habit can in turn impact the intention to recycle, independent of past recycling behaviour.

5.8.1.2.7 Recycling Habit

Another driver of recycling behaviour is the household's system trust in the local authority and the recycling programme. System or institutional trust is 'created and sustained by the continual, ongoing, and confirmatory experience of a social system's functioning. Residents with good relationships with the local government are more receptive to the recycling requirements' (Jesson et al., 2014) and shown to perform better (Xu et al., 2016). Commitment is undermined by mistrust. People's relationship with place and the local neighbourhood is also significant. As highlighted before, population mobility in big cities is increasing and so is the percentage of private renters and short- to medium-term tenants. These people often lack a direct service relationship with the local authority (or don't even know who that is), which represents another challenge for improving recycling behaviour (Jesson et al., 2014). Additionally, 'sense of community', referring to the level of social interaction, is regarded as increasingly important.

5.8.1.2.8 Economic Factors

Moreover, it is suggested that economic factors – perceived convenience and financial costs – strongly influence household recycling behaviour. People are said to be primarily utility maximizers motivated by costs and benefits. Lower socio-economic groups tend to spend less effort on recycling activities due to the simple fact that they have more pressing needs. On the other hand, financial incentives may be an effective driver for low-income households, e.g. benefiting from selling recyclables: It depends on the right 'stimulus'. Another related economic factor is the households' willingness to pay for the recycling service.

5.8.1.2.9 Political Background

Laws and regulations are influential external factors that increase the pressure on people to recycle and can influence moral norms. While in some areas the acceptance of legal norms is the most common reason for recycling, they appear less influential in others.

5.8.2 Good Practices to Improve the Recycling Performance of Urban Wastes

5.8.2.1 Consistent Laws and Policies

The legal requirement of household recycling in the cities with the highest reported rates varies significantly. "In most cases people do not recycle simply because they are not required to do so," said Tong Yen Wah, co-director of the Energy and Environmental Sustainability Solutions for Megacities programme at the NUS Environmental Research Institute. In a report for the European Commission the legal transpositions of separate collection schemes in the 28 EU capitals were reviewed. Regarding the setting of legislation, the introduction of mandatory separate collection systems for certain municipal waste fractions (waste paper, packaging, bio-waste) was recommended. Waste separation at the household level needs to become part of a new WM policy or by-law to enhance reuse. In Germany, which has a long legal recycling history, the so-called circular economy law obliges private households to separate their waste since 2015 with different regulations at the local level. In the US, Woodbury (New Jersey) became the first city to mandate recycling and promoted a curb-side pickup programme. The residents reacted by throwing trash on the lawn of then-mayor Don Sanderson, but within three months the city reached 85% compliance (Goodyear). As another 'recycling leader' worldwide, San Francisco adopted a Zero Waste Strategy – aiming to 'send zero waste to landfill' by 2020 (SF Environment, 2003). The local municipal 'Mandatory Recycling and Composting Ordinance' was passed in 2009, requiring all residents to separate their recyclables at source (SF Environment, 2009). Supported by organizations such as Zero Waste Europe (ZWE) and the Zero Waste International Alliance, 'zero waste' is presented as a holistic alternative solution for MSWMS (Connett, 2013). Cities worldwide have included this aspirational goal in their legislation and strategies to tackle waste problems, working towards sustainable development goals, optimum recycling and resource recovery simultaneously. Policies can set the base for social norms by increasing the visibility of recycling efforts to neighbours. Consistent recycling laws and policies should be implemented by all authority bodies, and their enforcement should be implemented according to the actual situation respectively (Jesson et al., 2014). Austria's former Environment Minister Nikolaus Berlakovich stated several factors as crucial to the success of recycling in Austria: The regulatory framework prescribing the requirements for a sustainable waste treatment, the implementation of economic incentives (producer responsibilities and landfill/incineration tax), regular waste controls and monitoring systems and lastly education and training programmes in combination with comprehensive information campaigns.

In a study based on 46 interviews with waste management experts from Bulgaria and Belgium the 'proper legal framework, control, and common and coordinated use of fiscal and eco-instruments' also turned out to be the top three factors when it comes to the successful implementation of policy instruments (*Gechev* et al., 2017).

5.8.2.2 Positive Image of Authority

To raise awareness, public relations and a positive image of the local WM authority play a very important role, particularly when a new separate collection system is implemented. Jesson et al. (2014) suggest that 'the persuasive influence of local authority messages and communications is enhanced by the existence of a strong and trusted relationship between people and their neighbourhood, and with their local council'.

In Hong Kong urbanites perceived the MSWMS as 'broken' or 'non-existent' and the results were low participation and system distrust. The development of strong leadership in local authorities can be improved through reliability (ability to effectively treat sorted waste), legitimacy (punitive measures) and effectiveness (dissemination of information about the programme benefits), such as strict refusal of unsorted waste by the communal collectors. Vienna, already enjoying a good image amongst citizens, uses the easy identification of all city collection vehicles through corporate design as a permanent positive advertisement for the authority. Noteworthy is the unusual popularity witnessed by Sweden's recycling brand Pantamera by turning Cuban song 'Guantanamera' into its official jingle and sponsoring a series of songs using the tune to go with their name meaning 'deposit more'. They even engaged Swedish musicians to make hit songs about recycling, resulting in ear-worms with lyrics 'recycle more'.

5.8.2.3 Collaboration with Community Groups

Another good practice is inviting press and interested stakeholders (associations, NGOs) to explain/demonstrate new steps in the collection system, even better if performed highly visibly or in events (e.g. public sorting of residual waste bin, demonstrating how much recyclables are still in there, etc.). The engagement of local trusted community groups to issue messages can strengthen their impact, help to build neighbourhood pride and improve local area quality (Jesson et al., 2014). A study from Uganda states that the involvement of active environmental organizations, women's clubs, church organizations and other associations is necessary. For areas with high cultural diversity the involvement of religious, faith and local ethnic community leaders is recommended. Ecological mobilization is reinforcing the perception that sorting wastes is what one 'ought' to do. 'The more people participate in recycling programs by being a member of an ecological association, the more the country's population at large is likely to engage in behaviors such as sorting and separating domestic wastes'. The involvement of volunteers to demonstrate and interact on a personal level – the use of an NGO as a 'broker' between residents and the local authority – was found to be a key element for recycling success and habit forming (Xu et al., 2016). The recognition of the informal sector (e.g. waste pickers) and inclusion in the MSWMS are also recommended.

5.8.2.4 Communication

Communication with residents is key to ensuring that they buy into new waste and recycling services and use them effectively. Persuasive communication has been found to produce attitude/

behaviour change and was already used during the 1980s to encourage recycling and 'manipulate' public commitment. The framing of messages in terms of content and emotional tone significantly influences how they are received. The use of humour is also recommended. A communication strategy is more likely to work if it addresses a specific audience, is worded in language familiar to that audience, addresses a specific problem commonly experienced by that audience, presents a solution to that problem and is directed to specific communities. Promotional recycling messages need to be customized and targeted in order to better link the different barriers faced by different segments of the population and maximize their impact. Content and message channels should both be highly contextually specific to the infrastructure and system situation (Jesson et al., 2014) as well as culturally sensitive. Communication is suggested to improve the acceptability of WM in the community. Increased acceptance leads to higher participation levels in communal waste activities which in turn improves communication as people have the same level of understanding. Communication strategy approaches can be based on mass information (e.g. television, radio, magazines and newspapers) or direct information (e.g. direct mailings, informative sessions, neighbourhood billboards, distribution of calendars with collection schedule). The success of an informational intervention depends on the interplay among content, media and format. Information needs to be permanently disseminated in order to help sustain a newly learned behaviour. Additionally, it is underlined that patience is required until the waste collection system 'takes root in society'.

5.9 CONCLUSION

With the rapid growth of China's population and the acceleration of industrialization, the output of urban solid waste is increasing. It brings serious harm to the ecological environment, human health and sustainable development of China. The management of solid waste in a scientific manner is one of the major issues faced by China. According to the different sources and hazards of solid waste, it is generally divided into domestic solid waste, industrial solid waste and hazardous solid waste. At present, incineration, landfilling, anaerobic digestion and composting are applied in solid waste treatment in China. However, there are problems existing in the disposal of solid waste such as imperfect laws and regulations, insufficient policy incentives, low residents' participation enthusiasm and so on. Based on these, it is necessary to continue implementing solid waste disposal polices, upgrade the industrial system, push forward economic and social reform and increase environmental education, which is beneficial for promoting the improvement of the solid waste management level.

Therefore, public awareness of segregating the MSW is mandatory, and for effective segregation government should make a plan to provide incentives and rewards to those who are doing their job effectively.

REFERENCES

Ajzen, I., 1991. The theory of planned behavior. *Organ. Behav. Hum. Decis. Process.* 50, 179–221.

Ando, A.W., Gosselin, A.Y., 2005. Recycling in multifamily dwellings: does convenience matter? *Econ. Inq.* 43, 426–438.

Awasthi, M.K., Pandey, A.K., Khan, J., Bundela, P.S., et al., 2014. Evaluation of thermophilic fungal consortium for organic municipal solid waste composting. *Bioresour. Technol.* 168, 214–221.

Barr, S., 2007. Factors influencing environmental attitudes and behaviors. *Environ. Behav.* 39, 435–473.

Bernstad, A., La Cour Jansen, J., Aspegren, A., 2013. Door-stepping as a strategy for improved food waste recycling behaviour-evaluation of a full-scale experiment. *Resour. Conserv. Recycl.* 73, 94–103.

Gechev, R., Dentchev, N., Bozhikin, I., 2017. Environmental fiscal policy and other instruments for sustainable and efficient municipal solid waste management: a cooperative study of Belgium and Bulgaria. *Waste Management* 65, I–III.

Connett, P., 2013. *The zero waste solution: untrashing the planet one community at a time.* Chelsea Green Publishing, White River Junction, VT, USA.

Chinese Environmental Yearbook Compiling Committee, 2011. *China environmental yearbook 2011*. China Environmental Science Press, Beijing, China.

Du, W.P., 2006. A study on greenhouse gas (methane) emissions in landfill of municipal solid waste (MSW). Master thesis, 14.

EC, 2015. *Closing the loop—an EU action plan for the circular economy, 2015 COM(2015) 614 final*. European Commission (EC), Brussels/BE. https://eur-lex.europa.eu/legal-content/EN/ALL/?uri=CELEX:52015DC0614.

EU, 2008. *Waste framework directive*. Directive 2008/98. European Parliament and the Council of the European Union (EU). http://data.europa.eu/eli/dir/2008/98/oj.

Ezeah, C., Roberts, C.L, 2012. Analysis of barriers and success factors affecting the adoption of sustainable management of municipal solid waste in Nigeria. *J. Environ. Manag.* 103, 9–14.

Goldstein, J., 2016. Just how "wicked" is Beijing's waste problem? A response to "The rise and fall of a "waste city" in the construction of an "urban circular economic system": the changing landscape of waste in Beijing" by Xin Tong and Dongyan Tao. *Resour. Conserv. Recycl.* 117, 171–182.

Jain, S., Jain, S., Wolf, I.T., et al., 2015. A comprehensive review on operating parameters and different pretreatment methodologies for anaerobic digestion of municipal solid waste. *Renewable Sustain. Energy Rev.* 52, 142–154.

Jesson, J.K., Pocock, R.L., Stone, I., 2014. *Barriers to recycling: a review of evidence since 2008*. Waste & Resources Action Programme (WRAP), Birmingham.

Jiao, X.D., Du, HZ., 2019. Study on the present situation, problems and environmental management countermeasures of urban waste disposal in China. *J. Recycl Resour Circular Econ.* 1(12), 18–22.

Kalyani, K.A, Pandey, K.K., 2014. Waste-to-energy status in India: a short review. *Renew. Sustain. Energy Rev.* 31, 113–120.

Kumar, S., 2010. Composting of municipal solid waste. *Crit. Rev. Biotechnol.* 31(2), 112–136.

Li, C.J., Huang, Y.Y., Harder, M.K., 2017. Incentives for food waste diversion: exploration of a long term successful Chinese city residential scheme. *J. Clean. Prod.* 156, 491–499.

Liu T., Awasthi M.K., Chen H., Duan Y., et al., 2019. Performance of black soldier fly larvae (Diptera: Stratiomyidae) for manure composting and production of cleaner compost. *J. Environ. Manage.* 251, 109593.

Miafodzyeva, S., Brandt, N., 2013. Recycling behaviour among householders: synthesizing determinants via a meta-analysis. *Waste Biomass Valori.* 4, 221–235.

Mosannenzadeh, F., Bisello, A., Vaccaro, R., et al., 2017. Smart energy city development: a story told by urban planners. *J. Cities.* 64, 54–65.

National Bureau of Statistics of People's Republic of China. 2016. *China Statistical Yearbook 2016*. China Statistics Press, Beijing, China.

NBSR 2016. *China statistical yearbook*. China Statistics Press, Beijing.

Recycling Magazin, 2018, April 17. bvse: recycling braucht verbesserte Mülltrennung. https://www.recyclingmagazin.de/2018/04/17/bvserecycling-braucht-verbesserte-muelltrennung.

Tang, S., Hao, P., Huang, X., 2016. Land conversion and urban settlement intentions of the rural population in China: a case study of suburban Nanjing. *Habitat Int.* 51, 149–158.

Tonglet, M., Phillips, P.S., Read, A.D., 2004. Using the theory of planned behaviour to investigate the determinants of recycling behaviour: a case study from Brixworth, UK. *Resour. Conserv. Recycl.* 41, 191–214.

Thomas, C., 2001. Public understanding and its effect on recycling performance in Hampshire and Milton Keynes. *Resour. Conserv. Recycl.* 32, 259-274.

Thomas, C., Sharp, V., 2013. Understanding the normalisation of recycling behaviour and its implications for other pro-environmental behaviours: a review of social norms and recycling. *Resour. Conserv. Recycl.* 79, 11-20.

Trivedi, A., Singh, A., Chauhan, A., 2015. Analysis of key factors for waste management in humanitarian response: an interpretive structural modelling approach. *Int J Disaster Risk Reduct.* 14, 527–535.

UN DESA, 2013. World economic and social survey 2013. United Nations Department of Economic and Social Affairs (UN DESA). https://www.un.org/en/development/desa/policy/wess/wess_current/wess2013/WESS2013.pdf accessed June 17, 2018.

Varotto, A., Spagnolli, A., 2017. Psychological strategies to promote household recycling. A systematic review with meta-analysis of validated field interventions. *J. Environ. Psychol.* 51, 168-188.

Wan, C., Shen, G.Q., Choi, S., 2017. Experiential and instrumental attitudes: interaction effect of attitude and subjective norm on recycling intention. *J. Environ. Psychol.* 50, 69–79.

Xu D.Y., Lin Z.Y., Gordon M.P.R., et al., 2016. Perceived key elements of a successful residential food waste sorting program in urban apartments: stakeholder views. *J. Clean. Prod.* 134, 362–370.

Yang, C., Yang, M., Yu, Q., 2012. An analytical study on the resource recycling potentials of urban and rural domestic waste in China. *Procedia Environ. Sci.* 16, 25–33.

Zhang, D.Q., Tan, S.Q., Gersberg, R.M., 2010. Municipal solid waste management in China: status, problems and challenges. *J. Environ. Manag.* 91, 1623–1633.

Zhao, P., Shen, Y., Ge, S., et al., 2014. Clean solid biofuel production from high moisture content waste biomass employing hydrothermal treatment. *Appl. Energy.* 131, 345–367.

Zhao, S.L., Huang, N.S., 2008. Development of comprehensive treatment of the waste plastics pollution. *J. Ecol. Environ.* 17, 2473–2381.

Zhou, B.Y., Sun, C.X., Yi, H.T., 2017. Solid waste disposal in chinese cities: an evaluation of local performance. *Sustainability.* 9, 2234.

6 Solid Waste Management Practices in India and China

Sustainability Issues and Opportunities

S. Sri Shalini, Kurian Joseph, Binghua Yan, O. P. Karthikeyan, K. Palanivelu and A. Ramachandran

CONTENTS

6.1 INTRODUCTION: BACKGROUND

The rapid increase in volume and types of solid and hazardous waste as a result of continuous economic growth, urbanization and industrialization is a serious concern for Asian cities, especially in China and India with about one-third of the world's population. India and China have become the fastest growing populations in the world in the past few decades, facing major challenges in achieving the United Nations (UN) Sustainable Development Goals (SDG) related to eradicating poverty (SDG 1), achieving greater equality (SDGs 5 and 10), combating hunger and malnutrition (SDG 2) and strengthening the coverage and quality of health and education systems (SDGs 3 and 4) (UN Population 2019).

The UN has also addressed waste management as a priority in achieving SDG and laid the stress on a shift from waste to resource management as part of responsible consumption and production. Although considerable efforts are being made by many Governments and other entities in tackling waste-related problems, there are still major gaps to be filled in this area. The solid waste practices in India and China are facing similar challenges with respect to the different elements of waste management. This chapter critically addresses the waste management practices in India and China, especially the gaps between solid waste generation and its treatment, assessment of available waste treatment methods, greenhouse gaseous emission inventory for solid waste and its climate change challenges, review of existing policies with solid waste management and based on the above outcomes a future recommendation for a sustainable holistic approach in solid waste management (SWM) from its generation, collection and treatment to its disposal are proposed.

6.2 DEMOGRAPHY, URBANIZATION AND WASTE GENERATION

China is the most populous country and the world's third largest nation by area (9,596,961 km^2), followed by India as the second most populous country in the world and the seventh largest nation by area (3,287,263 km^2). The land area of India is about 2,973,190 km^2, making it the third largest country in Asia, and the largest country in South Asia, and it is bounded by the Indian Ocean in the south, the Arabian Sea in the southwest and the Bay of Bengal in in the southeast; it shares land borders with Pakistan to the west, China, Nepal and Bhutan to the north and Bangladesh and Myanmar to the east. The land area of China is about 9,326,410 km^2, making it the second largest country in the world by land area and the second largest country in Asia, and it is a country in East Asia, adjacent to Vietnam, Laos, Myanmar, India, Bhutan, Nepal, Pakistan, Afghanistan, Tajikistan, Kyrgyzstan, Kazakhstan, Russia, Mongolia and North Korea.

The population growth and urbanization in these developing countries are the main cause for the billions of tonnes of solid waste generation. The population growth and population density projections of India and China are depicted in Figure 6.1. The Indian population has grown steadily (Figure 6.1a) to 1.38 billion people (2020) with a population density of 406 people per square kilometre (world rank 19th) at an annual growth rate of 0.99%. The current population of China is

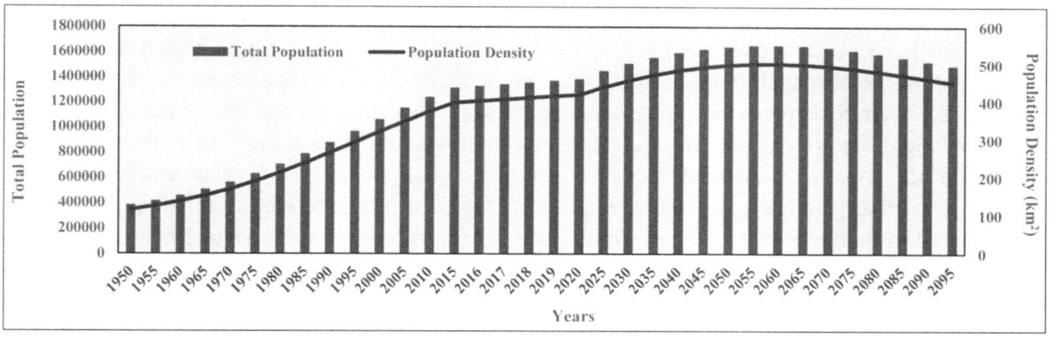

a) Indian population (1950 – 2095)

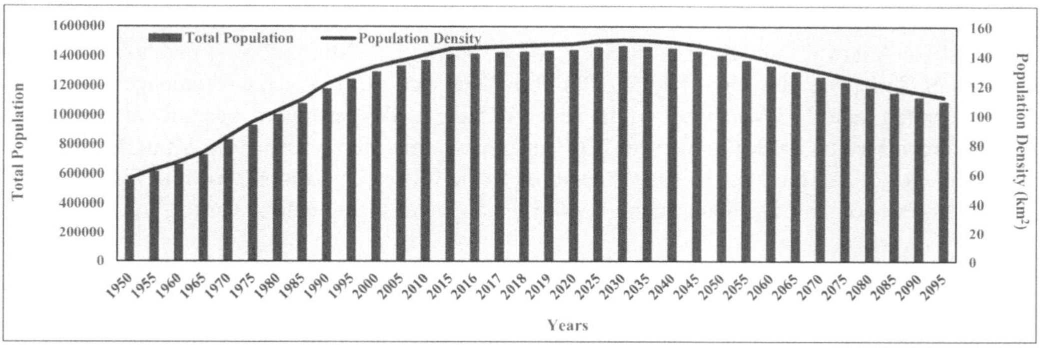

b) Chinese population (1950 – 2095)

FIGURE 6.1 Population growth and population density projections of India (a) and China (b). (a) Indian population (1950–2095). (b) Chinese population (1950–2095). (Source: Prepared by author (adapted from UN DESA 2018).)

around 1.4 billion people with a population density of 148 people per square kilometre (world rank 83rd) at an annual growth rate of 0.39%. Even though both the countries have a similar population in total, the population density difference highlights that India is 2.7 times denser than China. Most notably, the Indian population is predicted to exceed China, and it will soon become the world's most populous country in 2024. Whereas the Chinese population is expected to peak in 2028, then decrease in the succeeding years (Figure 6.1b). The capital city of India (New Delhi) is the world's second largest city with an agglomeration of 29 million, followed by Shanghai in China with 26 million. Recently, cities like Mumbai, India, and Beijing, the capital city of China, have close to 20 million inhabitants. Delhi is anticipated to grow significantly and turn out to be the world's most populous city in 2028 (UN DESA 2018). The population increase in these countries is in turn playing a significant part in urbanization and migration from rural to urban areas.

Despite the fact that India has the biggest rural population of 893 million and China has a rural population of 578 million, the urbanization rate is growing rapidly in these countries due to rural exodus in search of livelihood opportunities. By 2050, it is expected that India will have added 416 million urban dwellers and China will have an urban population of 255 million (UN DESA 2018). In 2018, 59.8% of the population of China lived in urban areas, which was a massive increase from 17.92% in 1978. Currently, India is at 34% urbanization rate, and it is expected to escalate to 50% by 2030 due to the population growth. The population residing in urban regions of India increased from 18% in 1961 to 31.2% in 2011 (Joshi and Ahmed 2016). Especially between the years 2001 and 2011, the urban population increased at a rate of 31.8%, i.e. 285 million to 377 million population, respectively, and 70% of this population resides in 366 cities (Singh et al. 2018).

Migration from rural areas to the urban areas has triggered an exponential increase in municipal solid waste (MSW) generation and changed the waste characteristics with respect to the economic status and living standards of people (Liu et al. 2019; Singh et al. 2018). India and specifically China both have imbalances in their per capita urban waste generation rates with respect to economic status, as enormous rural populations are poverty-stricken and deprived (Hoornweg and Bhada-Tata 2012). People in urban areas use nearly twice as many resources per capita than those people surviving in rural areas, and as a result urban areas generate more waste. In 2014, 377 million people resided in urban areas of India, generating 62 million tons of MSW per annum, and this is expected to increase to 165 million tons of waste by 2031 and 436 million tons by 2050 (Planning Commission Report 2014). MSW generation over the past decades in both countries is tremendous. The solid waste generation in different states and provinces of India and China and changes in their waste composition for the past two decades are shown in Figures 6.2 and 6.3, respectively. In India, MSW generation in different states has increased massively for the past twenty years (2000–2020) (Figure 6.2a). In 1999–2000, MSW generation was at 52,125 MT/day; this increased to 127,485 MT/day in 2010, 139,566 MT/day in 2015 and reached 147,613 MT/d in 2020 (Ahluwalia and Patel 2018; SBM 2020a; Singh et al. 2018). MSW generation in different provinces of China was also high (Figure 6.2b). Provinces such as Liaoning, Guangdong, Sichuan and Hubei generated more MSW in 1995. In the year 1980, the MSW generation was only around 30 Mt, whereas it increased to 100 Mt in the year 2000 and nearly amplified twice to 200 Mt in the year 2010 (Wang and Nie 2001; Zhang et al. 2010). According to the waste composition shown in Figure 6.3, the biodegradables or organic fraction is about 50–60%, mainly consisting of food wastes, and the quantity

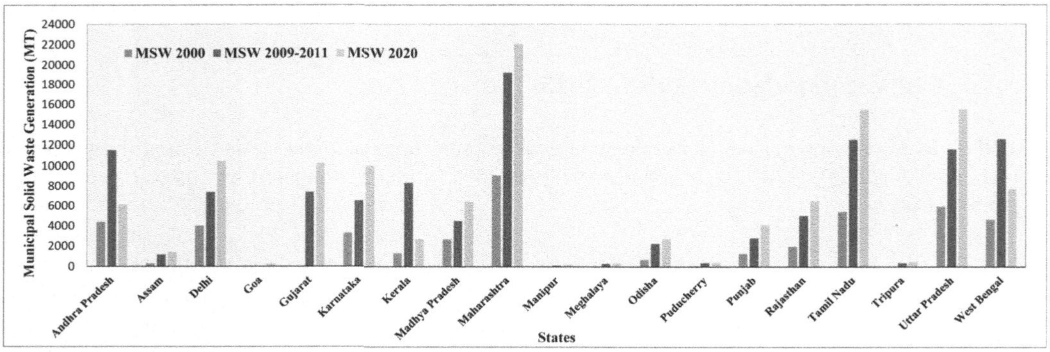

a. MSW generation in different Indian states

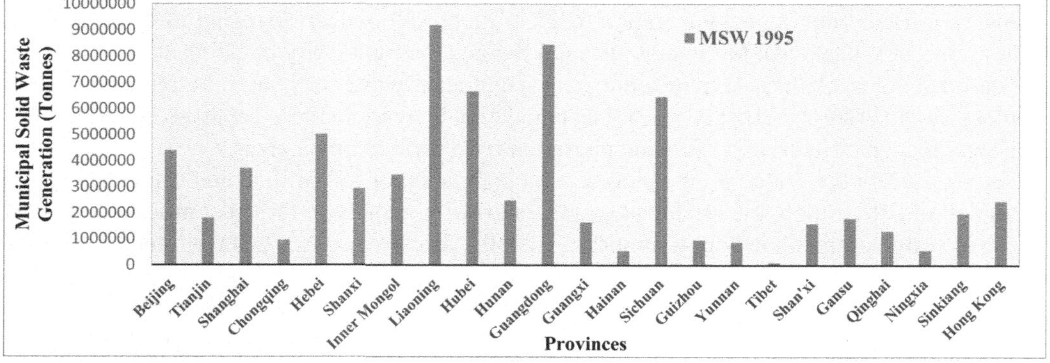

b. MSW generation in different Chinese provinces

FIGURE 6.2 MSW generation in different Indian states and Chinese provinces. (a) MSW generation in different Indian states. (b) MSW generation in different Chinese provinces. (Source: Prepared by author.)

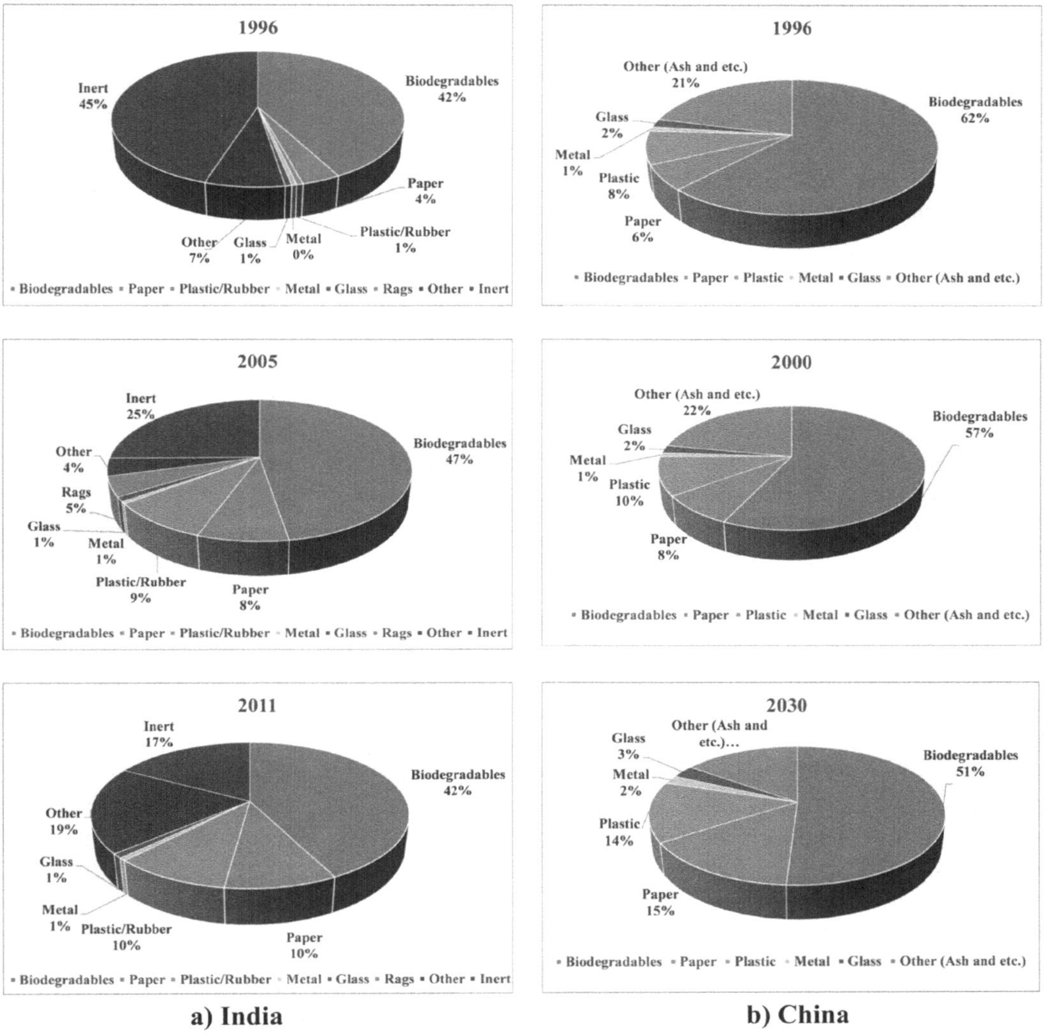

FIGURE 6.3 Changes in the waste composition in India and China for the past two decades. (Source: Prepared by author (adapted from Liu 2017; Planning Commission Report 2014).)

of inert waste is decreasing in both countries. The plastic waste was very low in the 1990s, whereas it is in an increasing trend now. The glass and metal composition is <4%.

Per capita MSW generation in major Indian cities ranges from 0.2 to 0.5 kg per day. MSW generation in bigger Chinese urban cities show much higher rates than the rural areas. The urban city of Beijing generates about 310 kg/capita/year, Shanghai 405 kg/capita/year, Chongqing 394 kg/capita/year in, Hangzhou 427 kg/capita/year and Hong Kong 485 kg/capita/year, whereas the rural areas of China generate a very low rate at 90.2 kg/capita/year (Dorn et al. 2010).

MSW management is a primary issue for many Indian cities due to the rising urban population and per capita waste generation rate, no proper public participation and unacceptable organizational and financial capacities of urban local bodies (Joseph et al., 2012). It is also the same in the case of China with its urban solid waste management. A high fraction (55%) of biodegradable waste ending up in dumpsites is a major constraint in urbanization, which could be avoided if urban areas are well connected with rural areas, where such waste can be utilized as animal feed or in the preparation of compost. The urban public service of solid waste management has become a major challenge. Solid waste will have a direct effect on the life and health of the people.

6.3 IMPACTS OF SOLID WASTE

Improper solid waste disposal will cause public health problems as well as water, soil and air pollution. Open dumping/landfills directly emit greenhouse gases (GHGs) into the atmosphere and affect the global temperature, resulting in an environmental threat. The dumping of solid wastes leads to substantial nutrient reductions (nitrate, potassium and phosphorus), reduces the ability to recycle or reuse waste and leads to high chances of groundwater pollution due to leachate generation (Kumar et al. 2017; Singh et al. 2018). The release of GHGs from landfills sometimes causes fire incidents (TERI 2019). The smoke from the open burning of MSW contains compounds of carcinogenic and hazardous nature such as polychlorinated dibenzo-p-dioxins and dibenzofurans as well as particulate matter, black carbon, carbon monoxide, carbon dioxide, nitrous oxide and other GHGs (Ferronato and Torretta 2019). The improper disposal of plastic wastes, usually litter in various parts of urban areas, creates a nasty landscape and chokes the drains (TERI 2019).

MSW open dumping generates leachate with high concentrations of heavy metals, BOD, COD, sulphate and ammonia which is released to the soil, polluting groundwater mainly used for drinking and household purposes. The risks concern the health of the population through direct and indirect (agriculture) intake. The open dumping of waste tires is an area of mosquito growth, especially in tropical areas, where dengue, malaria and yellow fever, among other diseases, are common. The breeding of flies and rats in the MSW dumps causes cholera, diarrhoea, dysentery, typhoid, hepatitis and plague. The impacts of improper waste management on public health causes higher chances of nose, skin, eye and throat infections, breathing problems, inflammation, bacterial infections, anaemia, reduced immunity, allergies, asthma and other infections (Ferronato and Torretta 2019; Kumar et al. 2017; Kurian Joseph 2014). The hazardous part of biomedical waste presents physical, chemical and/or microbiological risk to the general population and health care workers associated with the handling, treatment and disposal of waste (Datta et al. 2018). Open dumping of hazardous waste (HW) leaks into groundwater through drainage network, making it poisonous and unfit for drinking. The unscientific disposal of hazardous waste poses a high risk to the public health and environment. Methane gas from hazardous waste causes health diseases (lung infection, heart diseases and cancer) and pollutes surface or groundwater, making it unsuitable for any use. Improper burning of hazardous wastes and other untreated wastes releases toxic gases with dioxins, furans, mercury, heavy metals and carcinogens that cause air pollution and health issues, sometimes affecting future generations with reproductive abnormalities, physical deformities, permanent disorders and even deaths. The leaching of hazardous waste compounds into soil and land affects their quality. Working with unscientific hazardous waste handling practices leads to neurological disorders, skin diseases and genetic defects (Kothwale 2019; PIB 2016).

There are serious impacts on the health of informal sector workers and as a consequence a threat to their basic survival and socio-economic distress. During the collection of hazardous wastes and e-waste recycling, there are major impacts on informal sector workers' haemopoietic systems, nervous systems, renal systems, respiratory systems, dermal systems, metabolic systems and gastrointestinal systems, also causing cancer (Chaturvedi and Bhardwak 2013).

6.4 LEGAL FRAMEWORK FOR SWM IN INDIA AND CHINA

Both India and China have a legal framework for solid waste management in the handling and disposal of various solid wastes.

6.4.1 Framework for Solid Waste Management

The legal frameworks of the Government of India and the Government of China on solid waste management are summarized in Table 6.1, and the SWM institutional framework with the roles and responsibilities of different agencies is given in Figure 6.4.

TABLE 6.1
Legal Framework for SWM in India and China

S. No	Year	Rules, Policies, Schemes, Financial Plans, Technical Documents on Solid Waste Management	
		India	China
1.	**1974 and 1976**	**1976 –** Article 48A, directive principle, Part IV and Article 51 A(g), Part IVA, of the amendment of Constitution of India in 1976 (Waste Management mention in Environmental Policy)	**1974 –** Environment Protection Leading Group of the State Council was established The People's Republic of China started to carry out disposal of waste water, gases and solid waste
2.	**1986**	Environmental Protection Act 1986 Factories Act 1948 and its amendment in 1987 (Waste Management mention in Environmental Policy)	Environmental Protection Law (enacted on a trial basis in 1979)
3.	**1989 and 1991**	**1989** – The Hazardous Waste (Management & Handling) Rules	**1991** – Entry into force of the Basel Convention on the Control of Trans-boundary Movements of Hazardous Wastes and Their Disposal **1992** – City appearance and environmental sanitary management ordinance by State Council **1993** – Regulation regarding municipal residential solid waste by Ministry of Construction of PRC
4.	**1994, 1995 and 1996**	**1994–1995** – MSWM strategy paper by NEERI J.S. Bajaj Committee (The High Powered Committee on Urban Solid Waste Management)	**1996** – Law on Pollution Prevention Caused by Solid Waste
5.	**1998**	Bio-medical Waste (Management & Handling) Rules Supreme Court appointed Barman Committee	National Hazardous Waste Catalogue (Amended in 2008)
6.	**1999 and 2000**	**2000** – Municipal Solid Waste (Management & Handling) Rules CPHEEO Manual on MSW	**1999** – Administrative Measures for Hazardous Waste Transfer Manifests **2000** – Technical policies on the disposal of Domestic Waste and the Prevention of Pollution, Ministry of Construction of PRC
7.	**2001 and 2002**	Batteries (Management and Handling) Rules	**2001 –** 'Standard for pollution control on hazardous waste storage' (GB18597-2001) 'Standard for pollution control on hazardous waste incineration' (GB18484-2001) 'Standard for pollution control on the security landfill site for hazardous waste' (GB18598-2001) **2002** – Comments on promoting the industrialization of municipal waste water treatment and municipal solid waste treatment by State Planning Commission, the Ministry of Construction and State Environmental Protection Administration

(Continued)

TABLE 6.1 (CONTINUED)
Legal Framework for SWM in India and China

S. No	Year	Rules, Policies, Schemes, Financial Plans, Technical Documents on Solid Waste Management India	China
8.	**2003, 2004 and 2005**	**2005** – Report of the Technology Advisory Group on SWM JNNURM (2005–2012)—40 MSW projects costing Rs. 2,186 Cr sanctioned from a total of 65 cities covered UIDSSMT (2005–2012)—51 MSW projects costing Rs. 327 Cr sanctioned from a total of 632 cities covered 12th Finance Commission (2005–2010)—Rs. 2,500 Cr for 423 Class I cities	**2003** – Cleaner Production Promotion Law Hazardous Wastes and Medical Wastes Treatment Facility Construction Program Ordinances on Medical Waste Management Law for Environment Impact Assessment of PRC **2004** – Administrative measures on hazardous waste operation licenses Law on Pollution Prevention Caused by Solid Waste (Amended)
9.	**2006 and 2007**	**2006** – National Environment Policy (Waste Management mention in Environmental Policy) Strategy and Action Plan-Use of compost in cities **2007** – 11th Five-Year Plan (2007–2012)—Rs. 2,210 Cr for MSWM	**2007** – Identification standards for hazardous wastes: • General specifications (GB 5085.7-2007) • Identification for corrosivity (GB5085.1-2007) • Identification for extraction toxicity (GB5085.3-2007) • Identification for reactivity (GB5085.5-2007) • Identification for toxic substance content (GB5085.6-2007)
10.	**2008**	National Urban Sanitation Policy (NUSP) Service Level Benchmarks (SLBs) in MSWM Hazardous Waste (Management, Handling & Transboundary Movement) Rules National Mission on Sustainable Habitat (NAPCC)	Circular Economy Promotion Law
11.	**2010 and 2011**	**2010** – 13th Finance Commission (2010–2015)—Establishing standards for delivery of essential services Batteries (Management and Handling) Amendment Rules, 2010 **2011** – Plastic Waste (Management &Handling) Rules E-Waste (Management & Handling) Rules Draft Bio-medical Waste (Management & Handling) Rules	**2011** – Regulations on Management to Recovery and Treatment of Waste Electrical and Electronic Equipment
12.	**2012–2014 and 2016**	**2014** – Swachh Bharat Mission, October 2014 **2016** – Waste Management Rules, 2016 comprising Solid Waste Management Rules, Plastic Waste Management Rules, Bio-Medical Waste Management Rules, E-Waste Management Rules, Hazardous and Other Wastes (Management and Transboundary Movement) Rules, 2016 and Construction and Demolition Waste Management Rules, 2016	**2012** – The 12th-five-year Planning on Hazardous Waste Pollution Prevention **2013** – Circular Economy Development Strategy and Immediate Plan of Action (2013) establish sector specific development targets (for Five Year Plans)

(Continued)

TABLE 6.1 (CONTINUED)
Legal Framework for SWM in India and China

S. No	Year	Rules, Policies, Schemes, Financial Plans, Technical Documents on Solid Waste Management: India	China
		Technical documents on solid waste management under Swachh Bharat Mission (SBM) (Urban) Municipal Solid Waste Management Manual Draft Model Municipal Solid Waste (Management & Handling), Cleanliness and Sanitation RULES/BY-LAWS An Inclusive Swachh Bharat through the Integration of the Informal Recycling Sector: A Step by Step Guide	
13.	**2018 and 2019**	E-Waste Management Amendment Rules, 2018 National Policy on Electronics, 2019 Draft National Resource Efficiency Policy, 2019 **Technical documents on solid waste management under Swachh Bharat Mission (SBM) (Urban)** **2018** – Advisory on On-Site and Decentralised Composting of Municipal Organic Waste C&D waste ready reckoner Guidelines on usage of Refuse Derived Fuel in various industries Star ratings for garbage-free cities **2019** – Guidelines for disposal of legacy waste (old municipal solid waste), CPCB, MoEF	-

(Source: Adapted from CPCB 2019, Hoornweg et al. 2005; Liu 2017; MoEF & CC 2019; MoHUA 2018; MoUD 2016; TERI 2019.)

The Indian history of solid waste management, its evolution and legal framework with various rules and policies in SWM are summarized in detail in Table 6.1, and the institutional framework of SWM is shown in Figure 6.4a. The Ministry of Urban Development (MoUD), Government of India, issues policy guidelines from time to time. Waste management comes under the scope of MoUD and the Ministry of Environment, Forest and Climate Change (MoEF & CC). Pollution control boards at the central and state levels monitor compliance with service delivery benchmarks as set by MoUD. Solid waste management is one among the main missions of Swachh Bharat Mission (Urban) scheme.

The institutional framework for solid waste management in China is shown in Figure 6.4b, and the legal framework with existing policies and rules related to solid waste management is summarized in Table 6.1 with comparison to Indian policies. In China, the legislation framework for waste management is made up of the national law, the regulation promulgated by the State Council, the rule by Ministries and technology standards.

6.4.2 Definitions of Solid Waste

From the solid waste management rules of India and China, the definitions of various solid wastes are summarized below.

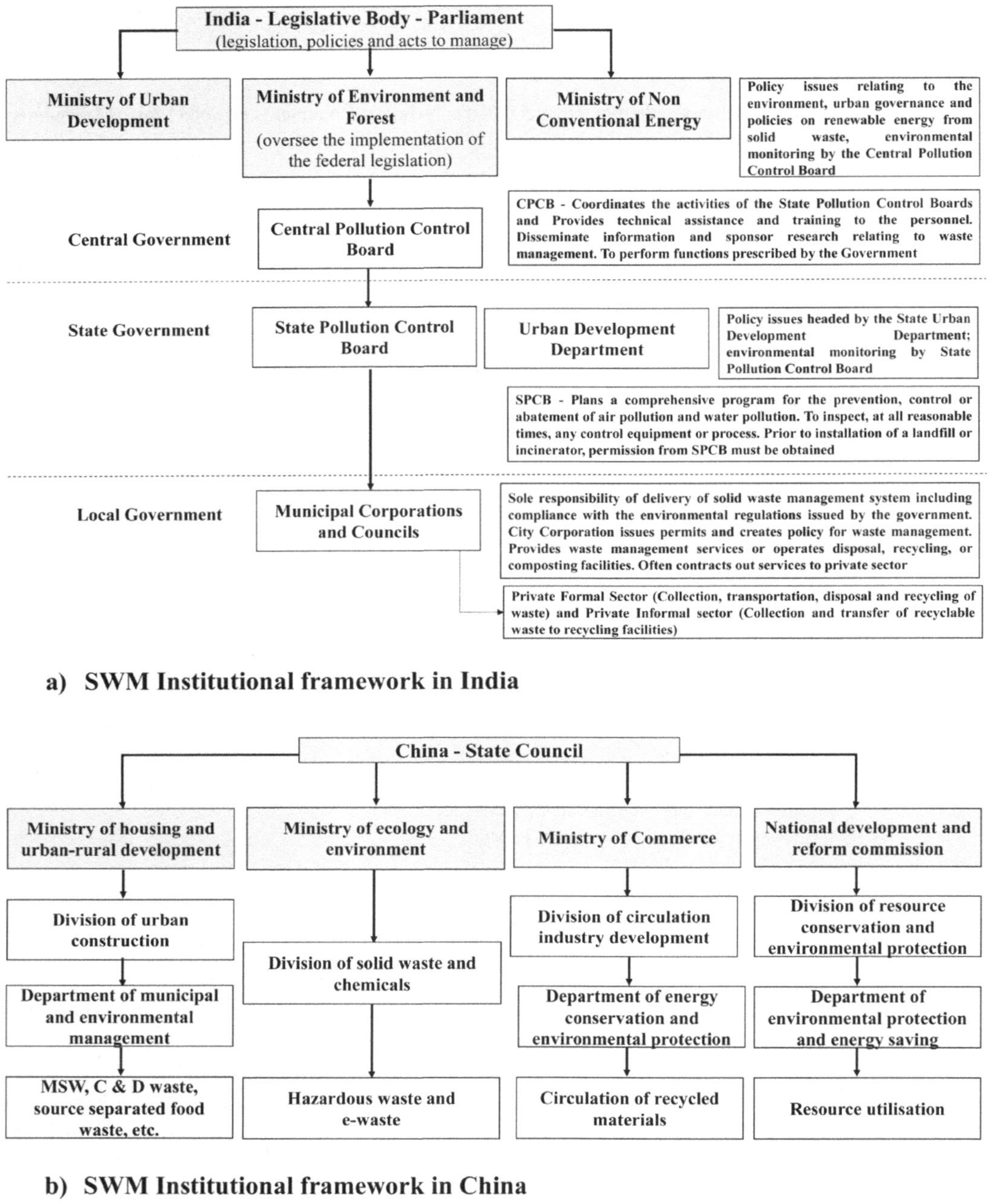

FIGURE 6.4 Solid waste management institutional framework in India and China. (a) SWM institutional framework in India. (b) SWM institutional framework in China. (Source: Prepared by author (Adapted from APO 2007; Wang et al. 2019)

6.4.2.1 India

According to the Waste Management Rules, 2016, and other related regulations (Table 6.1):

i. **Solid wastes** are solid or semi-solid domestic waste, sanitary waste, commercial waste, institutional waste, catering and market waste and other non-residential wastes, street sweepings, silt removed or collected from the surface drains, horticulture waste, agriculture and dairy waste, treated biomedical waste excluding industrial waste, biomedical waste and e-waste, battery waste and radio-active waste generated in the area under the local authorities

ii. **Hazardous wastes** are the wastes which by reason of characteristics such as physical, chemical, biological, reactive, toxic, flammable, explosive or corrosive cause danger or are likely to cause danger to health or the environment, whether alone or in contact with other wastes or substances, and shall include waste specified under column Schedule I, Schedule II and Schedule III of the Rules
iii. **Biomedical wastes** are the wastes generated during the diagnosis, treatment or immunization of human beings or animals or research activities pertaining thereto or in the production or testing of biological or in health camps
iv. **Electronic wastes** are electrical and electronic equipment, whole or in part, discarded as waste by the consumer or bulk consumer as well as rejects from manufacturing, refurbishment and repair processes
v. **Demolition and construction wastes** comprising of building materials, debris and rubble resulting from construction, re-modelling, repair and demolition of any civil structure
vi. **Plastic wastes** are any plastic discarded after use or after its intended use is over
vii. **Battery wastes** are used lead acid batteries after their intended use is over

6.4.2.2 China

According to the Law of the People's Republic of China on the Prevention and Control of Environmental Pollution by Solid Wastes (adopted in 1995, and amended in 2004) and related regulations (Table 6.1):

i. **Solid wastes** are articles and substances in solid or semi-solid state or gaseity in containers that are produced in the production, living and other activities and that have lost their original use values or are discarded or abandoned despite not having yet lost their use value, and articles and substances that are included in the management of solid wastes upon the strength of administrative regulations. 'Solid waste' is classified into three types: industrial solid waste (IW), municipal solid waste and hazardous waste
ii. **Municipal solid wastes** are solid waste discharged from everyday life or from services provided to everyday life as well as the solid waste that is regarded as municipal solid waste under laws and administrative regulations. It usually includes residential, institutional, commercial, street cleaning and non-process waste from industries. In some cases, construction and demolition waste is also included
iii. **Hazardous wastes** usually refers to industrial hazardous waste generated as a by-product of the manufacturing process, medical waste, small-scale generation of hazardous waste from households, institutions and commercial establishments, and occasionally small amounts of radioactive waste, e.g. smoke detectors and medical process waste

6.5 STATUS OF MUNICIPAL SOLID WASTE MANAGEMENT IN INDIA AND CHINA

Municipal solid waste management deals with control of the generation, collection, storage, transfer and transport, processing and disposal of MSW (Kurian Joseph 2014). The current status of solid waste management in India and China is discussed below.

6.5.1 Waste Generation and Quantity

The current quantity of waste generation in India is 147,613 MT/d (2020), and it is expected to reach 125 million tonnes by 2031 and 300 million tonnes per annum by 2051 (Ahluwalia and Patel 2018; SBM 2020a; Singh et al. 2018). The per capita generation of MSW in India varies from 200 to 700 g per day (Kurian Joseph 2014). In the last few decades, the amount of waste generated per capita has recorded an annual increase at a rate of 1% to 1.33% (Bhat et al. 2018).

MSW generation volume in China reached 234 million tons in 2016, which is an increase of 88% compared to that in 1996. Further, it is expected to increase to over 500 Mt by 2025 (Hoornweg and Bhada-Tata 2012; Liu et al. 2019; Zhang et al. 2010). Future MSW generation in China is predicted to increase about 2 times in the next 20 years (Wei et al. 2013). Liu et al. (2019) conducted a model study based on the panel data of 30 provinces in China, and showed that financial expenditure and consumption patterns have a significant effect on MSW generation and by 2021, the MSW volumes in Hubei and Sichuan province will reach more than ten thousand tons.

According to the World Bank (2012), MSW per capita rates in India and China are 0.34 kg/capita/day, and 1.02 kg/capita/day, respectively. In 2025, it will increase up to 0.70 kg/capita/day in India and 1.70 kg/capita/day in China (Hoornweg and Bhada-Tata 2012). MSW generation per capita in India ranges from approximately 0.17 kg per person per day in small towns to approximately 0.62 kg per person per day in cities (Kumar et al. 2017). A study by Ramachandra et al. (2018) showed that increases in family income and consumption rates led to more solid waste generation in Bengaluru, India. The changing life-style of people, their socio-economic status, cultural habits, climate, location, urban structure, density of population, extent of non-residential activities, food habits, gender, age, income, etc., act as major aspects in MSW waste generation rates, per capita rates and its varying composition (Bhat et al. 2018; Qu et al. 2009; Khan et al. 2016; Kurian Joseph 2014; Ramachandra et al. 2018).

6.5.2 Waste Composition

Based on the above social and environmental factors, the waste composition varies widely in India and China. MSW in India is approximately 40–60% compostable, 30–50% inert waste and 10–30% recyclable (Joshi and Ahmed 2016). China and India traditionally use coal as a household fuel source. The ash that is produced is very dense and tends to dominate the waste stream in terms of weight. Ash is included in the 'others' category and makes up 45% and 54% of India and China's waste composition, respectively (World Bank 1999). Khan et al. (2016) studied the Dhanbad municipality in Jharkand, India, and differentiated the population based on socio-economic parameters of education, occupation, income of the family and number of family members. The average MSW generation was 0.41 kg/capita/day, out of which 0.19 kg/capita/day was biodegradable and 0.22 kg/capita/day was non-biodegradable. Maximum waste was generated by the lower middle socio-economic group with average waste generation of 0.46 kg/capita/day. Plastic waste was maximum (15%) in the higher socio-economic group (HSEG) and minimum (1%) in the lower socio-economic group (LSEG). Food waste was the major component of generated waste in almost every socio-economic group with maximum (38%) in the case of HSEG and minimum (28%) in LSEG. A similar study was carried out in Beijing, China, by Qu et al. (2009) on the waste generation rate and its composition with varying socio-economic factors of family population, income, age and education. The per capita waste generation of six selected districts of Beijing, China, was 0.233 kg/capita/day with a composition of kitchen waste – 69.3% as the major component, paper/cardboard – 10.3%, plastics – 9.8%, textiles – 1.3%, metals – 0.8%, glass – 0.6% and other wastes – 2.7%. Families with a secondary education level produced less total household waste and kitchen waste than those with a primary or advanced education level. In contrast, families with a higher education level produced more waste paper and plastics. In conclusion middle-income families generate more household waste than low-income and high-income families.

6.5.3 Waste Segregation

The main issue of MSWM, especially in urban India and China, is the lack of any segregation of MSW and mixing up of other wastes with MSW. No proper segregation is carried out at the household or community bin level in either country. The absence of segregation leads to improper scientific disposal of waste (CPCB 2000). Sometimes the sorting of waste is carried out by informal sector workers in an unsafe manner.

6.5.4 Recovering and Recycling at Source

India and China strongly promote the practice of reusing and recycling materials as input production materials. The most common recyclable materials include paper, cardboard, glass, plastics, metals, etc. In India, unsegregated waste is dumped at community bins; its optimal recycling is not possible. However, rag-pickers usually sort out and take and sell recyclable material like plastics, glass, etc. The recycling rate of paper and paper-based products is at 27% in India. Plastic waste recycling is at 8 to 9%, of which about 60% is recycled, mostly by the informal sector. The polyethylene terephthalate (PET) plastic recycling rate is at 90%. A few states such as Tamil Nadu, Himachal Pradesh and other states regularly lay roads using plastics. In Tamil Nadu, 1,400 km of rural tar roads used plastic in 2003–2004, consuming about 1 tonne of waste per kilometre of single-lane road. In 2015, the Central Road Research Institute, India, mandated plastic roads for all National Highways up to 50 km from cities that have a population of over half a million (Ahluwalia and Patel 2018).

In China, recycling is under the Ministry of Commerce as resource management is considered an important economic activity. On the other hand, municipal solid waste management is under the Ministry of Housing and Urban-Rural Development (widely known as the Ministry of Construction) (Wilson et al. 2010). Liu et al. (2015) studied the recyclables in solid waste and the handling capacities of recycling centres. The study covered 500 families across 16 districts in Beijing, China. The total quantity of recyclable household solid waste was 1.805 million tons in 2013, and 72% of that was collected. The main generation categories were waste paper (24.4%), waste glass bottles (23.7%) and waste furniture (14.3%). They estimated there were about 5,700 permanent recycling centres and 22,800 curb-side recyclers in Beijing. China plans to recycle 35% of waste in major cities by the end of 2020, according to a National Development and Reform Commission report (BBC 2019). The full automation of recycling operations such as automated waste sorting using sensors and actuators was recently used (Gundupalli et al. 2017).

6.5.5 Waste Collection and Storage

Waste collection is primarily carried out by handcarts or tricycles with containers or bins, tricycles with hydraulic tipping containers, light commercial vehicles (mini trucks) with hydraulic tipping containers and four-wheeled mini trucks with international standard bins. The collected waste is stored in storage bins, maybe type of fixed bins on the ground (stationary) or movable bins (hauled). The fixed bins are more durable but their positions cannot be changed once they have been constructed, while the movable bins are flexible in transportation but lacking in durability (Bhat et al. 2018). Municipal authorities are responsible for providing the covered secondary storage of waste and also vehicles used for the transport of waste are covered. Both India and China regulate the waste treatment facilities through licensing systems. The waste is transferred from smaller collection vehicles to larger ones at transfer stations, and the waste is subsequently transported to disposal sites. The status of MSW collection, treatment and disposal in India and China is summarized in Table 6.2.

India permits only registered players (collectors, dismantlers and recyclers) in the waste management industry to operate with environmentally safe technologies. During the year 2011, about 127,486 TPD MSW was generated in the country, out of which 70%, or 89,334 TPD, was collected and only 12.45%, or 15,881 TPD MSW, was processed or treated (Joshi and Ahmed 2016). The collection efficiency of waste from households decreased to less than 60% in 2016, and only 15% of urban India's waste is processed (PIB 2016). Local bodies spend around Rs. 500–1,000 per

TABLE 6.2
MSW Collection, Treatment and Disposal in India and China

a) MSW Collection, Treatment and Disposal in Different States of India

S. No	States	Wards with 100% Door-to-door Collection	Wards with 100% Source Segregation	Total Wards	Total Waste Generation (MT/D)	Total Waste Processed (MT/D)	Total Waste Processing (%)
1.	Andhra Pradesh	3409	3300	3409	6141	3850	63%
2.	Andaman and Nicobar	24	23	24	90	86	95
3.	Arunachal Pradesh	75	11	75	181	0	0
4.	Assam	698	368	943	1432	759	53
5.	Bihar	3276	1107	3377	2272	1159	51
6.	Chandigarh	26	24	26	479	455	95
7.	Chhattisgarh	3217	3217	3217	1650	1485	90
8.	Daman and Diu	28	28	28	32	24	75
9.	Dadra and Nagar Haveli	15	15	15	55	55	100
10.	Delhi	294	59	294	10500	5775	55
11.	Goa	217	173	217	250	175	70
12.	Gujarat	1427	1187	1427	10274	8938	87
13.	Haryana	1401	935	1496	4783	2296	48
14.	Himachal Pradesh	490	490	497	377	294	78
15.	Jammu and Kashmir	809	137	1081	1489	238	16
16.	Jharkhand	897	752	932	2135	1281	60
17.	Karnataka	6464	3694	6464	10000	5400	54
18.	Kerala	3022	3536	3536	2696	1914	71
19.	Madhya Pradesh	7115	7005	7115	6424	5589	87
20.	Maharashtra	6590	6346	7322	22080	12806	58
21.	Manipur	270	196	306	174	101	58
22.	Meghalaya	27	27	114	268	10	4
23.	Mizoram	264	230	264	236	83	35

(*Continued*)

TABLE 6.2 (CONTINUED)
MSW Collection, Treatment and Disposal in India and China

a) MSW Collection, Treatment and Disposal in Different States of India

S. No	States	Wards with 100% Door-to-door Collection	Wards with 100% Source Segregation	Total Wards	Total Waste Generation (MT/D)	Total Waste Processed (MT/D)	Total Waste Processing (%)
24.	Nagaland	148	30	234	461	277	60
25.	Odisha	2009	1402	2024	2721	1306	48
26.	Puducherry	122	116	122	415	55	13
27.	Punjab	3064	2664	3213	4100	2501	61
28.	Rajasthan	5389	4419	5389	6500	4680	72
29.	Sikkim	53	50	53	89	62	70
30.	Tamil Nadu	12429	10891	12814	15437	10497	68
31.	Telangana	2020	1008	2112	8634	6735	78
32.	Tripura	277	243	310	450	239	53
33.	Uttar Pradesh	11872	8294	12007	15500	8990	58
34.	Uttarakhand	1170	669	1170	1589	731	46
35.	West Bengal	2527	558	2938	7700	700	9
	Total /Average	81135	63204	84475	147613	89545	60

b) MSW Collection, Treatment and Disposal in Some Cities of China

S. No	Provinces	MSW Generation (MSW + Manure Tonnes)	Collected/ Treated (Tonnes)	% of Total
1.	Beijing	6820000	845000	12.4
2.	Tianjin	1990000	1082000	54.4
3.	Shanghai	5880000	4481000	76.2
4.	Chongqing	2370000	2083000	87.9
5.	Harbin	2470000	2263000	91.6
6.	Shenyang	3550000	660000	18.6
7.	Xian	750000	220000	29.3
8.	Nanjng	1020000	832000	81.6
9.	Wuhan	2040000	1780000	87.3

(Source: Adapted from SBM 2020a; Wang and Nei 2001.)

tonne on SWM with 70% of this amount spent on collection and 20% spent on transport (Kumar et al. 2017). The percentage of MSW collection and treatment was increased in the following years, after the amendments of the MSW 2016 rules and their strict implementation. The current status of waste collection and treatment in different states of India is summarised in Table 6.2a. Solid waste management services are improved as the percentage of wards with 100% door-to-door collection of solid waste reached 96%, wards with 100% source segregation reached 74.8% and total waste processing is 60% under the operation of Swachh Bharat Mission (Urban). Better solid waste management services were found in states (12 nos) which attained 100% door-to-door collection in all their wards (Table 6.2a).

The current status of waste collection and treatment in selected cities of China are summarized in Table 6.2b. The highest MSW generation was in Beijing (in 2001) with 6,820,000 tonnes, whereas its collection or treatment of waste was lowest at 12.4% and the best services for the collection and treatment of MSW was in Harbin city with 91.6%, when compared to other selected Chinese cities. According to the EPA's statistics, in 2002 the total number of Government employees for general waste collection was 19,082, and there were 4,889 collection vehicle units (APO 2007). In 2003, 149 million tons of MSW were collected and transported by local Government, out of which 49% was dumped on appointed sites, 43% was disposed of in sanitary landfills, 3% was incinerated and 5% was composted (Liu 2017). MSW collection and treatment efficiency was improved in the following years. Municipal agencies collected and transported 426,450 tons/d of MSW in 2008, of which 82% was landfilled, 15% was incinerated and 3% was composted (Dorn et al. 2010). In 2012, 171 million tons of MSW were collected and transported by local Government; 15% was dumped on appointed sites, 62% was dumped in sanitary landfills, 21% was incinerated and 1% was composted (Liu 2017).

Both India and China were not able to collect 100% of the MSW generated, and as a consequence the disposal rate was also lower.

6.5.6 Waste Processing and Disposal

Different waste disposal methods were available for India and China such as landfilling, composting, waste to energy methods, pelletization and other new advanced treatment techniques.

6.5.6.1 Landfilling

Among the different disposal methods, landfilling is the most common method in India and China. The different landfilling methods in practise are open dumping and controlled dumping in sanitary landfills. Landfill bioreactor methods are very new in these countries. Both India and China have adopted unscientific disposal of MSW in almost every city, town and village (Dorn et al. 2010). Landfills should be mainly used for the disposal of non-biodegradables and non-recyclables, whereas the situation in India and China is different, as their primary disposal method is landfilling. The number and specifications of landfills in India and China are tabulated in Table 6.3.

From the Table 6.3, the top five most populated Indian cities with their number of landfills and current capacities are Mumbai with three landfills (40 ha), Delhi with three (66.4 ha), Bengaluru with two (40.7 ha), Chennai with two (465.5 ha) and Hyderabad with one (121.5 ha) (Swaminathan 2018). Even the most populous cities have fewer landfills. The number of landfills increased from 47 in 2011 to 59 constructed landfill sites in 2013; additionally 376 are under planning and at the implementation stage. Also 1,305 sites have been identified for future use (Joshi and Ahmed 2016). Indian cities with containment landfill sites include Mumbai, Kolkata, Chennai, Nashik, Vadodara, Jamshedpur, Allahabad, Amritsar, Rajkot, Shimla, Thiruvananthapuram and Dehradun (Kumar et al. 2017). In 2010, Bengaluru, India, the amount of waste disposed of at state of the art landfills was 1,364,188 tonnes/year, in which the amount disposed of at simple controlled disposal sites was 350,000 tonnes/year. Along with incineration, the total controlled disposal was at 65% and the remaining 15% of 209,875 tonnes/year were lost or illegally dumped. In Delhi, the amount of

TABLE 6.3
Landfills in India and China

	Landfills in India (as of 2011)				Large-Scale Landfills in China (1995)			
S. No.	States	Location/No. of Landfills	Area (Acre)	S. No.	Location	Total Capacity $\times 10^4$ m^3	Daily t/d	Service Period
1.	Andhra Pradesh	1 - Visakhaptnam	100.1	1.	Hangzhou Tianziling	600	2000	1991–2014
2.	Assam	1 – Guwahati	32.6	2.	Shanghai Laogang	1200	3000	1989–
3.	Chandigarh	1 - Chandigarh	44.5	3.	Guangzhou Licun	180	1200	1990–1997
4.	Chhattisgarh	1 – Raipur	36.1	4.	Guangzhou Datianshan	1700	200	1992–1999
5.	Delhi	3 – Delhi	164.1	5.	Xian Jiangxicun	4900	-	1995–2045
6.	Gujarat	5 – Surat; Ahmedabad; Vadodara; 2 – Rajkot	494.2; 207.6; 20; 3	6.	Shenzhen Xiaping	1400	1800	1997–
7.	Haryana	3 – Faridabad	5.9	7.	Baotou Qingshan	200	200	1990–2010
8.	Himachal Pradesh	1 – Shimla	1.5	8.	Beijing Asuwei	1070	2000	1995–2005
9.	Jammu & Kashmir	1 – Srinagar	75.1	9.	Beijing Liulitun	1060	1500	1996–2010
10.	Jharkhand	2 – Ranchi; Jamshedpur	37.1; 10.1	10.	Chongqing Changsheng	1850	1500	1999–2030
11.	Karnataka	2 – Bengaluru	100.6					
12.	Kerala	1 – Thiruvananthapuram	30					
13.	Madhya Pradesh	2 – Jabalpur; Indore	150.7; 147					
14.	Maharashtra	4 – Mumbai; 3 – Nasik	345.9; 85					
15.	Punjab	1 – Ludhiana	99.8					
16.	Rajasthan	3 – Jaipur	77.6					
17.	Tamil Nadu	2 – Chennai; 2 - Coimbatore -2; 1 - Madurai	1150.3; 721.5; 120.1					
18.	Telegana	1 – Hyderabad	300.2					
19.	Uttarakhand	1 – Dehradun	11.1					
20.	Uttar Pradesh	1- Kanpur; 2 - Meerut; 1 -Varanasi; 1-Agra; 1-Lucknow	61; 35.1; 4.9; 3.7; 3.5					
21.	West Bengal	2 – Kolkata; Asansol	61; 4.9					

(Source: Adapted from Ahluwalia and Patel 2018; CPCB 2017; Wang and Nie 2001.)

waste disposed of at simple controlled disposal sites was 1,810,035 tonnes/year, the percent of total controlled disposal was at 71% and the remaining 611,317 tonnes/year were lost or illegally dumped (Wilson et al. 2010). India will require 1,450 km^2 of land to dispose of MSW in a systematic manner in 2051 (Joshi and Ahmed 2016). Sanitary landfills should be the acceptable and recommended method for the ultimate disposal of MSW. But landfilling would continue to be an extensively accepted practice, even though metropolitan cities such as Delhi, Mumbai, Kolkata and Chennai have limited availability of land for waste disposal and designated landfill sites are running beyond their capacity. The old waste or mined waste or legacy waste in these landfills is creating immense environmental hazards, and the clearing of these wastes is the best option to reduce GHG emissions. Recently, CPCB has prepared guidelines for the disposal of legacy wastes (old municipal solid waste) (CPCB 2019). Research on the valorization of legacy waste is increasing significantly. A recent study by Singh and Chandel (2020) showed the effect of ageing on the characteristics of waste excavated from an Indian landfill and its potential valorization. The volatile matter, total and organic carbon and calorific value of excavated waste reduced with the age of waste. The excavated waste contained substantial amount of fine fraction (~45%) and it also showed the possible valorization routes for the excavated waste.

In China, landfills are among the main management options used in urban centres (Jia et al. 2018). In the 1990s, 160 illegal dumpsites were in use, as the waste generation increased massively and there was an unavailability of land area. In 2005, China's 660 cities had about 1,000 major landfills, and the land requirement for landfilling over the next 30 years (without incineration) is about 100,000 Ha (Hoornweg et al. 2005). The specifications of the large-scale landfills in China are tabulated in Table 6.3. The first landfill site in China for MSW was Tianziling landfill in Hangzhou city (Liu 2017) and also, it was the first site to install a leachate collection system. Forty-seven percent of landfills did not treat leachate, 10% of landfills discharged leachate into the sewage systems, while 20% of landfills used biochemical methods, 3% used membrane liners and 20% of landfills used 'other methods'. Chongqing landfill treated only a very small quantity of leachate before discharging into groundwater (Dorn et al. 2010). Usually, vertical wells are constructed in landfills to lower the leachate levels or pressures. However, recently horizontal well construction was inspected in Tianziling and Xingfeng landfills. In Tianziling landfill, the average leachate flow rates of two installed horizontal wells were 10.66 m^3/day and 3.93 m^3/day, respectively. In Xingfeng landfill, 12 horizontal wells were installed and after 157 days of drainage, a total volume of approximately 24,000 m^3 leachate had been discharged. Horizontal wells have been shown to be more effective in reducing leachate levels than vertical wells (Hu et al. 2020). The Jiangcungou landfill in Shaanxi Province, China, is one of the largest dumps, designed to take 2,500 tonnes/day, whereas it received 10,000 tonnes/day (the most of any landfill site in China) and serves over 8 million citizens (BBC 2019). In the next 50 years, China is likely facing a huge brownfield site clean-up cost (Hoornweg et al. 2005). Few of China's waste disposal sites are 'sanitary landfills' and currently, the Chinese Government prefers to develop more sanitary landfills (Havukainen et al. 2017).

A landfill life cycle assessment (LCA) study with and without energy recovery in Hangzhou, China, was carried out. It showed that environmental impacts decrease significantly in landfill with the utilization of biogas owing to the combined effects of emission reduction and electricity generation. Gas collection systems should be upgraded effectively (Dong et al. 2014). A study by Sri Shalini et al. (2010) in India operated different landfill environment conditions for treating MSW collected from Chennai landfills. It showed that MSW from different landfills methods were aerobically and anaerobically stable and results were comparable with compost. Further, the landfill bioreactor (LFBR) method gave better biological stability and higher methane content than other landfill operating conditions.

The future of landfill methods may be landfill bioreactors which are modified landfills with the addition of moisture and/or air injection in a controlled manner. The advantages of operating landfill as LFBR are an enhanced waste stabilization rate enabling efficient and effective use of landfill space, reduced leachate volume and an accelerated methane gas production rate. The operation of a landfill bioreactor loaded with MSW from Chennai landfills and nitrogen management was demonstrated by Sri Shalini and Kurian Joseph (2013; 2018). It revealed better removal of pollutants with enhanced biogas composition for future energy recovery.

6.5.6.2 Composting

The different composting technologies available are windrow, vermicomposting, decentralized composting (bin and box), aerated static pile composting, etc. It can also be classified as open or window, mechanical or closed, etc., depending upon operating conditions and design of plant. The most common composting methods are aerobic composting and vermicomposting. Indian and Chinese waste contains high percentages of organic content making it more suitable for biological processes like composting. Composting is beneficial in a SWM system, as it reduces organic waste to 25 to 30% of its initial weight. Composting should be an important waste management tool in India and China. The distribution of compost plants in India and China in different states/provinces with their operational details is given in Table 6.4.

In India, composting was widely encouraged, and early initiatives of the Government of India (GOI) regarding MSWM focused primarily on promoting the composting of urban MSW (CPCB 2017; Sharholy et al. 2008). Huge quantities of waste are treated by composting, where efficiency mainly depends on temperature (Nandan et al. 2017). As of 2017 in India (Table 6.4), 95 commercial-scale composting plants were installed with a capacity of 2.37 million tonnes/year, and total operational capacity is only about 14%, producing 0.33 million tonnes of compost/year (Ahluwalia and Patel 2018). Early in the 1975–1980 period, large-scale composting plants were installed in Indian cities like Bengaluru, Baroda, Mumbai, Kolkatta, Delhi, Jaipur, Kanpur and Indore, having a capacity of 150 to 300 tonne/day. But due to poor performance and no usefulness in soil enrichment, plants were shut down. Later in 1992, the first large-scale aerobic composting plant was set up in Mumbai with a capacity of 500 t/day of MSW developed by Excel Industries Ltd, and only 300 t/day capacity is being utilized. Another plant with 150 t/day capacity has been operated in Vijayawada, and over the years a number of other plants have been implemented in Delhi, Bengaluru, Ahmedabad, Hyderabad, Bhopal, Lucknow and Gwalior (Nandan et al. 2017). Indian cities such as Bengaluru, Chennai, Mumbai and Pune have very active community-based and decentralized composting schemes, by which sorted waste is turned into high-quality compost. These schemes have capacities of up to 1 ton per day. Several research studies are conducted for improving the compost properties for early maturity and stability (Rastogi et al. 2019; Sahu et al. 2020). The addition of bacterial consortia (*B. subtilis*, *B. tequilensis*, *B. venezuelans* and *B. amyloliquefaciens*) and cow dung to MSW composting facilitates early compost maturity compared to uninoculated treatments (Rastogi et al. 2019). The use of thermophilic ligno-cellulolytic microbial consortia (*Bacillus* spp., *Streptomyces* spp., *Rhizomucor pusillus* and *Aspergillus flavus*) for rapid composting was also observed. The key parameter for compost maturity is water-soluble carbon and organic nitrogen ratio, and dehydrogenase activity (DHA) is the significant factor for compost stability (Sahu et al. 2020).

The other Indian traditional indigenous method for MSW treatment is vermicomposting. There are around 138 vermicomposting plants in India (Planning Commission Report 2014). Vermicompost is an organic manure (bio-fertile) produced by the vermicast of earthworms feeding on biological waste material. Vermicomposting has been used in Hyderabad, Bengaluru, Mumbai and Faridabad. Experiments on developing household vermicomposting kits have also been conducted. However, the area required is larger, when compared to dry composting (Bhat et al. 2018). Developing countries like India offer large potential for the Clean Development Mechanism (CDM) because of their inherent dependence on fossil fuels for development. Many composting projects in India were operated under CDM. In 2003, the composting of MSW was 5% but in 2012, it was reduced to 1%. Early composting plants targeted the sale of fertilizer and compost; however the unsegregated MSW rendered any process unfeasible, leaving the output contaminated by plastics, ash and other

TABLE 6.4
Compost Plants in India and China

Indian Compost Plants (as of 2017)					Some of the China Compost Plants (as of 2000)			
S. No	States	No. of Plants	Installed Capacity (Tonnes/Year)	Operational Capacity (%)	S. No	Places	Type, Year Installed	Design Capacity (Tonnes/Day)
1.	Andaman and Nicobar Islands	1	90	-	**1.**	An'yang	Windrow	-
2.	Andhra Pradesh	2	2400	20	**2.**	Baise	Windrow	40–50
3.	Assam	1	15000	15	**3.**	Beijing Dongchun	-	100
4.	Chhattisgarh	1	1200	20	**4.**	Beijing Shejinshan	Bin (280.195 m^3×14 bins), 1991	1991 – 100 1994 – 300
5.	Daman and Diu	1	4050	-	**5.**	Beijing Nangong	Tunnel compost, 1995	400
6.	Delhi	4	168000	16.1	**6.**	Hangzhou	Trapezoid reactor	100
7.	Goa	1	1200	20	**7.**	Jixi	Pond (50 × 20 × 1.5 m^3)	100
8.	Gujarat	15	174300	19.5	**8.**	Nanping	Bin (25.6 × 29 m^3)	40
9.	Haryana	4	18600	15.3	**9.**	Shanghai	Bin (13.5 × 4.5 × 3.4 m^3)	300
10.	Karnataka	18	473400	10.1	**10.**	Tianjin	Windrow (100 × 30 × 3 m^3)	300
11.	Kerala	3	156000	20	**11.**	Tianjin Dagang	LD_{50-100} in-vessel	100
12.	Madhya Pradesh	1	36000	15	**12.**	Wuhan	Bin	100
13.	Maharashtra	13	488400	12.5	**13.**	Wuxi	Two step fermentation vessel	100
14.	Punjab	2	19200	15				
15.	Rajasthan	1	180000	15				
16.	Tamil Nadu	9	67680	15.8				
17.	Telangana	5	192000	15				
18.	Tripura	1	75000	6				
19.	Uttar Pradesh	7	124560	15.2				
20.	West Bengal	5	170400	15				
Total		95	2367480	14				

(Source: Adapted from Ahluwalia and Patel 2018; CPCB 2017; Wei et al. 2000.)

undesirable materials. The system ultimately collapsed as a result of an absent market for natural compost (Dorn et al. 2010).

Some of the compost plants in China along with their operational details are summarized in Table 6.4. Most MSW composting plants adopt a two-step fermentation. Tianjin Hexi compost plant and Wuxi sanitary pilot plant (100 tonnes MSW/d) were the earliest plants operating in one-step fermentation and two-step fermentation, respectively. In 1991, Beijing Dongchun windrow composting plant operated with 100 tonnes MSW/d, but it stopped running due to high cost. Another plant in 1991, Beijing Shejinshan treated 100 tonnes MSW/d, later expanded to 300 tonnes MSW/d in 1994 and produced compound fertilizer using MSW compost and some inorganic fertilizers. In 1995, Beijing Nangong composting plant was the first to develop an advanced tunnel composting system from Germany treating 400 tonnes MSW/d and was in operation in 1998. Most of the composting plants built in the 1980s are not in operation now, due to financial reasons, high costs, etc. (Wei et al. 2000). The number of composting plants decreased from 70 in 2003, to 7 in 2012 (Liu 2017). Several difficulties faced in compost plants were traditional and outdated technology; low- or poor-quality compost products due to mixed waste; heavy metals, ash, pieces of plastic and glass shards in the end-product not suitable for farmland applications and trouble in selling compost products (Hoornweg et al. 2005; Wang and Nei 2001). Many research studies were carried out on improving the quality of final composts (Zhu et al. 2020a, b). A study by Zhu et al. (2020) showed the core microorganisms (*Prosthecobacter*, *Paenalcaligenes* and *Solibacillus*) and key parameters (temperature, moisture and pH) responsible for regulating the dissolved organic matter stabilization for improving the compost stability.

6.5.6.3 Waste to Energy (WtE)

The different waste to energy technologies in India and China are incineration (thermal process), pyrolysis and gasification (thermo-chemical process), biomethanation (biochemical process) and refuse-derived fuel (mechanical and thermal process).

6.5.6.3.1 Incineration (Thermal Process)

Incineration is a thermal combustion process of drying and burning waste with the end-products of CO_2, H_2O in vapour form and ash with a large amount of heat. The operation temperature ranges between 980 to 2000°C (Sharholy et al. 2008). Incineration reduces the final MSW roughly up to 75% (Nandan et al. 2017). This process is not widely practised in India, whereas China is implementing many new incineration plants throughout the country.

The numbers of incineration plants in India and China are given in Table 6.5. The incineration plants are distributed in the states of Delhi, Telangana, Tamil Nadu, Madhya Pradesh and Himachal Pradesh (Ahluwalia and Patel 2018; CPCB 2017; Kumar et al. 2017). This shows that the incineration plants for MSW treatment in India are very limited as the MSW composition is not preferable. Indian MSW contains high concentrations of moisture content and organic content, about 40–60% each, high inert content with 30–50% and lower calorific value of 800–1100 kcal/kg, not suitable for incineration (Sharholy et al. 2008). The incineration plant at Timarpur, Delhi, was the first plant built in 1987 with 300 t/d developed by Miljotecknik volunteer, Denmark. It is currently not in operation due to its bad performance. Another incineration plant at Okhla, South Delhi, had air emission problems that processes 1,950 tonnes of unsegregated MSW/d, produces 16 MW of electricity and 30% of MSW as residual ash and char which is disposed of in Okhla landfill. Further, in 2017 two more incineration plants at Ghazipur with 1,300 TPD and at Narela-Bawana with 2,000 TPD were in operation (Nandan et al. 2017). An incineration plant in Trombay, Mumbai, at BARC is in operation for treating institutional waste (Sharholy et al. 2008). Recently, the incinerability of MSW generated in different Indian cities across the country was mapped using an incinerability index or *i*-Index. This is helpful for evaluating the feasibility of WtE projects in the country. *i*-Index reflects the influence of regional, topographic and climatic factors on the incinerability of MSW (Sebastian et al. 2020).

TABLE 6.5
Waste-to-Energy Plants in India and China

S. No	State	Place	Developer	Capacity (TPD)	Power MW
		WtE Plants in India			
1.	Delhi	Okhla	Jindal	1950	16.0
2.		Ghazipur	IL&FS	1300	14.0
3.		Bawana	Ramky	2000	24.0
4.	Telangana	Hyderabad	Ramky	2400	20.0
5.			IL&FS	1000	11.0
6.	Tamil Nadu	Chennai	Essel	300	2.9
7.	Madhya Pradesh	Jabalpur	Essel	600	9
8.	Himachal Pradesh	Shimla	Elephant Energy	70	1.75
		WtE Plants in China			
1.	Shenzhen	Guangdong	Mitsubishi, Japan	450	-
2.	Pudong	Shanghai	Alstrom, France	1000	17
3.	Ningbo	Zhejiang	Novel, Germany	1050	-
4.	Longgang	Guangdong	Richway, Canada	300	-
5.	Hangzhou	Zhejiang	Zhejiang University	800	12
6.	Shaoxing	Zheijiang	Chinese Academy of Social Sciences	400	-
7.	Haerbin	Heilongjiang	Ebara, Japan	200	-
8.	Laogang	Shanghai	Onyx	1000	-
9.	Changzhou	Jiangsu	JFE Holidays, Japan	450	-
10.	Minhang	Shanghai	-	3000	-
11.	Chongqing	Sichuan	-	1000	-
12.	Suzhou	Jiangsu	Everbright International	1000	-
13.	Huizhou	-	Enersave, Singapore	600	12

(Source: Adapted from Ahluwalia and Patel 2018; CPCB 2017; Kumar et al. 2017; Whitworth 2005.)

In China, the common MSW disposal method is landfill similar to India, but recently the trend is moving towards incineration. This is because of the unavailability of land space in China, secondary pollution from illegal dumping (in rural areas) and to handle the huge quantities of MSW, forcing the country to prefer the incineration method, even though it causes risk to public health from toxic emissions (Wong 2019). Recently, several research studies were conducted on MSW incineration emissions and control strategies (Ma et al. 2020; Xu et al. 2020). A new scheme was proposed, namely a leader-follower optimized approach to carbon-economy and environmental equilibrium in incineration power plants that includes a carbon allowance allocation scheme (IPP-CAAS) for tackling carbon emissions in the incineration industry (Xu et al. 2020). Even the EPA has adopted a strategy favouring incineration as the primary treatment method for MSW, with landfill as a supplement. China's WtE incineration industry is expected to grow enormously in the coming years due to the benefits of recovering energy from discarded MSW and producing electricity and/or steam heating, identifying as a source of renewable energy supply in China and also reduction in GHG emissions (Cheng et al. 2010). An incineration plant with waste to energy in Hangzhou, China, was compared with landfills with or without energy recovery by a life cycle assessment study. It showed that incineration is preferable to landfill, but toxicity-related impacts also need to be improved, especially heavy metals and dioxin emissions (Dong et al. 2014). The numbers of incineration plants in China are given in Table 6.5. In 2003, 47 incineration plants were operated, later it increased to 74 in 2010 and reached 138 plants in 2012 using three main technologies (Liu 2017). Stoker grate and fluidized bed furnaces hold an approximate 40% market share each, with rotary kiln technology accounting for the final 20%. Stoker grate incineration plants have a daily capacity of 1,000–1,500 tonnes/d. Fluidized bed furnaces have a smaller capacity ranging between 100 and 500 t/d (Dorn et al. 2010). The fluidized bed combustion is more appropriate than foreign developed technologies for Chinese MSW, as it has high moisture content (Whitworth 2005). Developing countries like China prefer incineration projects under CDM. In 2013, the amount of incinerated MSW in Hangzhou (one of the most developed cities in China) represented 56% of total MSW. MSW incineration is primarily performed via a process of co-incineration with coal because MSW has a low heating value (Havukainen et al. 2017). The first incineration plant in China was built in Shenzheng city, Guangdong province (Liu 2017). Recently, a new incineration plant was opened in November 2019 and Xi'an city waste was treated; an additional four plants are expected to open by 2020. Together, they are expected to be able to process 12,750 tonnes of rubbish per day (BBC 2019). MSW incineration in China is also currently blamed for substandard emissions due to a lack of operating experience and reliable supervisory measures, deficient funding for compliance with the emission standard and elemental MSW compositions (Lu et al. 2017). China still has a waste composition that is barely autogenic (a high enough heating value to burn on its own). Chinese cities would have to use supplemental fuel in order to burn their solid waste, and thus there would be no net energy generation to offset the high costs of incineration (Hoornweg et al. 2005).

6.5.6.3.2 Pyrolysis and Gasification (Thermo-Chemical Process)

Pyrolysis is a process of thermochemical decomposition of waste biomass operated at higher temperatures, usually at 300 to 700°C, to produce solid biochar, liquid bio-oil and syngas in an inert atmosphere. Pyrolysis is an irreversible reaction, and its scale-up is more flexible than incineration units. The different operating conditions of the pyrolysis process convert the MSW into useful products of biochar, bio-oil and syngas (Sri Shalini et al. 2020). MSW pyrolysis is becoming popular recently, and several thermal or kinetic studies have been conducted (Chhabra et al. 2020; Vuppaladadiyam et al. 2019). The slow pyrolysis of unsegregated MSW was studied using thermogravimetry coupled with gas phase FTIR, differential scanning calorimetry, synchrotron-based infrared microspectroscopy and DRIFT-FTIR. This showed that the rubber component of MSW when co-pyrolysed with biomass and plastic is expected to improve the stability of pyrolysis oil. The gas yields during pyrolysis of various combinations of MSW components at different temperatures are used to identify the temperature range for MSW pyrolysis, accounting for the effect of interactions on the volatiles evolution (Chhabra et al. 2020). The gasification process is the conversion of a biomass material at atmospheric elevated temperatures above 800°C with a fast heating rate in a gaseous medium like oxygen, air, nitrogen, carbon dioxide and steam that results in a gaseous mixture fuel generally known as synthesis gas (syngas) (Sri Shalini et al. 2020). The gasification plants in India have two different designs of gasifiers. They are the NERIFIER gasification unit at Nohar, Hanungarh, Rajasthan by Narvreet Energy Research and Information (NERI) and the TERI gasification unit installed at Gaul Pahari campus, New Delhi, by Tata Energy Research Institute (TERI). The NERIFIER gasification unit operated for agro-wastes, saw mill dust and forest wastes with a feeding capacity of 50–150 kg/h, yield efficiency of 70–80% and 25% of fuel gas recycled back into the system (for gasification) and the remaining recovered (for power generation) (Sharholy et al. 2008). China's waste biomass is currently adopting treatment using advanced gasification and liquefaction techniques for obtaining environmentally conscious products (Liu 2017).

6.5.6.3.3 Biomethanation (Bio-chemical Process)

The advanced method for the biochemical conversion of biodegradable waste is anaerobic decomposition or biomethanation. It is the process of conversion of organic matter into stable, inert residue by microorganisms in an oxygen-free environment. It yields methane-rich biogas which can be used for electricity and cooking, and inert residue which can be used as manure (Nandan et al.

2017). In India, the four biomethanation plants are in the states of Bengaluru, Karnataka (Nobel Exchange), with an installed capacity of 250 TPD and output of manure 25 TPD; Pune, Maharashtra (Nobel Exchange), with a capacity of 300 TPD and output of Bio-CNG 4TPD and manure 7.5 TPD; Solapur, Maharashtra (Organic Recyclers), with a capacity of 400 TPD and output of electricity 3 MW and manure 60 TPD and, lastly, Chennai, Tamil Nadu (Ramky), with a capacity of 30 TPD and output of electricity 0.26 MW and manure 3 TPD (Ahluwalia and Patel 2018). China's contribution to the biogas industry using waste feedstock is highlighted in Zheng et al. (2020). The different policies and subsidies for biogas-to-biomethane (BNG) are also detailed.

Many research studies were conducted in India and China, to potentially increase the biogas production from the organic fraction of MSW by mechanical, thermal, physical, chemical, biological and/or combined treatment technologies (Bala et al. 2019; Dasgupta and Chandel 2020; Panigrahi and Dubey 2019; Xu et al. 2016). Among the chemical treatment methods, the alkali pretreatment of MSW was evaluated by many researchers for biogas production. The alkali pretreatment of the organic fraction of MSW increased the biogas yield as well as decreasing the digestion period. The cumulative biogas generation increased by 19.6–34.8% post-NaOH pretreatment. The organic fraction of MSW at pH 10 with NaOH had a maximum biogas yield of 407 mL/gVS with 68.9% methane content (Dasgupta and Chandel 2020). In another study, the biogas production after pretreatment of the organic fraction of MSW with 0.1–0.9 N NaOH concentrations gave biogas in the range of 370–465 mL/gVS, respectively (Bala et al. 2019). The physical treatment of high-pressure extruding pretreatment of organic fraction of MSW gave methane production of 674 mL CH_4/gVS at 40 MPa pressure (Xu et al. 2016).

6.5.6.3.4 Refuse-Derived Fuel (RDF)/Pelletization (Mechanical and Thermal Process)

Refuse-derived fuel is a mechanical and thermal process to produce an improved solid fuel or pellets from MSW. Pulverization and drying methods are involved. The RDF method is used as a source of fuel in coal power plants, cement kiln industry, plasma arc, gasification modules, pyrolysis plants, etc. MSW properties such as particle size and moisture content have a major role in the RDF pelletization process. It is shown that smaller RDF particles (4-mm) consumed higher pelletization energy and produced pellets with lower durability and hardness than pellets from larger particles. The large-size particles (such as paper) (ground with 6-mm grinder screen) interfered with the free flowing to the pellet die. A 20% moisture content of MSW was shown to be the optimum moisture content for the RDF pelletization process (Rezaei et al. 2020). In India, as of 2017, there are seven RDF plants: Chandigarh (Jaypem) with 500 TPD (RDF-300 TPD), Surat, Gujarat (Hanjer) with 500 TPD (RDF – 125 TPD), Bengaluru, Karnataka (MSGP & KCDC) with 500 TPD & 200 TPD, Kochi, Kerala (Kochi MC) with 400 TPD (RDF – 100 TPD), Pune, Maharashtra (Rochem) with 400 TPD (RDF – 250 TPD), Navi Mumbai, Maharashtra (Pyrocrat) with 300 TPD and Jaipur, Rajasthan (Vikram Cements) with 500 TPD (RDF – 150 TPD) (Ahluwalia and Patel 2018). In China, an LCA programme study showed that RDF production and incineration could improve Hangzhou's MSW management global warming potential from −33% to 0%, the acidification potential from −90% to 34% and the eutrophication potential from −1,200% to 350% in comparison to the co-incineration of MSW with coal (Havukainen et al. 2017).

The status of other solid wastes generated in India and China is as follows. In India at present, 62 million tonnes of waste generated annually contain 5.6 million tonnes of plastic waste, 0.17 million tonnes of biomedical waste, 7.90 million tonnes of hazardous waste and 15 lakh tonnes of e-waste (PIB 2016; Ray and Rahman 2016).

6.6 STATUS OF OTHER WASTES IN INDIA AND CHINA

6.6.1 Construction and Demolition Waste (C&D)

In the past, India generated around 12–15 million tonnes of C&D waste per annum. This has increased to nearly 100 million tonnes as estimated by Ministry of Housing and Urban Affairs,

Government of India (MoHUA) recently (MoHUA 2018). Wastes and rejects are also generated during the construction of new buildings. New construction and repair may lead to waste generation in the range of 45–50 kg/m^2. However, demolition can lead to waste generation of 425 kg/m^2. The estimated consumption of construction materials in India is very high (TERI 2019). Currently, in India the number of operating C&D waste processing plants is extremely small when compared to other countries. In China, 30 to 40% of the total waste is construction and demolition waste. It is commonly disposed of in landfills. The recycling rate of construction and demolition waste is very low, at an average of 5% (Huang et al. 2018a). The construction and demolition waste was quantified in a construction site of Shenzhen city, South China. The study showed a waste quantity of 3.275 to 8.791 kg/m^2 consisting of miscellaneous waste, timber for formwork and falsework and concrete as three largest components amongst the generated waste (Lu et al. 2011).

6.6.2 Hazardous Waste (HW)

Both India and China have specific legislation with respect to hazardous waste management, and it is fully implemented (as given in Table 6.1). In India, there are 18 categories of HW, 50 constituents listed are hazardous, and 38 processes are listed as generating hazardous wastes. As per a 2017 estimate, annually 7.46 million MT of hazardous waste is generated from 43,936 industries in the country. Of this, 3.41 million MT is landfillable waste (46%), 0.69 million MT is incinerable (9%) and 3.35 million MT is recyclable hazardous waste (45%). The annual increase of HW generation is 2 to 5%, and 10–15% of industrial waste generated is hazardous in nature. Around 88 hazardous waste incinerators and 220 recycling facilities exist in the country, along with 2 engineered landfills (both in the State of Gujarat) (Karthikeyan et al. 2018). There are only 40 common hazardous waste treatment, storage and disposal facilities (TSDFs) available in 17 states/UTs. The few facilities in Gujarat, Maharashtra, and Andhra Pradesh are not sufficient to handle the entire quantity of HW in India. Many states like Karnataka, Kerala, Punjab and Orissa don't have such a disposal mechanism for hazardous waste. Hazardous waste such as lead acid battery scraps, used oil, waste oil, spent catalyst, etc., and other waste such as waste tyres, paper waste, metal scrap, etc., are used as raw material by the industries involved in the recycling of such waste and as supplementary resources for material and energy recovery, to avoid incineration and landfilling. There are about 1,080 registered recyclers, 47 cement plants permitted for co-processing and about 108 industries permitted for the utilization of hazardous waste (PIB 2016). The state of Gujarat leads the HW generation, amounting to 28.8%. This is due to the large number of HW generating facilities (7,751) in the state. Coal ash from thermal power stations accounts for more than 70% of all industrial hazardous waste (Karthikeyan et al. 2018).

In China, the industrial hazardous waste generation in 2005 was about 11.62 million tons. This amounts to the industrial solid waste volume of around 1.1%. On average 43.4% of industrial hazardous waste was recycled, 33.0% was stored, 23.0% was securely disposed and 0.6% was discharged improperly. There were 177 formal treatment and disposal centres for industrial hazardous waste management in 2004. The reported quantity of industrial hazardous waste disposed in these centres was only 416,000 tons, 65% of which was landfilled and 35% incinerated. The quantity of waste alkali and acid ranked the first among industrial hazardous waste categories, accounting for 30.9%. And 39% of industrial hazardous waste was generated from the raw chemical materials and chemical products industry sectors. Southwest China had the maximum generation of industrial hazardous waste with 40% (Duan et al. 2008). According to the *China Statistical Yearbook*, total industrial hazardous waste generated in 2013 was 31.6 million tonnes. There are 49 categories of hazardous waste and 498 types of waste hazardous chemicals in China (Liu et al. 2017).

6.6.3 Plastic Waste and Coastal Plastic Waste

Plastic waste is growing, and its recycling industry is also growing simultaneously. In India approximately 8.33 metric tons of plastic items are consumed every year. According to a CPCB survey, plastic contributes to 8% of the total solid waste, with Delhi delivering the maximum quantity followed by Kolkata and Ahmedabad. Large variation in the estimates of recycling of plastic wastes exists. Around 28 to 60% of the total plastic waste is being recycled (TERI, 2018). For materials such as PET bottles, the recycling rates are around 60%. Usually, plastic packaging is collected or recycled or processed via thermal recovery. CPCB estimates reveal that India generates around 15,342 tonnes of plastic waste per day (TERI 2019). According to the National Resource Efficiency Policy, 2019 (Draft), the targets set to handle plastic wastes are 100% recycling and reuse rate of PET plastic by 2025, 100% recycling of PET plastic and 75% recycling and reuse rate of other plastic packaging materials by 2030 and a ban on the disposal of recyclable waste (plastics, metals, glass, paper, cardboard and 159 types of biodegradable waste) in landfills by 2025 (MoEF & CC 2019). India has enacted bans on single-use plastic carry bags with a view to prevent flooding resulting from clogged drainage systems, maintain clean cityscapes and reduce waste at source. Plastic waste also affects the marine environment. Most of the generated plastics end up in oceans, making up more than 80% of all marine debris (Franzen 2017; Schimdt et al. 2017; Xu and Stanway 2019). The plastic fragments into micro-plastics which significantly pollute the oceans and rivers.

The microplastics load in Indian and Chinese rivers is given in Table 6.6. The load of marine plastic debris, microplastics and macroplastics that these rivers transport into the sea is studied by Schmidt et al. (2017) using models and summarized in Table 6.6. Ten rivers carry 90% of the total plastic pollution into the oceans; among them, the four rivers are in China, the Yangtze, Yellow, Hai and Pearl rivers, and two of them are in India, the Indus and Ganges rivers (Franzen 2017).

China is the world's largest producer and exporter of plastic. Recently, it has discarded about 200 million m^3 of waste into its coastal waters. This illegal disposal in coastal waters in the year 2018 was marked as the year with the highest amount of waste that was discarded in comparison to a ten-year period. It was also 27% more when compared to 2017. China quantified the floating waste in its surface waters in 2017, revealing 24 kg of waste per 1000 m^2 containing approximately 89% plastic content. Even in seabed waste, the plastic waste composition was highest. China is planning to make around 30% of its coastal waters completely off-limits to development as part of a national 'ecological red line' scheme (Xu and Stanway 2019). China accounts for 30% of the world's total plastic products. In 2017, China took in seven million tonnes of plastic rubbish from other countries of Europe, Japan and the US – and 27 million tonnes of waste paper. Later, China banned the import of waste as it is facing difficulties in handling its own waste (BBC 2019). India and China both can opt for bio-based and biodegradable plastics completely instead of plastics from chemicals.

6.6.4 Biomedical Waste

Biomedical waste (BMW) is any waste produced during the diagnosis, treatment or immunization of humans or animal research activities pertaining thereto or in the production or testing of biological or in health camps. The amount of waste generated during treatment, diagnosis, immunization and tests per day is enormous. In a country like India only 1 to 1.5% of the total amount of solid waste generated in a city is biomedical waste, of which 10–15% is considered infectious (Tiwari and Kadu 2013). Only about 10–25% of BMW is hazardous, and the remaining 75–95% is non-hazardous (Datta et al. 2018). The total quantity of BMW generation in the country was approximately 517 TPD (by CPCB) in 2016 and 550 TPD in 2018 (by the Associated Chambers of Commerce and Industry of India and Velocity). It will increase to 775.5 TPD by 2022. According to CPCB in 2017, there are 199 common biomedical waste treatment facilities (CBWTFs) in operation and 23 are under construction. BMW amendments will soon phase out chlorinated plastic bags (excluding blood bags) and gloves; establish a barcode system for both generators and operators; institute global positioning systems (GPS) in vehicles of CBWTF and ensure on-site pretreatment of non-sanitary wastewater through disinfection or sterilization as prescribed in the WHO Blue Book, 2014 (Henam and Shrivastav 2019). The number of health care facilities (HCFs) using CBMWTFs is 131,837, and approximately 21,870 HCFs have their own treatment facilities on-site (Datta et al.

TABLE 6.6
Micro-Plastics Load in India and China Rivers

S. No	River	Ranking the Higher Plastic Waste in Rivers	Receiving Sea	Catchment Area (km^2)	MMPW Generation per Capita (kg/d)	Population (Million)	Population Density	Micro-Plastic Load Model 1 (Tons/ Year)	Micro-Plastic Load Model 2 (Tons/ Year)	Macro-Plastic Load (Tons/ Year)
1.	Yangtze river (Chang Jiang)	1	East China Sea	1907295	0.092	504	264	1469481	85440	69282
2.	Indus	2	Arabian Sea	854106	0.069	191	224	164332	12378	11977
3.	Yellow river (Huang He)	3	Yellow Sea	761437	0.092	122	160	124249	9678	9561
4.	Hai He	4	Yellow Sea	211489	0.092	103	486	91858	7434	7515
5.	Ganges (Meghna, Bramaputra)	6	Bay of Bengal	1571571	0.013	621	395	72845	6039	6230
6.	Pearl river (Zhujiang)	7	South China Sea	388705	0.092	75	193	52958	4577	4823

MMPW – Mismanaged plastic waste generation in the river catchment.
Microplastic (particles <5mm); macroplastic (particles >5mm).
(Source: Adapted from Schmidt et al. 2017)

2018). The general methods employed for the disposal of biomedical waste are landfill, incineration, autoclaving and recycling – reuse (Tiwari and Kadu 2013).

In China, 650,000 tonnes of medical waste are generated annually. Approximately, only 10% was soundly managed (Duan et al. 2008). The State Council promulgated the Regulation on the Control of Medical Waste (Order No. 380) as China's first legislation to address medical waste management. Hazardous medical waste is divided into five categories as infectious waste, pathological waste, sharps, pharmaceutical waste and chemical waste (according to Regulation 287) (Gao et al. 2018). In China, there is no policy on health care waste management (HCWM); however, different guides and standards on the management of HCW from generation to packing, transport and treatment have been issued by the Ministry of Health and Ministry of Environmental Protection (MEP): (i) Pollution Control Standard for Hazardous Wastes Incineration (GB18484-2001), (ii) Technical Standard for Medical Waste Transport Vehicle (GB 19217-2003), (iii) Technical Standard for Medical Care Incinerator (GB 19218-2003) and (iv) Technical Specification for Centralized Disposal of Medical Waste (Trail). There is no national HCWM strategy and national action plan for HCWM. However, a National Plan for Construction of Hazardous and Health-care Waste Treatment Facilities exists (WHO 2015). The main methods of disposing of this medical waste are landfills, solidification, recycling and incineration. The medical waste generated in 261 large and medium-sized cities was 547,500 tons, and disposal capacity was 542,100 tons in 2013. The medical waste disposal rate in most cities has reached 100% (Liu 2017).

6.6.4.1 COVID-19 and Waste Management

With the coronavirus disease (COVID-19) pandemic continuing to spread in the world, Governments are urged to treat waste management, including of medical, household and other hazardous waste, in order to minimize possible secondary impacts upon health and the environment. During such an outbreak, many types of additional medical and hazardous waste are generated, including infected masks, gloves and other protective equipment, together with a higher volume of non-infected items of the same nature. Several guidelines have been released by the UN (the UN Basel Convention's Technical Guidelines on the Environmentally Sound Management of Biomedical and Healthcare Wastes) and WHO ('Safe management of wastes from health-care activities: a summary', World Health Organization) for biomedical waste management (UN 2002; WHO 2017). COVID-19 leaves China with mountains of medical waste especially in Wuhan, the city at the centre of the COVID-19 outbreak. The city is producing more than 240 tonnes a day (compared with 40 tonnes before the epidemic) forcing authorities to deploy mobile treatment facilities. The Central Government has deployed 46 mobile medical waste treatment facilitates to the city, and built a new plant with a capacity of 30 tonnes within 15 days (Zuo 2020). Several regulations and notices with respect to the management of medical wastes from COVID-19 have been released by the Chinese Government (BCRC 2020). Recently, India through CPCB has also released a separate guideline for the handling, treatment and disposal of biomedical waste generated during the treatment/diagnosis/quarantine of COVID-19 patients (CPCB 2020). The medical wastes from COVID-19 are separately collected in yellow bags, further disinfected and incinerated.

6.6.5 Agricultural Waste

Globally, 998 million tonnes of agricultural waste are produced every year. The highest agriculture production in India was sugar cane and rice, and that in China is maize. According to the World Bank, in 2012 the agricultural waste generation in India was 0.105 kg/capita/day and China about 0.153 kg/capita/day. The percentage of agriculture GDP relative to the total GDP of the country in 2013 was around 17.95% for India and 0.01% for China (Hong Kong SAR, China – 0.06%). The total biomass generation (sugarcane biomass) in India was 95,536,000 tonnes in 2013, and in China (maize waste) it was more than 244,386,028 tonnes/year. The biodiesel production will increase to 1 to 1.5 billion litres. Some of the policies and main programmes in India related to agriculture waste are the National Biogas and Manure Management Programme (NBMMP), the National Biomass Policy (1970s), the Fuel

Policy Committee (FPC) (1974), the Working Group on Energy Policy (WGEP) (1979), the National Programme on Bagasse Based Cogeneration (1994), Biomass Policies under the Ninth Plan (1997 to 2002) and the National Biofuel Policy of India adopted in December 2009. In China, the Renewable Energy Law of the People's Republic of China was issued in 2005 and it's the basis for biomass energy development policy. There is potential to reduce methane (CH_4) emission from rice fields in China and India by 26% according to the Asian Development Bank (ADB), as the effective utilization of agriculture and biomass waste contributed to greenhouse gas emissions mitigation (UNCRD 2015).

6.6.6 E- Waste

In 2014 India and China were among the top six countries generating nearly half of all global e-waste. China is the second largest producer of e-waste with 4.4 kg per capita in 2014, and India is the fifth largest producer in the world with 1.3 kg per capita (Balde et al. 2015; TERI 2019). Both India and China have implemented policies for e-waste management (Table 6.1). India generates around 2 million tonnes of e-waste per year, and it is rising at a rate of 5% per annum (TERI 2019). Between 2006 and 2012, India's domestic e-waste generation increased by an estimated factor of eight (Chaturvedi and Bhardwak 2013). According to ASSOCHAM India (2017), the amount of e-waste is growing at a compound annual growth rate (CAGR) of 30% and will reach more than 5 million tonnes per year by end of this decade. Maharashtra ranks highest in terms of e-waste generation, followed by Delhi (21.20%) and Bengaluru (10.10%). The disposal/recycling of e-waste by formal activities was only less than 9% in 2013–2014. The Ministry of Electronics and Information Technology has developed indigenous technology at C-MET and Central Institute of Plastics Engineering & Technology (CIPET) for the recovery of precious metals and plastics from e-waste, respectively (TERI 2019). Bengaluru city generates 18,000 tons of waste electrical and electronic equipment (WEEE or e-waste) per year, thousands of which are landfilled illegally (Ferronato and Torretta 2019). The National Policy on Electronics (2019) promotes research, innovation and support to industry for green processes and sustainable e-waste management.

China is not only the largest producer and consumer of electrical and electronic equipment, but also the country most seriously polluted from illegal e-waste importation and informal recycling (Zeng et al. 2017). In 2013, e-waste generated was about 109.8 million tonnes and only 38% was disposed of/recycled by formal facilities (Liu et al. 2017). The material and the financial flows of e-waste in China had their own specific characteristics. Nearly 60% of the generated e-wastes were sold to private individual collectors and passed into informal recycling processes. More than 90% of Chinese citizens are reluctant to pay for the recycling of their e-waste (Liu 2006).

6.6.6.1 Extended Producer Responsibility (EPR)

The extended producer responsibility provides an overall framework for linking formal systems of production to the informal system of recycling. The concept uses a product life-cycle approach that puts the responsibility on the producer. National e-waste legislation (covering 'take-back' of domestic waste) exists both in India and China (Balde et al. 2015). India has regulations enjoining consumers to return their used batteries, and manufacturers to take back their products. In China, EPR has been introduced but is not well defined. China policy-makers enjoin producers to take responsibility for their products throughout the entire product life cycle – from design to spent-product management, the principle is incorporated into its environmental law.

6.7 ENGAGEMENT WITH KEY STAKEHOLDERS (INFORMAL SECTOR, PPP AND PSP)

The roles and responsibilities of different stakeholders such as the public, municipalities, city planners, NGOs/social workers, corporations, etc., in sustainable solid waste management are explained in detail by Kurian Joseph (2006). Sustainable MSW management from segregation and collection to disposal requires support from public-private partnerships (PPPs), private sector participation (PSP) and also from the informal sector (for recycling). Ensuring the financial viability of MSWM systems through revenue generation and encouraging PSP and PPPs is important. Private participation in the provision of MSW services is not new to India (it has been in practise since 1985) or China.

A public-private partnership is a contract between the Government and a private sector actor in funding that provides the technical, financial and risk assurance of monitoring a project, improving and providing a quality in living for the public for any project to implement successfully. Usually, conventional PPPs have public sector and private sector (formal) aspects. PPP services and types of PPP projects or models being implemented by different cities in India for MSWM are (PPP services: cities/projects in India with PPP models):

- Door-door collection (PPP services): Bengaluru, Ahmadabad, Nagpur, Dumdum, Gandhinagar Jaipur, North and Delhi (PPM models)
- Sweeping streets: Hyderabad and Surat
- Storage and transportation: Surat, Ahmadabad, Mumbai and Delhi
- Integrated treatment and disposal: Delhi, Bengaluru, Coimbatore, Kolkata, Ahmadabad and Chennai
- Integrated primary collection, street sweeping, storage and transportation: Chennai
- Integrated MSWM (complete value chain): Guwahati and Hyderabad

(Bommareddy and Asadi 2017)

In China, many waste-to-energy plants have been established mainly by the PPP model. Around 200 WtE plants are operated through PPP, i.e. by the private sector, accounting for about 80% (Huang et al. 2018a).

The non-conventional PPPs have public sector and private sector (informal) aspects. Recycling has potential for non-conventional PPPs as recycling or valorization is a private trading activity undertaken by the informal sector. The status of the informal sector in recycling services plays a major role in India and China. The amounts of recyclables informally picked out of the municipal solid waste stream are unknown, as informal waste workers do not record their activities (Linzner and Salhofer 2014). Informal recycling is a new and expanding low-cost recycling practice in managing WEEE or e-waste. Informal e-waste recycling is not only associated with serious environmental and health impacts, but also the supply deficiency of formal recyclers and the safety problems of remanufactured electronic products (Chi et al. 2011). A recent study of 6 Indian cities found that waste pickers recovered approximately 20% of waste, with 80,000 people involved in recycling approximately 3 million tonnes of waste. It is estimated that every tonne of recyclable material collected saved the ULB approximately INR 24,500 per annum and avoided the emission of 721 kg CO_2 per annum (Kumar et al. 2017). There are 173,832 and 40,000 informal sector workers in Delhi and Bengaluru, respectively, collecting 5 and 6 tonnes per worker per year. The informal sector workers contributed to 1.3% (117 informal workers per km^2) and 0.5% (50 informal workers per km^2) of the total population in Delhi and Bengaluru, respectively (Wilson et al. 2010). They earn their living from e-waste management, collection, dismantling and metal extraction, including many children (Chaturvedi and Bhardwak 2013). In Beijing, China, the informal collectors, namely waste pickers, waste merchants and middle men, on average collect 16,311 and 890 kg of recyclables per day and capita respectively, and achieve turn-over levels of 1200, 2500 and 5250 CNY per month, respectively (Steuer et al. 2017). The number of informal waste pickers in Beijing, China, is 150,000 people (Ferronato and Torretta 2019). The informal sector is active in the collection, processing and trading of recyclable materials in urban China. Formal waste management organizations have established pilot schemes for the source separation of recyclables, but this strategy is still in its infancy. A secondary data study showed that approximately 0.56–0.93% of the urban population or 3.3–5.6 million people are involved in informal waste collection and recycling activities in urban China. This is equivalent to estimated informal recycling rates of approximately 17–38 w/w% of the municipal solid waste generated (Linzner and Salhofer 2014). Investment arrangements between local Governments and the informal sector are potentially good areas for PPPs in SWM that would be nonconventional in nature.

6.8 GREENHOUSE GAS EMISSIONS FROM WASTE

The methane emissions from waste are higher in India and China. The emission of CH_4 is directly related to the amount of waste generated from the households or the society in a region. MSW generated by households is considered the third largest anthropogenic source of methane emissions, constituting 11% of all global CH_4 emissions (Singh et al. 2018). A World Bank report revealed the methane emissions from India and China in 2005 from post-consumer MSW disposal, GHG emissions (CO_2, CH_4, N_2O) and % methane from disposal sites relative to total GHG emissions were about 14 $MtCO_2e$, 1,210 $MtCO_2e$ and 1.1% for India and 45 $MtCO_2e$, 3,650 $MtCO_2e$ and 1.2% for China, respectively (Hoornweg and Bhada-Tata 2012). The different case studies of waste emission from landfills in India and China are as follows.

6.8.1 Case Studies of Waste Emission from Landfills in India

Landfills rank third in terms of greenhouse gas emissions in India (Swaminathan 2018). Methane emission from landfill is about 3 to 19% of anthropogenic sources in the world. India, one among the world's largest emitters of CH_4 from landfills, produces about 16 tons CO_2eq/yr. CH_4 constitutes about 29% of the total GHG emissions in India which is nearly twice the worldwide average of 15% (Singh et al. 2016). Another study estimating the net increase in methane emissions from MSW landfills in India using the IPCC default method, modified triangular method and first order decay method showed net annual emission of CH_4 increased from 404 Gg in 1999–2000 to 990 Gg in 2011 and reached 1,084 Gg in 2015, respectively (Singh et al. 2018).

A study by Singh et al. (2016) revealed the methane emissions from three landfills in Delhi that produce about 8,360 tons of MSW/d in 2012, of which 87% is collected (7,273 TPD) and only 28% (2,000 TPD) is treated. In the end, it is disposed of at the three Delhi landfill sites, viz. Bhalswa, Ghazipur and Okhla landfills. Methane emissions from these Bhalswa, Ghazipur and Okhla landfills for the year 2011–2012 were 91.23 Gg/yr, 3,845.20 Gg/yr and 77.42 Gg/yr by the default method and 61.97 Gg/yr, 2,611.99 Gg/yr and 52.60 Gg/yr by the first order decay method, respectively.

Several methane emission studies have been carried out in two landfills in Chennai. Perungudi landfill has been in operation since 1987; the total area used for dumping is around 99 ha of the total 3,177 ha with 2,200 to 2,400 MT of MSW dumped per day. Kodungaiyur landfill has been in operation since 1980; the total area of this site is about 804 ha, of which about 127 ha acres are used to date for dumping and 2,100 to 2,300 MT of MSW is dumped per day. Karthikeyan et al. (2012) studied the methane emission potential of Chennai landfills using TNO, LandGEM and zero-order models. The total calculated methane emissions from Chennai landfills were 33×10^6 m^3, 36×10^6 m^3 and 52×10^4 m^3 for the applied TNO, LandGEM and zero-order models, respectively. Another study showed CH_4 emission rates for Chennai landfills in 2015–2016 using closed flux chamber as 8.8 to 11.3 Gg/yr and 6.1 to 9.1 Gg/yr in Kodungaiyur and Perungudi, respectively. The CH_4 emission rates of Chennai landfills using the IPCC default method for the period of 1981–2050 were in the range of 25.8 to 124.88 Gg/yr (an increased value of 88%) and 60 to 266 Gg/yr (an increased value of 79%) in Kodungaiyur and Perungudi, respectively (Srinivasan et al. 2020).

6.8.2 Case Studies of Waste Emission from Landfills in China

In China, CH_4 emissions from MSW waste treatments in 2005 were 3.824 million tonnes, of which the CH_4 emissions from solid waste disposals (landfills) were 2.204 million tonnes, accounting for 57.62%.

The greenhouse gas emissions from waste treatments were 111 Mt CO_2eq, CO_2 emissions from the incineration of fossilized material-originated wastes were 2.658 Mt and emissions from organic waste incinerations were 3.927 Mt (excluded in the total) (Liu 2017). Zhao et al. (2009) showed GHG emissions from Tianjin's MSW management system account for 467.34 Mg CO_2 eq/yr depending on the treatment of MSW collected in the central districts in 2006, and the key issue is landfill gas (LFG) released. The integrated MSW management scenario, combining different improvement options, shows the highest GHG mitigation potential. Greenhouse gas emissions from different municipal solid waste management scenarios in China, based on carbon and energy flow analysis, were studied by Liu et al. (2017). Poor landfill gas collection efficiency and low carbon storage caused landfilling with flaring and landfilling with biogas recovery scenarios showed the highest amounts of GHG emissions of 192 and 117 kg CO_2 eq/t, respectively. Incineration had the best energy recovery rate (19%), and, by grid emissions substitution, led to a substantial decrease in net GHG emissions (−124 kg CO_2 eq/t). The scenarios with biological treatment of the organic fractions after sorting, such as composting or anaerobic digestion (AD), did not lead to the outstanding GHG reductions (emissions of 32 and −36 kg CO_2 eq/t, respectively) expected. A GHG accounting model study in six cities of China showed that MSW incineration in most of the cities was the source of GHGs, with emissions of 25–207 kg CO_2 eq t^{-1} rw. Within all process stages, the emission of fossil CO_2 from the combustion of MSW was the main contributor (111–254 kg CO_2 eq t^{-1} rw), while the substitution of electricity reduced the GHG emissions by 150–247 kg CO_2 eq t^{-1} rw (Yang et al. 2012).

6.9 CHALLENGES: LESSONS LEARNT FROM INDIAN AND CHINESE WASTE MANAGEMENT PERSPECTIVES

The level of waste management delivery in India and China does not meet the compliance and standards as mentioned in their respective waste management rules and regulations (as shown in Table 6.1). Numerous challenges exist in segregation, collection, waste processing and disposal practices as described in Table 6.7. MSW management services have many difficulties in their implementation. The handling of other wastes such as C&D wastes, e-wastes, hazardous, biomedical and agricultural wastes has various complications with respect to collection and disposal. The lessons learnt from Indian and Chinese waste management perspectives show that lack of integration is the key deficiency in the existing systems. So, proper strategies should be framed in order to address the challenges in the system.

6.10 POSITIVE SIGNS

Even though India and China have faced many problems in handling and treating waste, some of the good initiatives of these Government support effective waste management. The schemes of 'smart cities' in India and 'zero-waste cities' in China support sustainable solid waste management in the countries.

In India, Swachh Bharat Mission or Swachh Bharat Abhiyan (SBA) or the Clean India Mission campaign launched on 2 October 2014 to eliminate open defecation and improve SWM in urban and rural areas. Under SBM, the various schemes implemented are Swachh Bharat Mission for Urban Areas, Swachh Bharat Mission (Gramin), Swachh Vidyalaya Abhiyan, Rashtriya Swachhata Kosh and other schemes of Swachh Sarvekshan annual cleanliness survey (MSW services), star rating of garbage-free cities, Swachhata Hi Seva (single-use plastic free), etc. As of SBM (2020b) urban, 100% door-to-door collection and source segregation was achieved in 81,535 and 64,730 wards, respectively, total waste processing is 65%, WtE is 61 MW and waste to compost is 3,182,435 MT. National sanitation improved from 38.7% in 2014 to 90% in 2018.

6.10.1 Smart Cities in India

The Smart Cities Mission in India aims for urban planning with sustainable solid waste management. The Indian Government has initiated the National Smart Cities Mission, an urban renewal

TABLE 6.7
Major Challenges of Solid Waste Management in India and China

S. No	Solid Wastes and SWM Services	Challenges
1.	Urbanization	Rapid urbanization in India and China is directly linked with higher waste generation rates. The Indian and Chinese Governments through urban public services are not able to manage and treat the produced MSW effectively.
2.	Segregation	No proper segregation of waste. Segregation of hazardous and non-hazardous wastes or biodegradable and non-biodegradable is not practised.
3.	Collection	One hundred percent collection efficiency is not reached in all wards. Littering of MSW on the streets and clogging the drains, contaminating the aquifers and polluting the environment.
4.	Transportation	The frequent transportation of waste to treatment sites is not possible due to lack of trucks or other vehicles. Also the trucks are not closed or covered during transportation.
5.	Treatment	The waste sent to treatment plants is lower in both the countries. The numbers of compost plants or WtE plants in India and incineration plants in China were not meeting the requirement of the present quantity of waste generated. Hence, treatment efficiency is insufficient.
6.	Disposal	Inadequate disposal facilities. Mixed waste ends up in the disposal plants. The quality of compost is poor and not feasible to be used in agricultural fields. Incineration of mixed waste not effective. In many cases, the treatment plants were shut down due to mixed waste problems. Landfills with uncontrolled dumping that don't comply with regulations. No scientific disposal was adopted.
7.	Recycling	The gap between the urban service and recycling service is huge. The advantage of informal sectors is that they handle large volume of wastes for recycling, whereas the disadvantage of informal sector is improper handling of wastes that harm the environment and health. Informal workers are poorest.
8.	C&D wastes	No proper disposal sites and lack of collection. Mix-up with MSW impacting its processing.
9.	Public	No complete awareness about solid waste management and services, so segregation and collection of wastes are difficult. The not in my backyard concept (NIMBYism) exists.
10.	E-waste	Ineffective implementation of current policies. Price of e-waste in the informal market is higher than formal recycler.
11.	Stakeholders	Lack of coordination among stakeholders, disturbing the SWM chain and unskilled stakeholders.
12.	Hazardous waste	Untrained waste collectors, no segregation and hence recycling of waste is not possible, lack of infrastructure for scientific disposal and recycling, burning of waste causing threat to environment and health. Number of incineration plants to dispose of hazardous waste is very low.
13.	Biomedical waste	Improper collection of waste, low number of treatment facilities and high cost for treatment.
14.	Agricultural waste	Lower quantity of collection and treatment of the agricultural waste.
15.	Governance	Limitations of proper and strict governance, no stringent compliance and regulation and lack of technical and financial resources.

(Source: Prepared by author)

and retrofitting programme to develop smart cities across the country, making them citizen friendly and sustainable. There are smart solutions for waste management, water and energy management, e-Governance and citizen services, urban mobility and others (MoUD 2015). The main purpose of smart waste management is to help to reduce waste, create waste-to-energy sources and also help to keep the environment clean and neat (Mokale 2019). As of April 2020, 99 cities have been added to the Smart Cities Mission and these 'Smart Cities' in India are already considering the issue of solid waste management and recycling.

6.10.2 Zero-Waste Cities in China

To achieve sustainable urban development in China, the recycling of MSW waste, harmless disposal and waste-to-energy technologies are all important issues to be addressed. In January 2019, the State Council, China's Cabinet, issued a work plan for the construction of zero-waste cities. According to the plan, a zero-waste city is an advanced urban development and management model that aims to promote green design, production and living, minimize the amount of waste produced, strengthen recycling programmes and ensure that waste released into the environment is harmless (Wang et al. 2019). According to the Ministry of Ecology and Environment, China, the regions of strategic importance, such as Beijing-Tianjin-Hebei (JJJ region), the Yangtze River economic belt and the Greater Bay Area, will be considered first in the pilot city selection due to hotspot areas.

6.11 WAY FORWARD: INTEGRATED WASTE MANAGEMENT

Based on the current practices and challenges existing in the waste management systems of India and China, the following recommendations are proposed for a sustainable solid waste management system. The recommendations are sub-divided into different categories based on the concepts and roles of the target groups.

6.11.1 Implementation of Waste Minimization Strategies

- Adoption of waste prevention and waste reduction
- Implementation of integrated waste management
- Use waste hierarchy of the '4 Rs' concept (reduce, reuse, recycle and recover)
- Application of circular economy
- Effective and sustainable solid waste management practices
- Waste recovery of valuable materials

6.11.2 Roles of Government Agencies

- Strict monitoring and enforcement of existing regulations and standards
- Ban of single-use plastic items
- Strictly follow polluter pays principle
- Subsidies for carbon reduction technologies
- Implementation of incentives and subsidies for informal sector workers to reduce improper recycling and formalization of informal sector
- Review and make amendments of policies based on current waste practices
- Creation of new policies for waste management and recycling
- Punishments and fines for illegal dumping of wastes
- Incentives for buying recycled products
- Policymakers to focus more attention on e-waste, micro-plastics and marine debris, etc.

6.11.3 Responsibilities of Different Stakeholders

- ULBs should have consistency in data collection of waste quantities and composition
- Stakeholders should know their roles and responsibilities, and should be well equipped with waste-related information

- Door-to-door collection by municipal authorities to reach 100%
- Source segregation of hazardous/non-hazardous and biodegradable/non-biodegradable compounds
- Separate bins for waste collection (bio-degradable/recyclable/hazardous), kept in appropriate places
- Regular street sweeping with coverage of all areas
- Improve household waste segregation at source
- Transportation of waste must be in closed containers/trucks
- Treatment of bio-degradable waste to be composted or used in waste-to-energy
- Stringent monitoring and standards followed in waste services
- Control of waste (mostly inert) disposed in landfills
- Better interaction and co-operation with multi-stakeholders (informal, NGOs, ULBs, private sectors, Government sectors, etc.)
- Implementation of home composting of organic wastes
- Encourage PPP and PSP in MSWM services
- Expand extended producer responsibility
- Strengthen urban public services for waste collection, treatment and disposal
- Special collection trucks for hazardous waste collection and disposal
- Increase the waste collection frequency based on (MSW quantity) demand
- Prevention of plastics entering into marine systems

6.11.4 Strengthening the Waste Treatment Techniques

- Implementation of decentralization facilities for waste processing
- Efficient use of methane gas from landfills as an energy source
- Adopt sustainable and cost-effective treatment techniques
- Improve landfill operations and reduction of wastes to landfills
- Co-processing of solid wastes
- Opt for bio-based and biodegradable plastics as alternatives for plastics
- Old dumpsite remediation and reuse of wastes for useful purposes
- Implementation of new treatment methods for special wastes
- Proper waste treatment adoption based on waste characteristics
- Avoid mixed wastes as feed to incinerators and compost plants
- Usage of smart networks for monitoring and regulating waste collection services until disposal
- Improvement of on-site waste disposal methods
- Up-gradation of waste treatment methods from traditional practices
- Production of sustainable materials from recycled wastes
- Reuse of construction and demolition waste in other processes
- Better control emissions from incinerators
- Proper and scientific disposal of hazardous waste
- Predictions of waste quantity using models
- Mapping of waste distribution and characteristics
- Increase the common biomedical waste treatment facilities

6.11.5 Waste Management Finance

- Better management of waste finance
- Frequent waste audits
- Corporate social responsibility funds for waste management
- Urban local bodies to spend money efficiently on innovative waste management systems

6.11.6 Capacity Building and Awareness

- Creating public awareness about solid waste management
- Behavioural changes towards solid waste generation, collection and disposal
- Skilled training for municipal workers and informal workers

6.12 CONCLUSION

India and China have substantial similarities and dissimilarities in waste management practices. Both countries have several gaps in their waste management services, which can be addressed and managed by an integrated solid waste management approach. Managing the solid waste sector has the potential for numerous business opportunities and job creation. The waste market including segregation, collection, waste processing and disposal creates jobs in source-separation schemes of MSW, recycling of materials, sorting and processing of recyclables, management of compost plants, etc. Recycling industries, material-recovering industries and the waste-to-energy market have huge business opportunities. Encouraging sustainable solid waste management is a move towards the green economy in the nation.

ACKNOWLEDGEMENTS

The corresponding author (Dr S. Sri Shalini) gratefully acknowledges the financial support provided by the Department of Science and Technology (DST), Ministry of Science and Technology, Government of India under Women Scientists Scheme-A (WOS-A) (Grant no. SR/WOS-A/EA-37/2018).

REFERENCES

Ahluwalia, I.J., and U. Patel. 2018. Solid waste management in India: an assessment of resource recovery and environmental impact. ICRIER working paper 356. http://icrier.org/pdf/Working_Paper_356.pdf (Accessed on March 20, 2020).

APO. 2007. Solid waste management: issues and challenges in Asia. Report of the APO survey on solid wastes management 2004–05, APO, Japan. https://www.apo-tokyo.org/00e-books/IS-22_SolidWasteMgt/IS-22_SolidWasteMgt.pdf (Accessed on March 15, 2020).

Bala, R., G.K. Gupta, B.V. Dasgupta, et al. 2019. Pretreatment optimisation and kinetics of batch anaerobic digestion of liquidised OFMSW treated with NaOH: models verifcation with experimental data. *Journal of Environmental Management* 237:313–321.

Balde, C.P., R. Kuehr, K. Blumenthal, et al. 2015. The global e-waste monitor—2014. United Nations University, IAS-SCYCLE, Bonn, Germany. https://i.unu.edu/media/unu.edu/news/52624/UNU-1st Global-E-Waste-Monitor-2014-small.pdf (Accessed on March 12, 2020).

BBC. 2019. A rubbish story: China's mega-dump full 25 years ahead of schedule. https://www.bbc.com/news/world-asia-50429119 (Accessed on March 17, 2020).

BCRC. 2020. Basel convention regional centre for Asia and the Pacific, Beijing, United Nations. http://bcrc.tsinghua.edu.cn/en/col/1257152450718/index.html (Accessed on April 1, 2020).

Bhat, R.A., S.A. Dar, D.A. Dar, et al. 2018. Municipal solid waste generation and current scenario of its management in India. *International Journal of Advance Research in Science and Engineering* 7(2):419–433.

Bommareddy, S.S.R., and S.S. Asadi. 2017. Solid waste management through public private partnership model. *International Journal of Civil Engineering and Technology* 8(1):263–272.

Chaturvedi, B., and S. Bhardwak. 2013. Learning to re-e-cycle: what working with e-waste has taught us. Chintan Environmental Research and Action Group, New Delhi. https://www.chintan-india.org/sites/default/files/2019-07/chintan-study-learning-to-re-e-cycle.pdf (Accessed on March 18, 2020).

Cheng, H., and Y. Hu. 2010. Municipal solid waste (MSW) as a renewable source of energy: current and future practices in China. *Bioresource Technology* 101(11):3816–3824.

Chhabra, V., K. Bambery, S. Bhattacharya, et al. 2020. Thermal and in situ infrared analysis to characterise the slow pyrolysis of mixed municipal solid waste (MSW) and its components. *Renewable Energy* 148:388–401.

Chi, X., M. Streicher-Porte, M.Y.L. Wang, et al. 2011. Informal electronic waste recycling: a sector review with special focus on China. *Waste Management* 31(4):731–742.

CPCB. 2000. Status of solid waste generation, collection, treatment and disposal in metro-cities. Series: CUPS/46/1999-2000, MoEF & CC, Government of India.

CPCB. 2017. Annual reports of CPCB on SWM status. MoEF & CC. http://164.100.107.13/MSW_AnnualReport_2015-16.pdf (Accessed on March 5, 2020).

CPCB. 2019. Guidelines for disposal of legacy wastes (old municipal solid waste). MoEF & CC. http://www.toxicslink.org/docs/LegacyWasteBiomining_guidelines_29.04.2019.pdf (Accessed on March 17, 2020).

CPCB. 2020. Guidelines for handling, treatment and disposal of waste generated during treatment/diagnosis/quarantine of COVID-19 patients. MoEF & CC. https://ncdc.gov.in/WriteReadData/l892s/63948609501585568987.pdf (Accessed on March 17, 2020).

Dasgupta, A., and M.K. Chandel. 2020. Enhancement of biogas production from organic fraction of municipal solid waste using alkali pretreatment. *Journal of Material Cycles and Waste Management* 22:757–767.

Datta, P., G.K. Mohi, and J. Chander. 2018. Biomedical waste management in India: critical appraisal. *Journal of Laboratory Physicians* 10(1):6–14.

Dong, J., Y. Chi, D. Zou, et al. 2014. Comparison of municipal solid waste treatment technologies from a life cycle perspective in China. *Waste Management & Research* 32(1):13–23.

Dorn, T., M. Nelles, S. Flamme. 2010. Circular economy in China. In: ISWA Conference proceedings—urban development and sustainalility—a major challenge for waste management in the 21st century. International Solid Waste Association, Hamburg. https://www.iswa.org/uploads/tx_iswaknowledgebase/Dorn.pdf (Accessed on March 18, 2020).

Duan, H., Q. Huang, Q. Wang, et al. 2008. Hazardous waste generation and management in China: a review. *Journal of Hazardous Materials* 158(2–3):221–227.

Ferronato, N., and V. Torretta. 2019. Waste mismanagement in developing countries: a review of global issues. *International Journal of Environmental Research and Public Health* 16(6):1060.

Franzen, H. 2017. Almost all plastic in the ocean comes from just 10 rivers, Deutsche Welle. https://www.dw.com/en/almost-all-plastic-in-the-ocean-comes-from-just-10-rivers/a-41581484 (Accessed on March 10, 2020).

Gao, Q., Y. Shi, D. Mo, et al. 2018. Medical waste management in three areas of rural China. *PLoS ONE* 13(7):e0200889.

Gundupalli, S.P., S. Hait, A. Thakur. 2017. A review on automated sorting of source-separated municipal solid waste for recycling. *Waste Management* 60:56–74.

Havukainen, J., M. Zhan, J. Dong, et al. 2017. Environmental impact assessment of municipal solid waste management incorporating mechanical treatment of waste and incineration in Hangzhou, China. *Journal of Cleaner Production* 141:453–461.

Henam, S., and R. Shrivastav. 2019. Biomedical waste management in India: still a looming concern. Down to Earth. https://www.downtoearth.org.in/blog/health/biomedical-waste-management-in-india-still-a-looming-concern-63896 (Accessed on March 16, 2020).

Hoornweg, D., and P. Bhada-Tata. 2012. What a waste: a global review of solid waste management. World Bank. https://openknowledge.worldbank.org/handle/10986/17388 (Accessed on March 25, 2020).

Hoornweg, D., L. Philip, and C. Manisha. 2005. Waste management in China: issues and recommendations (English). Urban development working papers; no. 9. Washington, DC: World Bank. http://documents.worldbank.org/curated/en/237151468025135801/Waste-management-in-China-issues-and-recommendations (Accessed on February 5, 2020).

Hu, J., H. Ke, L.T. Zhan, et al. 2020. Installation and performance of horizontal wells for dewatering at municipal solid waste landfills in China. *Waste Management* 103:159–168.

Huang, B., X. Wang, H. Kua, et al. 2018a. Construction and demolition waste management in China through the 3R principle. *Resources, Conservation and Recycling* 129:36–44.

Huang, J., S. Liu, A. Plaza, et al. 2018b. Creating an enabling environment for public–private partnerships in waste-to-energy projects. ADB sustainable development working paper series. https://www.adb.org/sites/default/files/publication/471811/sdwp-058-ppp-waste-energy-projects.pdf (Accessed on March 17, 2020).

Jia, X., S. Wang, Z. Li, et al. 2018. Pinch analysis of GHG mitigation strategies for municipal solid waste management: a case study on Qingdao City. *Journal of Cleaner Production* 174(10):933–944.

Joseph, K. 2006. Stakeholder participation for sustainable waste management. *Habitat International* 30:863–871.

Joseph, K. 2014. Municipal solid waste management in India. In *Municipal Solid Waste Management in Asia and the Pacific Islands*, ed. A. Pariatamby, and M. Tanaka, Environmental Science and Engineering, Springer-Verlag, Singapore.

Joseph, K., S. Rajendiran, R. Senthilnathan, et al. 2012. Integrated approach to solid waste management in Chennai: an Indian metro city. *Journal of Material Cycles and Waste Management* 14:75–84.

Joshi, R., and S. Ahmed. 2016. Status and challenges of municipal solid waste management in India: a review. *Cogent Environmental Science* 2:1139434.

Karthikeyan, L., V.M. Suresh, V. Krishnan, et al. 2018. The management of hazardous solid waste in India: an overview. *Environments* 5:103.

Karthikeyan, O.P., K. Joseph, and S. Esakku. 2012. Methane emission potential of open dumps in Chennai: a case study. *International Journal of Environmental Technology and Management* 15, 3/4/5/6:291–304.

Khan, D., A. Kumar, and S.R. Samadder. 2016. Impact of socioeconomic status on municipal solid waste generation rate. *Waste Management* 49:15–25.

Kothwale, S. 2019. The hazard of hazardous waste management in India: a legal overview, legal services india.com. http://www.legalservicesindia.com/law/article/1246/14/The-Hazard-of-Hazardous-Waste-Management-in-India-A-Legal-Overview (Accessed on February 12, 2020).

Kumar, S., S.R. Smith, G. Fowler, et al. 2017. Challenges and opportunities associated with waste management in India. *Royal Society Open Science* 4(3):160764.

Linzner, R., and S. Salhofer. 2014. Municipal solid waste recycling and the significance of informal sector in urban China. *Waste Management & Research* 32(9):896–907.

Liu, C. 2017. *Country Chapter on the People's Republic of China, State of the 3Rs in Asia and the Pacific*. UN Centre for Regional Development, Japan.

Liu, J., Q. Li, W. Gu, et al. 2019. The impact of consumption patterns on the generation of municipal solid waste in China: evidences from provincial data. *International Journal of Environmental Research and Public Health* 16:1717.

Liu, T., Y. Wu, X. Tian, et al. 2015. Urban household solid waste generation and collection in Beijing, China. *Resources, Conservation & Recycling* 104:31–37.

Liu, X. 2006. Electrical and electronic waste management in China: progress and the, barriers to overcome. *Waste Management & Research* 24(1):92–101.

Liu, Y., W. Sun, and Liu, J. 2017. Greenhouse gas emissions from different municipal solid waste management scenarios in China: based on carbon and energy flow analysis. *Waste Management* 68:653–661.

Lu, J., S. Zhang, J. Hai, et al. 2017. Status and perspectives of municipal solid waste incineration in China: a comparison with developed regions. *Waste Management* 69:170–186.

Lu, W., H. Yuan, J. Li, et al. 2011. An empirical investigation of construction and demolition waste generation rates in Shenzhen city, South China. *Waste Management* 31(4):680–687.

Ma, Y., X. Lin, and Z. Chen. 2020. Influence factors and mass balance of memory effect on PCDD/F emissions from the full-scale municipal solid waste incineration in China. *Chemosphere* 239:124614.

MoEF & CC. 2019. Report on the national resource efficiency policy, 2019 (Draft)—charting a resource efficient future for sustainable development, Ministry of Environment, Forest and Climate Change, Government of India. http://moef.gov.in/wp-content/uploads/2019/07/Draft-National-Resourc.pdf (Accessed on April 9, 2020).

MoHUA. 2018. Utilisation of recycled produce of construction & demolition waste a ready reckoner. Building Materials and Technology Promotion Council, Ministry of Housing & Urban Affairs Government of India. http://164.100.228.143:8080/sbm/content/writereaddata/C&D%20Waste_Ready_Reckoner_BMTPC_SBM.pdf (Accessed on March 15, 2020).

Mokale, P. 2019. Smart waste management under smart city mission—its implementation and ground realities. *International Journal of Innovative Technology and Exploring Engineering* 8(12):3095–3013.

MoUD. 2015. Mission statement and guidelines—smart cities. Ministry of Urban Development, Government of India. http://164.100.161.224/upload/uploadfiles/files/SmartCityGuidelines(1).pdf (Accessed on March 17, 2020).

MoUD. 2016. SBM municipal solid waste management manual, Part-II, CPHEEO, MoUD. http://cpheeo.gov.in/upload/uploadfiles/files/Part2.pdf (Accessed on March 5, 2020).

Nandan, A., B.P. Yadav, S. Baksi, et al. 2017. Recent scenario of solid waste management in India. *World Scientific News* 66:56–74.

Panigrahi, S., and B.K. Dubey. 2019. A critical review on operating parameters and strategies to improve the biogas yield from anaerobic digestion of organic fraction of municipal solid waste. *Renewable Energy* 143:779–797.

PIB. 2016. New hazardous waste rules will ensure resource recovery and disposal of hazardous waste in environmentally sound manner. Press Information Bureau, MoEF&CC, Government of India. https://pib.gov.in/newsite/PrintRelease.aspx?relid=138521 (Accessed on March 28, 2020).

Planning Commission Report. 2014. Reports of the task force on waste to energy (Vol-I). (in the context of Integrated MSW management, Planning Commission, Government of India. http://planningcommission.nic.in/reports/genrep/rep_wte1205.pdf (Accessed on February 15, 2020).

Qu, X.Y., Z.S. Li, X.Y. Xie, et al. 2009. Survey of composition and generation rate of household wastes in Beijing, China. *Waste Management* 29(10):2618–2624.

Ramachandra, T., H. Bharath, G. Kulkarni, et al. 2018. Municipal solid waste: generation, composition and GHG emissions in Bangalore, India. *Renewable and Sustainable Energy Reviews* 82:1122–1136.

Rastogi, M., M. Nandal, and L. Nain. 2019. Additive effect of cow dung slurry and cellulolytic bacterial inoculation on humic fractions during composting of municipal solid waste. *International Journal of Recycling of Organic Waste in Agriculture* 8:325–332.

Ray, M., and M. Rahman. 2016. An overview of legal framework for waste management system in india with special allusion to SWM rules. *International Journal of Interdisciplinary and Multidisciplinary Studies* 4(1):13–19.

Rezaei, H., F. Yazdanpanah, C.J. Lim, et al. 2020. Pelletization properties of refuse-derived fuel—Effects of particle size and moisture content. *Fuel Processing Technology* 205:106437.

Sahu, A., M.C. Manna, S. Bhattachariya, et al. 2020. Dynamics of maturity and stability indices during decomposition of biodegradable city waste using rapo-compost technology. *Applied Soil Ecology* 155:103670.

SBM. 2020a. State-wise status of implementation of various components under Swachh Bharat Mission, upto January 2020. Government of India. http://swachhbharaturban.gov.in/writereaddata/Statewise_status_of_implementation.pdf (Acessed on April 2, 2020).

SBM. 2020b. Swachh Bharat Mission urban. http://swachhbharaturban.gov.in/ (Accessed on April 2, 2020).

Schmidt, C., T. Krauth, and S. Wagner. 2017. Export of plastic debris by rivers into the sea. *Environmental Science & Technology* 51(21):12246–12253.

Sebastian, R.M., D. Kumar, and B. Alappat. 2020. Mapping incinerability of municipal solid waste in Indian subcontinent. *Clean Technologies and Environmental Policy* 22:91–104.

Sharholy, M., K. Ahmad, G. Mahmood, et al. 2008. Municipal solid waste management in Indian cities—A review. *Waste Management* 28(2):459–467.

Singh, A, and M.K. Chandel. 2020. Effect of ageing on waste characteristics excavated from an Indian dump-site and its potential valorisation. *Process Safety and Environmental Protection* 134:24–35.

Singh, C.K., A. Kumar, and S.S. Roy. 2018. Quantitative analysis of the methane gas emissions from municipal solid waste in India. *Nature Scientific Reports* 8:2913.

Singh, S.K., G. Anunay, G. Rohit, et al. 2016. Greenhouse gas emissions from landfills: a case of NCT of Delhi, India. *Climatol Weather Forecasting* 4:157.

Sri Shalini, S., and Kurian Joseph. 2013. Start-up of the SHARON and ANAMMOX process in landfill bioreactors using aerobic and anaerobic ammonium oxidising biomass. *Bioresource Technology* 149:474–485.

Sri Shalini, S., and Kurian Joseph. 2018. Combined SHARON and ANAMMOX processes for ammoniacal nitrogen stabilization in landfill bioreactors. *Bioresource Technology* 250:723–732.

Sri Shalini S., K. Palanivelu, A. Ramachandran et al. 2020. Biochar from biomass waste as a renewable carbon material for climate change mitigation in reducing greenhouse gas emissions—a review. *Biomass Conversion and Biorefinery* https://doi.org/10.1007/s13399-020-00604-5

Sri Shalini, S., O.P. Karthikeyan, and Kurian Joseph. 2010. Biological stability of municipal solid waste under simulated landfill conditions. *Bioresource Technology* 101:845–852.

Srinivasan, P., R. Andimuthu, S.N. Ahamed Ibrahim, et al. 2020. Methane emission from municipal solid waste dumpsites: a case study of Chennai city in India. *Advances in Environmental Research* 9(2):97–107.

Steuer, B., R. Ramusch, F. Part, et al. 2017. Analysis of the value chain and network structure of informal waste recycling in Beijing, China. *Resources, Conservation & Recycling* 117:137–150.

Swaminathan, M. 2018. How can India's waste problem see a systemic change?, *Economic & Political Weekly* 53(f 16) ISSN (Online)—2349–8846. https://www.epw.in/engage/article/institutional-framework-implementing-solid-waste-management-india-macro-analysis (Accessed on March 20, 2020).

TERI. 2018. Fact sheet on plastic waste in India. Energy and Resources Institute, New Delhi, India. https://www.teriin.org/sites/default/files/files/factsheet.pdf (Accessed on March 12, 2020).

TERI. 2019. Reference report for national resource efficiency policy for India. Prepared for MoEF & CC, 2019. Energy and Resources Institute, New Delhi, India. https://www.teriin.org/sites/default/files/2019-04/National-Policy-Report.pdf (Accessed on March 9, 2020).

Tiwari, A.V., and P.A. Kadu. 2013. Biomedical waste management practices in India—a review. *International Journal of Current Engineering and Technology* 3(5)2030–2033.

UN. 2002. Technical guidelines on the environmentally sound management of biomedical and healthcare wastes. United Nations. USA. https://digitallibrary.un.org/record/475506?ln=en (Accessed on March 20, 2020).

UN DESA. 2018. Revision of world urbanization prospects, population division of the United Nations Department of Economic and Social Affairs, USA. https://www.un.org/development/desa/publications/2018-revision-of-world-urbanization-prospects.html (Accessed on March 1, 2020).

UN Population. 2019. World population prospects 2019: highlights, population division of United Nations, Department of Economic and Social Affairs, USA. https://www.un.org/en/sections/issues-depth/population/index.html (Accessed on March 2, 2020).

UNCRD. 2015. Sixth regional 3R forum in Asia and the Pacific, circular economic utilisation of agriculture and biomass waste—a potential opportunity for Asia and the Pacific, Maldives. United Nations entre for regional development, Japan. https://www.uncrd.or.jp/content/documents/2674Front%20cover%20for%20Background%20paper-%20PR-3-%20Prof.%20Agamuthu_merged.pdf (Accessed on March 10, 2020).

Vuppaladadiyam, A.K., M. Zhao, M.Z. Memon, A.F. Soomro, et al. 2019. Solid waste as a renewable source of energy: a comparative study on thermal and kinetic behavior of three organic solid wastes. *Energy & Fuels* 33(5):4378–4388.

Wang, H., and Y. Nie. 2001. Municipal solid waste characteristics and management in China. *Journal of the Air & Waste Management Association* 51(2):250–263.

Wang, X., J. Chen, M. Gu, et al. 2019. Working towards zero waste in China's cities: status, challenges and opportunities. Proceedings of ISWA World congress, Circular Economy, What are you doing. Bilbao, Spain, pp 137–138. https://www.iswa2019.org/wp-content/uploads/2019/11/PROCEEDINGS2019.pdf (Accessed on March 16, 2020).

Wei, Y., Y. Xue, J. Yin, et al. 2013. Prediction of municipal solid waste generation in china by multiple linear regression method. *International Journal of Computers and Applications* 35(3):136–140.

Wei, Y-S., Y-B. Fan, M-J. Wang, et al. 2000. Composting and compost application in China. *Resources, Conservation and Recycling* 30:277–300.

Whitworth, A.P. 2005. Thermal breakdown characteristics of municipal solid waste components in varying oxygen environment and municipal solid waste management in China. Master's Thesis, Department of Earth and Environmental Engineering, Columbia University in the City of New York, USA.

WHO. 2015. Regional office for the Western Pacific. Status of health-care waste management in selected countries of the Western Pacific Region. Manila: World Health Organization. Regional Office for the Western Pacific. https://apps.who.int/iris/handle/10665/208230 (Accessed on March 17, 2020).

WHO. 2017. Safe management of wastes from health-care activities: a summary. World Health Organization, Switzerland. https://apps.who.int/iris/handle/10665/259491 (Accessed on February 17, 2020).

Wilson, D.C., L. Rodic, A. Scheinberg, et al. 2010. Comparative analysis of solid waste management in cities around the world. In proceedings waste 2010: waste and resource management, Warwickshire, UK. https://edepot.wur.nl/169398 (Accessed on March 2, 2020).

Wong, S. 2019. Amount of disposed waste in China 1990–2018. https://www.statista.com/statistics/279117/amount-of-disposed-garbage-in-china/ (Accessed on February 6, 2020).

World Bank. 1999. What a waste: solid waste management in Asia. The World Bank Report. Washington, DC.

Xu, J., M. Zhu, and S. Zhao. 2020. Leader-follower optimized approach for carbon-economy equilibrium in the municipal solid waste (MSW) incineration industry. Environmental Science and Pollution Research. https://doi.org/10.1007/s11356-020-09076-6 (Accessed on July 3, 2020).

Xu, M., and D. Stanway. 2019. China's ocean waste surges 27% in 2018. Ministry, Environment, Reuters, October 29, 2019. https://www.reuters.com/article/us-china-pollution-oceans/chinas-ocean-waste-surges-27-in-2018-ministry-idUSKBN1X80FL (Accessed on March 10, 2020).

Xu, S., X. Kong, J. Liu, et al. 2016. Effects of high-pressure extruding pretreatment on MSW upgrading and hydrolysis enhancement. *Waste Management* 58:81–89.

Yang, N., H. Zhang, M. Chen, et al. 2012. Greenhouse gas emissions from MSW incineration in China: impacts of waste characteristics and energy recovery. *Waste Management* 32(12):2552–2560.

Zeng, X., H. Duan, F. Wang, et al. 2017. Examining environmental management of e-waste: China's experience and lessons. *Renewable & Sustainable Energy Reviews* 72:1076–1082.

Zhang, D.Q., S.K. Tan, and R.M. Gersberg. 2010. Municipal solid waste management in China: status, problems and challenges. *Journal of Environmental Management* 91:1623–1633.

Zhao, W., E.V.D. Voet, Y. Zhang, et al. 2009. Life cycle assessment of municipal solid waste management with regard to greenhouse gas emissions: case study of Tianjin, China. *Science of the Total Environment* 407(5):1517–1526.

Zheng, L., J. Chen, M. Zhao, et al. 2020. What could China give to and take from other countries in terms of the development of the biogas industry? *Sustainability* 12:1490.

Zhu, L., Z. Wei, T. Yang, et al. 2020a. Core microorganisms promote the transformation of DOM fractions with different molecular weights to improve the stability during composting. *Bioresource Technology* 299:122575.

Zhu, Q., G. Li, Z. Jiang, et al. 2020b. Investigating the variation of dissolved organic matters and the evolution of autotrophic microbial community in composting with organic and inorganic carbon sources. *Bioresource Technology* 304:123013.

Zuo, M. 2020. Coronavirus leaves China with mountains of medical waste. *South China Morning Post*, March 12, 2020. https://www.scmp.com/news/china/society/article/3074722/coronavirus-leaves-china-mountains-medical-waste (Accessed on April 2, 2020).

Part II

Trends and Strategies with Case Studies

7 Waste Management in Russia
Technological Advances

D. Pavlova and Y. Milshina

CONTENTS

7.1 INTRODUCTION

Nowadays in Russia there are about 15,000 legal, authorized waste disposal facilities (landfills), which together occupy an area of about 4 million hectares (Government of the Russian Federation, 2018). This territory is annually expanding by 300,000–400,000 hectares due to the increasing amount of solid municipal waste (MSW). From 1998 to 2013 this indicator increased from 30 million tons per year to 60 million tons per year. In 2013 the amount of MSW generated annually by the average Russian exceeded 400 kg, while in 2000 it was only 220 kg (Ministry of Natural Resources and Ecology of the Russian Federation, 2013). According to the Federal Service for Supervision of Natural Resources (Rosprirodnadzor), the active use of packaging materials is one of the main reasons for the constantly increasing amount of MSW and expanding landfill areas (Council, 2019).

In this regard, the Presidential Council for Civic Society and Human Rights under the President of the Russian Federation on the development of civil society and human rights (2019) emphasizes that the prevention of the constantly increasing volume of MSW in Russia is one of the key national priorities according to the hierarchy implemented in 2015 (Federal Law of 06.24.1998 N 89-FZ). This hierarchy of waste management in Russia corresponds to the basic principles integrated into the waste management system of the European Union, described in the Directive 2008/98/EC, and includes the following sequence of actions:

1. Maximum use of raw materials and prevention of waste
2. Reduction of the volume of waste generated and its hazard class

3. Treatment, recycling and disposal of waste

Experts emphasize that the maximum attention should be paid to the first stage of this hierarchy. While the key indicators aimed at the increase of total capacities for MSW treatment and recovery are already integrated into the Russian National Project 'Ecology', the majority of the regional governments still include such indicators as the rise of the volume of the MSW in the strategic plans. They do not take special measures to prevent it or to stimulate the manufacturers to maximize the use of resources.

Regarding the last stage of this hierarchy, it should be mentioned that only 7–8% of MSW is recycled now in Russia, while more than 90% of that waste is still landfilled (Council, 2019). The Ministry of Natural Resources and Ecology of the Russian Federation explains this problem in the same way as Rosprirodnadzor: The key reason for landfilling the MSW is the rapid growth of consumption and, consequently, the volume of packaging materials.

However, the rapid growth of volume of packaging materials is not the only problem of the Russian WMS: It is also characterized by the extremely low share of waste being collected for further recycling. For example, the share of paper waste being collected for recycling was only 30% in 2018, while the same share of polyethylene waste – 20%, polypropylene waste – 17%, waste from polyvinyl chloride – 12%, waste from polystyrene – 12%, waste from thermoplastics – 12%, glassware waste – 7.5% (Government of the Russian Federation, 2018). This means that at least 9 million tons of wastepaper, 2 million tons of plastic and 0.5 million tons of glass are sent to landfills every year. In this way, Minpromtorg highlights that a considerable proportion of the above-mentioned components of MSW, which are used by consumers for a short period of time as packages, fall out of the economic cycle and damage the environment.

Thus, the uncontrolled growth in the use of disposable consumer goods, especially packaging materials, as well as their low recoverability from the general volume of MSW in Russia, causes cluttering and plastic pollution of territories (Council, 2019). Apart from having a negative impact on the environment and human health, it also damages such economic activities as tourism, fishing and shipping. The disposable plastic products are the main source of pollution of water and coastal areas, accounting for up to 50% of marine litter.

Another important problem of the Russian WMS is the constantly increasing amount of food waste. It accounts for 24–35% of the total volume of MSW (Government of the Russian Federation, 2018). Moscow alone sends annually 2 million tons of food waste to landfills, which is almost a quarter of the total volume of MSW generated within the city. According to the Association of Retail Companies, up to 700,000 tons of food products do not find consumers in supermarkets annually, which leads to the formation of at least 5% of all food waste in Russia (TASS, 2019). At the same time, the majority of the expired food products are actually destroyed at MSW landfills, which leads to the formation of landfill gas, which has a negative impact on the environment and public health. The problem is compounded by the fact that the present tax regulation system in Russia makes it cheaper for the companies to landfill the food waste than to transfer it, for example, to charitable organizations: The amount of tax deductions that an organization must make under the Tax Code of the Russian Federation as part of the transfer of goods to charity can reach 40%, while the cost of landfilling one ton of garbage in Moscow is only 600–800 rubles.

For these reasons, the social tension continues to grow in Russia. A significant increase in complaints and social protests, caused by the violations of citizens' rights to protect their health, sanitary and epidemiological welfare, was recorded. At the beginning of 2019, large-scale 'anti-garbage' protests were held in 30 regions of Russia.

As the 'garbage' crisis escalates, the involvement of civil society in the prevention of waste generation also rises. Russians have a high level of concern about the current situation in the field of waste management. Moreover, they consider landfills (16%) as one of the most acute environmental problems after air pollution (22%) (VTsIOM, 2019). In addition, up to a half of the Russians believe that the problem of landfills can be solved by reducing the amount of disposable packaging in supermarkets (Kommersant, 2018). Furthermore, up to a third of Russians are ready to stop using disposable containers and packaging if stores and catering outlets provide convenient reusable alternatives (Kommersant, 2018).

Hence, a special role for the retail industry in preventing waste generation and maximizing the use of raw materials should be noted. As they are a link between producers and consumers, retailers are able to form a sustainable model of non-waste sales. Worldwide retail experience confirms the effectiveness of waste prevention through the reduction of disposable packaging and the sale of expired food products. The international practices of food retailers to reduce turnover and reject disposable plastic goods, containers and packaging in favour of reusable alternatives have already been investigated by Russian scholars (NRU HSE, 2018).

Still, the innovative technologies that may be used by the consumers, retailers, producers and public utilities providers in order to prevent waste generation, particularly the increasing volume of wasted packages and food, are ignored by scholars. For this reason, this chapter aims to bring light to the opportunities for Russian WMS improvement by means of technological innovations. This chapter focuses mostly on the MSW generated by consumers, but not the hazardous waste generated by industry, which is discussed elsewhere in the book.

The chapter is organized in the following way: The first section contains an overview of the studies that were previously conducted in this area, mostly by the Russian scholars who investigated innovative waste management technologies and practices as well as legislative tools at the international and national levels. The second section is dedicated to the description of the methodology used in this study, particularly the semantic analysis of research papers collected in worldwide citation databases (Web of Science or WoS, Scopus, etc.). The third section represents the results of the sematic analysis – a semantic map that shows the technologies connected with the global waste management system. This section also demonstrates a number of Russian cases that illustrate the development and implementation of the innovative waste management technologies. In the final section the opportunities for the introduction and wide adoption of these technologies are discussed in the context of the prevailing management concepts and recent legislative initiatives. After that, some concluding remarks are drawn.

7.2 LITERATURE REVIEW

This literature review contains 22 studies dedicated to the investigation of the Russian WMS and conducted mostly by Russian scholars. These studies are differentiated by the geographical scope (foreign countries/Russia/Russian regions) and subject focus (legislative framework/management concepts, practices and tools/innovative technologies) of the research. First, the studies that explored and compared the foreign and Russian WMSs are discussed. Next, the studies that analysed the Russian WMS only are provided. Finally, the studies that investigated the WMSs in the particular regions of Russia are mentioned.

7.2.1 Comparison of Foreign and Russian WMSs

7.2.1.1 Legislative Framework

Starting with the studies which compared the foreign and Russian WMSs with a focus on legal issues, the research conducted by Smetanina et al. (2014) should be mentioned. They explored the environmental management tools, including the waste management ones, existing in the developed countries and suitable for Russia. In particular, the scholars highlighted the green taxes programmes in the USA and Europe, especially waste and environment levies for fertilizers and pesticides in the agricultural field, 'command and control tools' (courts, police, fines) for waste management in the European Union, environmental permits for industrial installations implemented by the IPPC Directive (Directive 2008/1/EC), the European Pollutant Emission Register, grant programmes for the development of low-carbon energy and energy efficiency technologies (e.g. Farm Waste Treatment Facilities in the UK) and the ISO 14001, which addresses cost reduction of waste management.

Moreover, Ghosh et al. (2016) examined a certain type of waste management in BRICS countries – the management of electrical and electronic equipment waste. The study is based on extensive literature review and interviews with stakeholders. The authors provided brief statistical information on the volume of such waste being generated in Russia as well as the legislation framework that regulates its management. The scholars highlighted that there is no enacted legislation for the recycling and management of electronic waste. Also, they uncovered the fact that despite the fact that most of such waste goes to landfills, some part of it undergoes dismantling and metal recovery. Still, detailed information on this issue is unavailable.

In addition, Plyaskina and Kharitonova (2016) conducted a comparative analysis between the Russian MSW management system and the same systems of the economically developed countries. By means of statistical analysis, the researchers assessed the influence of these systems on the effectiveness of MSW utilization. Moreover, the researchers analysed the formation of the circular economy in Russia and identified the solutions for waste sorting problems: Selective component collection of waste by the population followed by automated sorting, selective fraction collection of waste by the population followed by manual or automated sorting, factory automated sorting. In the final part of the study, the scholars identified the administrative barriers (frequent reorganizations, high risks of non-compliance and investment risks, low quality of housing and communal services) in the Russian MSW management system as well as economic mechanisms (creation of regional waste operators, introduction of 'garbage tax' differentiated for sorted and non-sorted waste, licensing system for waste transporting companies) that stimulate its development on the basis of the document analysis of the official legislation.

7.2.1.2 Management Concepts, Practices and Tools

Continuing on with the studies, which compared the foreign and Russian WMSs with a focus on management concepts, practices and tools, the research of Pakhomova et al. (2016) should be noticed. They conducted a comparative analysis of the WMSs existing in Russia and the EU. In particular, the scholars compared the liability standards (legislation, disposal rates for individual product groups, environmental fees, fines) for Russian and European producers regarding the waste management schemes. In this way, the authors emphasized an important role for remanufacturing and the concept of extended producer responsibility, which are widely spread in Europe. Moreover, the researchers conducted a case study of the automobile industry, especially the contribution of automobile remanufacturing and utilization (done by such companies as Nissan, Mercedes-Benz, Mitsubishi, Volvo, Renault, Rolls-Royce) to the reduction of the amount of resources used, energy consumed and pollution gases emitted. Apart from the ecological effects, the scholars mentioned such economic benefits of automobile remanufacturing and utilization as the decrease of cost of the final product and increase of employment. In the end, the researchers strongly suggest implementing similar practices into the Russian WMS and provide some practical recommendations for it.

Furthermore, Verenikina and Gorokhova (2016) explored the opportunities for the implementation of resource management practices into the Russian WMS on the basis of foreign experience (USA, EU). In this way, the researchers showed the differentiation of traditional and resource management contracts by their payment schemes, system of motivation, type of interaction between the client and contractor and range of services available. In the end, the scholars uncovered the advantages (services differentiation, potential increase of the market share, risks diversification, rise of customers loyalty, cost minimization) and disadvantages (low awareness of business and population, risks in compensation models, problem of diminishing returns for service providers) of the resource management approach.

Finally, Kolotyrin (2016), while analysing the foreign practices of ecological risk reduction in the waste management field, highlighted an important role of risk diversification and insurance. According to the researcher, the use of environmental insurance can prevent adverse environmental effects and violations. Moreover, the scholar argued that the diversification of risks within public-private partnerships can increase the efficiency of the WMS. In this way, Kolotyrin (2016) suggested to implement these instruments into the Russian WMS in order to increase its attractiveness for investors and reduce the ecological risks.

In addition, Likhacheva and Sovetov (2017) examined the governance problems of the modern Russian WMS. On the basis of the statistical data analysis, document analysis of the Russian waste management legislation and literature review of foreign official reports related to this issue the researchers analysed the experience of MSW management in Russia and abroad. As a result, the scholars highlighted the problem of the blurred responsibilities of the national governmental authorities for managing the MSW and suggested the use of target-oriented tools in order to increase the efficiency of the Russian WMS. Moreover, the researchers emphasized the necessity to establish an integrated system of MSW management at the regional and municipal levels.

7.2.1.3 Innovative Technologies

Concluding this part of comparison of the foreign and Russian WMSs, the only study which focuses on innovative technologies, particularly waste recycling, can be provided. Alekseyev (2014) investigated the recycling sector in Finland as an innovative industry segment that could be also developed in Russia (according to the study, in Finland 45% of waste is recycled, while in Russia – only 5%). For this purpose, the scholar, first, constructed a typical logistics scheme of waste recycling. After that, Alekseyev (2014) analysed the structure of allocated and processed waste fractions in Finland. Next, the researcher conducted a case study of a Finnish waste recycling company 'Kuusakoski Oy' based on its economic indicators. As a result, Alekseyev (2014) emphasized several multiplier effects created by the recycling industry in Finland: Reduction of the amount of non-renewable natural resources used, technological development, employment increase, environment protection, growth of small and medium enterprises. In this way, the scholar demonstrated the investment attractiveness of the recycling industry.

7.2.2 Analysis of Russian WMS

7.2.2.1 Legislative Framework

Regarding the studies that investigate the Russian WMS in general with a focus on legislative regulations, the following papers should be considered. For example, Panina (2015) examined the legal regulation of waste management in Russia. In particular, the researcher studied the problems of the implementation of control and supervisory activities. In this way, the author highlighted the inconsistency of powers distributed among the federal, regional and local government authorities related to the Russian WMS. In the end, the scholar emphasized a strong need to give more power to local government bodies, which should be responsible for ecological control at the municipal level so as to create a favourable environment for the local population.

Moreover, Prisyach and Shvetsova (2018) explored the elements of innovative scenarios of WMS in Russia. In particular, the researchers conducted a document analysis of legislative regulations related to the practice of waste management in Russia and statistical analysis of the development of environmental management strategy within the country. Next, the scholars identified eco-technical parks as a key innovative component of the development of Russian technologies for waste treatment. Finally, the authors suggested a formula for the calculation of the amount of equipment for waste processing, utilization and disposal required by these eco-technical parks.

In addition, Putintseva and Chekalin (2018) examined the Russian WMS as a factor for improving national energy efficiency. On the basis of the analysis of the regulatory documents, the scholars investigated the stages of the reformation of the WMS in Russia as well as formation of the circular economy. Moreover, after analysing the statistical data, the scholars drew a conclusion about the inefficiency of waste incineration in Russia and identified the problems of separate waste collection and sorting. In this regard, Putintseva and Chekalin (2018) as well as Prisyach and Shvetsova (2018) highlighted an important role for eco-technical parks, which should not only process the waste, but produce the secondary raw material as well in order to increase the rate of return on investment in waste.

7.2.2.2 Management Concepts, Practices and Tools

The following studies examined the general Russian WMS in terms of the emerging management concepts, practices and tools. For instance, Kirillova and Musinova (2018) noticed a need for the development of a single economic-organizational mechanism of MSW management in Russia. In this regard, the scholars identified the components that might be integrated in this mechanism. The organizational mechanisms include separate collection and sorting of MSW, construction of recycling plants, modern landfills, etc. The economic mechanisms, in turn, contain soft loans and credits for the domestic producers of the equipment for MSW processing, preferential taxation for the Russian companies that implement the principles of circular economy, state subsidies for the realization of projects in the field of waste disposal and utilization and public-private partnerships in the waste management field. Moreover, the researchers demonstrated the interconnection of these mechanisms with the legal (standards, licenses, fines, etc.) and informational (education, promotion, public control, etc.) ones. As a result, the researchers concluded that such a mechanism should increase the efficiency of the Russian MSW management system.

Furthermore, Fedotkina et al. (2019) studied drivers and barriers for the development of an efficient national WMS in the context of the formation of the circular economy in Russia. For this purpose, the authors conducted a literature review of the general circular economy concept as well as particular successful factors and barriers connected with its implementation in the USSR and Russia at the first stage. At the next stage, the researchers made a review of the regulatory documents (state laws, development programmes and initiatives) which focus on the waste processing issues in Russia, especially the distribution of roles among the stakeholders of the Russian WMS. At the final stage, the scholars conducted a case study of best available techniques in the leading Russian companies, which actively implement the principles of circular economy and waste management within the country. As a result, this study provides a theoretical framework for the global movement toward the circular economy as well as the current waste management situation, circular economy and waste management reforms in Russia. What is more important, this study represents a number of best available techniques in the field of circular economy used by the leading Russian companies from different sectors of the economy: PJSC SIBUR – petrochemical industry, PJSC Gazprom – oil and gas industry, Russian Railways (RZD Holding) – transport services, Evraz – mining industry, OPTICOM Ltd. – consumables and household goods for business. Among these techniques are waste reduction and maximum use of raw resources at the first stages of production, waste disposal and recycling at the final stages of production and separate solid waste collection at the companies' office buildings, administrative and domestic premises. To achieve higher efficiency in these techniques, according to the authors of the study, the companies plan full re-equipment and modernization of the technological processes, launch of promotional and educational initiatives and exchange of best practices with the other companies by making mutually beneficial partnerships with them. In the end, the researchers demonstrate a full description of the roles and responsibilities of WMS stakeholders in Russia (generator, collector, disposer, processor, educator, promoter etc.).

7.2.3 Analysis of Regional WMSs in Russia

7.2.3.1 Legislative Framework

Regarding the studies that analysed the regional WMSs in Russia with a focus on legal issues, the following studies should be mentioned. First, Andreassen (2016) explored waste management in Russia as a part of research dedicated to the investigation of sustainable energy development in the Arctic region. In this way, the scholar examined the Russian Arctic energy industry and its politico-economic context as well as sustainability priorities, which include not only the prevention of pollution and waste, but transport modernization, environmental security, emergency and coordination systems preparedness, governance and international partnerships. For this purpose, the author used a documentary search for the relevant policy documents (presidential decrees, governmental orders, etc.) and news releases freely available online on the websites of media agencies focused on the Arctic. After that, the content of these documents was analysed manually by the scholar. As a result, the author represented a framework for the sustainability concept in Russia, identified the sustainability-related priorities in the Arctic energy industry and discovered the media position.

Moreover, Prianikova et al. (2018) investigated the sustainable development of the Arctic zone of Russia with a special focus on the waste processing system. In order to establish a system of steady management in the field of ecology in the Arctic zone, the researchers developed a multilevel mathematical model for the guaranteed achievement of the management goal, which includes several indicators such as demographic activity, power security, interest of people in development of the economy, development of the real sector of the economy, security of jobs, compliance of the population of security and compliance of development of the economy of security. On this basis, they suggested the whole concept of public administration of Arctic zone.

7.2.3.2 Management Concepts, Practices and Tools

The management concepts, practices and tools were much less examined by the Russian scholars at the regional level compared to the international and national ones. The only study which can be provided in this sub-section is the research of Selezneva (2016), who analysed the current environment for the development of source-separated WMS in Saint Petersburg and estimated the possibility of its implementation. By means of conducting a literature review of the existing legislative framework related to the regional WMS and official documents and research reports of public organizations operating in the field of source-separated waste management promotion the author identified the key challenges to the establishment of such a system in Saint Petersburg. Moreover, Selezneva (2016) interviewed a number of main stakeholders involved in the regional WMS in order to suggest the most suitable way for the development of a source-separated waste management system in the city and launched a public opinion poll so as to assess the social environmental consciousness of citizens. As a result, the researcher uncovered a number of main driving forces for the transition to a separate waste collection system in Saint Petersburg: Economic forces (new taxation system, extended producer responsibility, ecological dues), management ones (infrastructure development, law development and enforcement, data availability on waste treatment) and social ones ('pay-as-you-go' throw principle for households, environmental education, local community involvement). Moreover, the scholar developed four scenarios for separate waste collection differentiated by number of containers (six/three/two/one), separated fractions accordingly (biowaste, paper, metals, glass, plastics, mixed waste/biowaste, recyclables, mixed waste/recyclables, mixed waste/mixed waste) and treatment facilities required (mechanical biological treatment/material recovery facilities/ landfill).

7.2.3.3 Innovative Technologies

However, the innovative technologies were much more investigated by the Russian scholars at the regional level rather than international or national ones. For example, Rodionov and Nakata (2011) developed a design of an optimal waste utilization system and demonstrated its applicability to a WMS in Saint Petersburg. For this purpose, the scholars used a linear programming model that evaluates the proposed WMS with respect to its energy, economic and environmental impacts, particularly energy generated from waste, system net cost and the carbon dioxide emissions. As a result, the authors of the study argued that 'the proposed MSW system with energy recovery from waste along with a high level of material recovery has energy, environmental and economic benefits compared to the conventional treatment system' (Rodionov and Nakata, 2011). This system is based on such conversion technologies as material recovery, compositing, anaerobic digestion system, incineration, RDF production and landfilling with LFG recovery. Thus, it covers a wide range of types of waste generated (plastic, glass, textile, organic, etc.) as well as a wide range of stakeholders who can benefit from this conversion (industrial sector, agriculture sector, electric network, district heating system).

Moreover, Kaazke et al. (2013) conducted an environmental evaluation of waste treatment scenarios for the towns Khanty-Mansiysk and Surgut, where the current methods of waste disposal have recently reached their limits. The scholars aimed their study at the identification of the influence of waste composition structure and logistics routes on the life cycle assessment results, assessment of the current option for the treatment of MSW and its comparison with the proposed scenarios. The waste management scenarios included such options as landfilling, aerobic mechanical biological treatment, anaerobic mechanical biological treatment, incineration, landfilling and recycling, aerobic mechanical biological treatment and recycling, anaerobic mechanical biological treatment and recycling and incineration and recycling. They were evaluated by means of the life cycle assessment model. The results of the study demonstrated that the options with the lowest environmental impact were 'incineration with recycling' and 'anaerobic mechanical biological treatment with recycling' (Kaazke et al., 2013). Still, as the researchers noticed, there are great logistical challenges in transferring the recyclables.

Furthermore, Tulokhonova and Ulanova (2013) studied the WMS in Irkutsk and assessed four MSW management scenarios by means of the same life cycle assessment model. These scenarios included such options for waste processing and disposal as landfilling, recycling, compositing and aerobic mechanical biological treatment. They were compared with respect to their ecological, economic and social impacts. The results of the analysis showed that the development of environmentally sustainable and socially friendly WMS increases the cost of handling waste. The best solution for the Irkutsk WMS that was chosen by the authors of the study is based on 'the separate collection and reprocessing of recyclables in combination with an aerobic mechanical biological pre-treatment of the residual waste before landfilling' (Tulokhonova and Ulanova, 2013). The cost of this solution is 3.6 higher than the existing system, as was noticed by the researchers.

Starostina et al. (2014) also investigated waste management in the Irkutsk Region (Siberia) by means of environmental assessment of current practice focusing on landfilling. For this purpose, the researchers used various official reports from the regional authorities as well as studies of the State Technical University in Irkutsk and supplementary literature to collect the data about WMS in that region. After that, they employed the mass and substance flow model STAN to assess the composition of the regional MSW. Next, the scholars used the EASETECH model to conduct a life cycle assessment of the Irkutsk WMS. The environmental impact assessment was made by the researchers in accordance with the EU-JRC's recommendations. As a result, the authors represented the composition of the structure of the MSW in Irkutsk and a model of a new and modern landfill which could replace the old one since it is going to run out of capacity in a few years, with a higher performance due to collection of landfill gas and its utilization for electricity production.

Later Starostina et al. (2018) examined the same problem of landfilling in Irkutsk and on the basis of the previous research on this topic and built seven scenarios for the development of WMS in this region. These scenarios are based on the regional waste composition structure (generated by both households and industry) and consider different options such as material recycling, collection and transportation of waste and recyclables, incineration of residual waste, mechanical biological pre-treatment, bio-drying mechanical biological treatment, landfilling and energy utilization. After conducting the life cycle assessment modelling as well as sensitivity analysis of each scenario, the authors concluded that incineration with efficient energy recovery, electricity recovery for the grid and steam or heat recovery should be the key element of the environmentally sound WMS in Irkutsk. At the same time, the commercial and institutional waste has great potential for recycling, according to the authors, since it contains larger quantities of recyclables (compared to household waste) and its collection is easier to administer. As the researchers notice, despite the fact that transport distances for recyclables affect the overall environmental savings, they still do not make the recycling questionable.

In addition, Zhuchkov and Tupikova (2017) studied the problems of environmental pollution in Russia in the context of the constantly increasing volumes of solid municipal and industrial waste. The researchers explored the methods of reformation of the Russian WMS as well as the regional WMS in the Primorsky Krai and city of Vladivostok. In particular, the scholars analysed the effectiveness of the existing legislative framework and experience of introduction of modern eco-friendly waste recycling technologies. Regarding these technologies, Zhuchkov and Tupikova (2017) highlighted the great potential for mechanical waste processing and production of useful secondary resources from ash waste in Russia. Finally, they suggested coordination measures for the creation of a sustainable environment in the country.

The key points of the studies described above are summarized in Table 7.1.

According to Table 7.1, the following conclusions can be drawn.

- First, while comparing the foreign and Russian WMSs, the domestic scholars paid more attention to the legal issues and management concepts, practices and tools rather than innovative technologies

TABLE 7.1
Summary of the Key Points of the Literature Review

Scope \ Focus	Legislative Framework	Management Concepts, Practices and Tools	Innovative Technologies
Russia and foreign countries	Taxes and grant programmes, command and control tools, administrative barriers	Concept of extended producer responsibility, resource management practices, ecological risks diversification and insurance tools, target-oriented tools	Waste recycling, remanufacturing
Russia	Control and supervisory activities, eco-technical parks	Organizational and economic mechanisms, corporate practices of waste reduction, disposal and recycling, promotional and educational activities, exchange of best practices	-
Regions of Russia	Sustainability-related priorities and public administration in the Arctic zone	System of separate waste collection in Saint Petersburg, including 'pay-as-you-go' throw principle for households	Compositing, material recovery, mechanical biological pre-treatment, bio-drying mechanical biological treatment, aerobic mechanical biological treatment, anaerobic mechanical biological treatment, incineration with heat and energy recovery, recycling and production of useful secondary resources from the ash waste, landfilling with LFG recovery, collection of landfill gas and its utilization for electricity production, RDF production

Source: Compiled by the authors.

- Second, while analysing the Russian WMS in general, the domestic scholars examined only the legislative regulations and managerial aspects, but not the innovative technologies
- Third, while investigating the regional WMSs in Russia, the domestic scholars paid little attention to the legal and managerial issues, but widely discussed the opportunities for the implementation of the innovative technologies for waste recycling and disposal in different regions of Russia (Irkutsk, Surgut etc.)
- Finally, the majority of the studies mentioned above employ traditional research methods such as statistical and document analysis, literature review and modelling

For these reasons, this study is relatively novel for this field since it is aimed to uncover the innovative technologies, or technological advances, in the global WMS (which could be further integrated into the Russian one) by means of semantic analysis, which has never previously been used by Russian scholars in this field. Moreover, this study provides a number of cases of the development of innovative technologies of MSW management technologies in Russia that correspond to the global agenda in this field. The methodology of this study is described in detail in the following section.

7.3 METHODOLOGY

The methodology of this study is based on semantic analysis conducted by means of the intellectual system of big data analysis Intelligent Foresight Analytics (iFORA) developed by the researchers of the Institute for Statistical Studies and Economics of Knowledge (ISSEK) of the National Research University Higher School of Economics (HSE University).

The system of intellectual big data analysis iFORA processes large volumes of unstructured text data using machine learning and computer linguistics algorithms (Kuzminov et al., 2018). Thus, the system allows the detection of complex semantic concepts (trends, technologies, markets, products, relevant topics of scientific research), assessment of the dynamics of their development and identification of their relationships.

At the moment the iFORA database contains about 350 million documents, including scientific articles and patents, market analytics and professional media, legislation documents and consultant reports, which ensure comprehensive and relevant analytical results. For the purposes of this study, the texts of the scientific articles published over the past ten years (2009–2019) were used. These texts were obtained from the worldwide leading scientific citation databases and journals (CROSREF, Microsoft Academic Graph (MAG), Web of Science (WoS), Semantic Scholar, Springer Nature, etc.).

The processing of large volumes of text data by means of iFORA was carried out in five stages: Primary processing of natural language, syntactic-semantic analysis of texts, thematic modelling, clustering (classification) and identification of semantic patterns.

The natural language processing with the use of iFORA included the following operations:

- Segmentation, which means separation of the texts into sentences
- Tokenization, that is, breaking down the sentences into individual words
- Lemmatization of words, which means their reduction to normal form
- Stemming, or identification of the stem of the word
- Parsing of the text, including the allocation of syntactic links in the sentence, the definition of the properties and functions of words, as well as the analysis of triplets of relationships 'subject-action-object'
- Morphological analysis of words (identification of part of speech, case, declension, etc.)
- Construction of vector representations of words and documents in order to assess their semantic proximity
- Statistical analysis, graph construction, clusterization

The iFORA methodology combines two main algorithms: Identification of compatibility – the joint use of terms (topics) in sentences, as well as a vector representation of the identified terms by assigning to each term a unique numerical vector obtained by analysing the context of the use of the term in texts. Terms used in a similar context have the same numerical vectors and are closely located in the vector space.

The first algorithm clusters the identified terms or gathers them into similar thematic categories. The result of this analysis is visualized as a semantic map (Figure 7.1). Nodes (dots) on semantic maps indicate individual topics and their names, while the colours indicate clusters of topics. The sizes of the dots and their signatures on the semantic map demonstrate the significance of the identified topics: The larger the point and signature – the more significant and popular the corresponding topic in the analysed texts. The proximity of the dots of different colours in space indicates:

- Convergence of directions (topics are started to be used in the same context, which means that different directions are beginning to merge into one) or
- Divergence of directions (topics which traditionally were used in the same context recently began to differ)

The second algorithm calculates the frequency of occurrence of terms in the analysed texts in the selected period (in this study – in 2009–2019). The result of this analysis is visualized as a trend map (Figure 7.2). Its main purpose is to identify the trends and classify them by significance (frequency of occurrence of terms) and dynamics (growth rate of frequency of occurrence of terms). The trend map is divided into four quadrants. The topics in the upper right quadrant of the trend map are the leading trends: They have high indicators of significance and dynamics, which means that they set the current agenda for scientific and technological development. These trends are steadily growing in popularity. The topics in the upper left quadrant are the mature trends: They are widespread, but their popularity is stable and has almost stopped growing in recent years. The topics in the lower right quadrant are the emerging trends: Their significance is relatively low but has been growing quite rapidly in recent years. This means that these trends with a high probability will determine the scientific and technological development in the long run. The topics in the lower left quadrant are the niche trends: They are not popular and dynamic at a global level. Still, they should not be underestimated since they might be potentially important in the long run ('weak signals'). The trend map, thus, classifies the identified topics into the leading, mature, emerging and niche trends by their significance and dynamics of development.

The reliability of the iFORA methodology as well as robustness of its analytical results was tested and approved at the international level, particularly during the scientific and practical workshop 'Semantic Analysis for the Purpose of Innovation Policy', organized by the Organization for Economic Cooperation and Development (OECD, 2018).

The results of iFORA analysis are accompanied by a number of case studies of the development of innovative technologies of MSW management in Russia. These case studies were selected on the basis of the hottest research areas discovered by means of the semantic analysis. In this way, they demonstrate the development of leading and mature, emerging and niche trends in the Russian MSW management field. Each case is described in terms of the technology developed or applied, its advancement compared to the other technologies existing, effects on the economy, society and environment, stages of development, including the intellectual property rights obtained, and potential areas of application, especially prospects for commercialization.

7.4 RESULTS

The general results of the study are represented by a semantic map (Figure 7.1) and a trend map (Figure 7.2), which reflect the identified clusters of topics, their significance and dynamics,

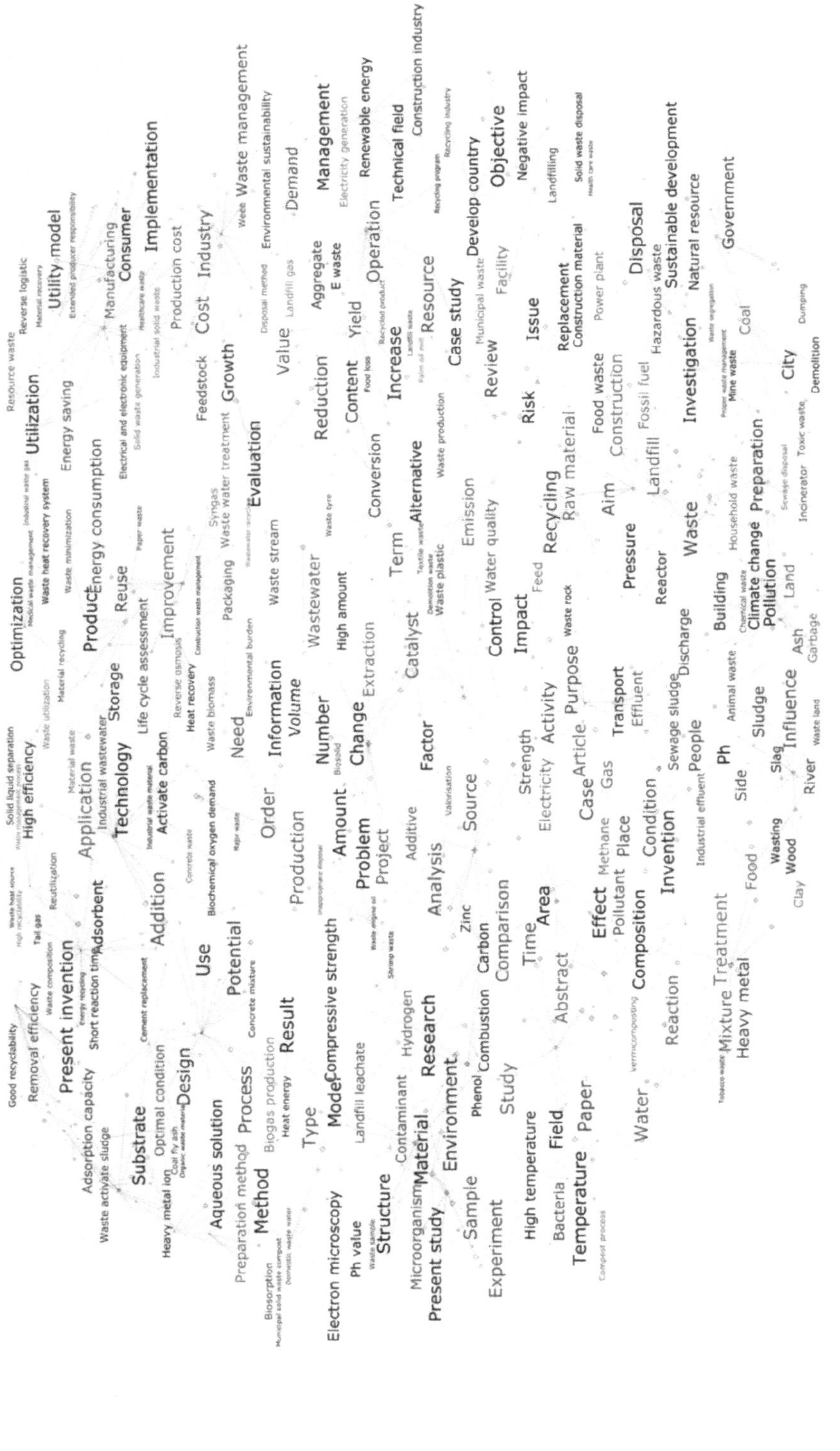

FIGURE 7.1 Semantic map on waste recycling, disposal and landfilling. (Source: Made by the authors on the basis of the openly available published materials with the use of the developed text-mining instruments.)

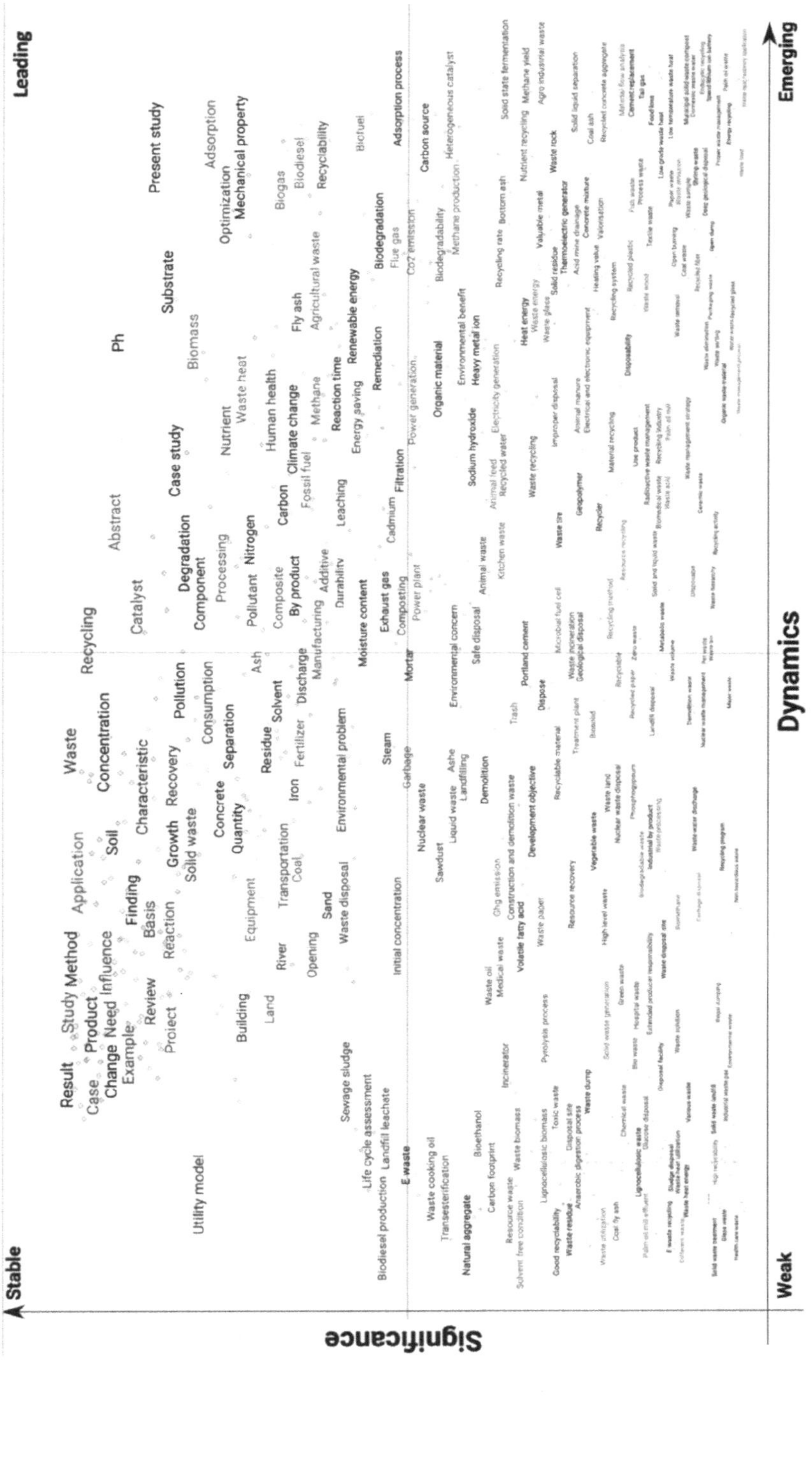

FIGURE 7.2 Trend map on waste recycling, disposal, landfilling. (Source: Made by the authors on the basis of the openly available published materials with the use of the developed text-mining instruments.)

constructed on the basis of the research papers published in 2009–2019 by means of iFORA according to the methodology described in the previous section.

These results were obtained after making a relevant search query, which was 'waste recycling, disposal, landfilling'. These stages of waste management were selected with respect to the key priorities of the Russian waste management policy described in the National Project 'Ecology' (Roscongress, 2019). For this search query we identified the most relevant keywords listed in Table 7.2. The relevance indicator was calculated by counting the query normalization and coordination factors as well as the sum of the weights for each term in the query for document (Compose, 2016). So, it shows the extent to which the word found in the document matches the search query.

Based on the list of the most relevant keywords mentioned above (Table 7.2) regarding waste recycling, disposal and landfilling, the semantic map (Figure 7.1) and trend map (Figure 7.2) were constructed.

The semantic map (Figure 7.1) demonstrates a number of clusters of topics related to 'waste recycling, disposal and landfilling' and mentioned in the research papers published worldwide in 2009–2019. The largest clusters of topics are listed in Table 7.3.

According to Table 7.3, the largest cluster of topics (Cluster 1) represents the general terms used in research (e.g. aim, work, analysis, etc.). Cluster 2, Cluster 3 and Cluster 4, in turn, include a number of very specific terms related to the technologies of waste processing and utilization (e.g. anaerobic digestion, sewage disposal, waste heat recovery), which are a focus of this study. At the same time, Cluster 5, Cluster 6 and Cluster 7 involve the terms that characterize the background of the study (e.g. environmental sustainability, huge waste, greenhouse gas). The dynamics and significance of these clusters and terms are illustrated in Figure 7.2.

The trend map (Figure 7.2) demonstrates the same clusters of topics according to their significance (frequency of occurrence of terms) and dynamics (growth rate of frequency of occurrence of terms). These indicators of the clusters mentioned above are represented in Table 7.4.

According to Table 7.4, the most significant clusters of topics, apart from Cluster 1, which represents the general research terms, are Cluster 3 and Cluster 6. Cluster 3 includes a number of terms related to wastewater recycling and treatment, while Cluster 6 involves the terms that describe the global problem of the increasing amounts of waste. Less significant clusters are Cluster 4, which illustrates a number of issues related to the generation of energy from waste, and Cluster 7, which covers the terms that describe the problem of global warming, climate change and greenhouse, especially carbon dioxide, emissions. The least significant clusters of topics, according to Table 7.4, are Cluster 2, which represents the terms related to the production of renewable energy from waste (biomass, biogas, biofuel, etc.), and Cluster 5, which includes the topics of environmental protection and sustainability.

Regarding the dynamics of these clusters, it should be noticed that Cluster 2, while being less significant than the others, is the most dynamic one. This means that the production of biomass, biofuel and biogas from waste is one of the most actively investigated topics by scholars nowadays. Figure 7.2 approves this assumption, since these terms are allocated at the upper right quadrant (leading trends). Other actively investigated topics represented on the trend map are the problems of global warming and climate change (Cluster 7) and issues of environmental protection and sustainability (Cluster 5). However, the terms from these clusters are widely spread across all four quadrants of the trend map (Figure 7.2) and, thus, cannot be attributed to any types of trends (leading, emerging, stable, etc.). Less dynamically discussed, but still highly significant topics in the scientific literature related to global waste management, according to the Table 7.4, are the wastewater treatment and recycling issues as well as technologies for power generation from waste. These terms are spread mainly across the upper and bottom right quadrants of the trend map (Figure 7.2), which means that they represent the leading and emerging trends.

To analyse the particular innovative technologies in the waste management field, which are represented on the semantic map and trend map, the intellectual big data analysis conducted by iFORA was complemented by a comparative study of multiple cases conducted by the authors. In this

TABLE 7.2
Keywords Used for the Semantic Analysis Sorted by Relevance

№	Keyword	Relevance
1.	Waste	417.79
2.	Recycling	163.59
3.	Disposal	122.95
4.	Landfill	69.98
5.	Solid waste	55.06
6.	Wastewater	48.88
7.	Waste management	44.94
8.	Waste heat	40.40
9.	Food waste	37.76
10.	Waste material	36.91
11.	Municipal solid waste	34.07
12.	Organic waste	32.25
13.	Recyclability	27.36
14.	Industrial waste	27.33
15.	Agricultural waste	27.17
16.	Waste product	18.53
17.	Wastewater	17.41
18.	E waste	15.95
19.	Solid waste management	15.71
20.	Anaerobic digestion	15.53
21.	Biogas	15.25
22.	Leachate	15.04
23.	Recycle	14.76
24.	Plastic waste	14.23
25.	Environment	14.20
26.	Invention	13.98
27.	Waste disposal	13.88
28.	Water	13.27
29.	Reuse	13.21
30.	Waste generation	13.19
31.	Waste cooking oil	13.03
32.	Recycled aggregate	12.43
33.	Landfill leachate	12.32
34.	Waste heat recovery	12.28
35.	Recycled material	12.17
36.	Raw material	12.11
37.	Present invention	12.02
38.	Study	11.99
39.	Sludge	11.67
40.	Result	11.41
41.	Heavy metal	10.93
42.	Environmental impact	10.80
43.	Environmental pollution	10.65
44.	Construction waste	10.64
45.	Electronic waste	10.61
46.	Waste stream	10.46

(Continued)

TABLE 7.2 (CONTINUED)
Keywords Used for the Semantic Analysis Sorted by Relevance

№	Keyword	Relevance
47.	Pollution	10.18
48.	Production	10.03
49.	Hazardous waste	9.98
50.	Incineration	9.97
51.	Recycling process	9.81
52.	Waste gas	9.65
53.	Disposable income	9.21
54.	Energy waste	9.16
55.	Biogas production	8.95
56.	Waste water treatment	8.73
57.	Radioactive waste	8.47
58.	Compost	8.29
59.	Household waste	8.25
60.	Use	8.12
61.	Landfill site	8.04
62.	Energy	7.90
63.	Comprehensive strength	7.80
64.	Wasting	7.66
65.	Circular economy	7.52
66.	Catalyst	7.39
67.	Process	7.37
68.	Fly ash	7.16
69.	Waste treatment	7.10
70.	Compositing	7.08
71.	Landfilling	6.95
72.	Kitchen waste	6.89
73.	Construction and demolition waste	6.86
74.	Natural aggregate	6.81
75.	Waste plastic	6.77
76.	Plurality	6.69
77.	Waste activate sludge	6.62
78.	Product	6.60
79.	Biodiesel	6.58
80.	Cement	6.57
81.	Soil	6.54
82.	Municipal waste	6.48
83.	Muscle wasting	6.44
84.	Final disposal	6.35
85.	Effluent	6.35
86.	Resource waste	6.25
87.	Recycling rate	6.24
88.	Sewage sludge	6.20
89.	Nuclear waste	6.20
90.	Domestic waste	6.16
91.	Material	6.16
92.	Waste collection	6.14
93.	Liquid waste	6.13

(*Continued*)

TABLE 7.2 (CONTINUED)
Keywords Used for the Semantic Analysis Sorted by Relevance

№	Keyword	Relevance
94.	Ash	6.06
95.	Waste reduction	6.03
96.	Paper	6.01
97.	Waste biomass	5.99
98.	Biomass waste	5.96
99.	Order	5.94
100.	Agro industrial waste	5.88

Source: Calculated by the authors on the basis of the openly available published materials with the use of the developed text-mining instruments.

TABLE 7.3
The Largest Clusters of Topics Represented on the Semantic Map

Colour	Cluster	Number of Terms	Terms
	Cluster 1	49 terms	Order, addition, resource, work, time, advantage, system, source, analysis, need, aim, term, presence, world, purpose, influence, activity, value, end, type, case, way, possibility, form, lot, challenge, mean, condition, generation, combination, example, comparison, strength, place, level, people, same time, top, respect, benefit, future, side, effort, opportunity, practice, basis, ability, opening, other hand
	Cluster 2	36 terms	Municipal solid waste, agricultural waste, anaerobic digestion, biogas, biogas production, waste gas, biodiesel, municipal waste, biomass, fuel, landfill gas, biochar, fossil fuel, biofuel, emission, wood waste, gasification, greenhouse gas emission, flue gas, methane production, biogas plant, anaerobic digester, coal, methane, waste wood, waste vegetable oil, agricultural residue, co2 emission, renewable resource, alternative fuel, syngas, biomethane, co2, Ghg emission, ethanol, cement kiln
	Cluster 3	22 terms	Waste water, wastewater, solid waste management, water, waste water treatment, effluent, recycled water, wastewater treatment, sewage, water recycling, groundwater, water quality, digester, water resource, sewage disposal, surface water, water treatment, wastewater recycling, water, waste, treatment plant, treat waste water, reverse osmosis
	Cluster 4	11 terms	Waste heat, waste heat recovery, energy, waste to energy, energy recovery, electricity, power generation, energy production, energy source, electricity generation, power plant
	Cluster 5	11 terms	Environment, environmental impact, sustainability, environmental protection, environmental problem, natural resource, resource recovery, environmental issue, environmental performance, environmental sustainability, environmental concern
	Cluster 6	6 terms	Amount, large amount, huge waste, quantity, huge amount, high amount
	Cluster 7	6 terms	Carbon dioxide, greenhouse gas, carbon, global warming, climate change, atmosphere

Source: Made by the authors on the basis of the openly available published materials with the use of the developed text-mining instruments.

TABLE 7.4
Dynamics and Significance Indicators of the Clusters of Topics Represented on the Trend Map

Colours	Cluster	Significance	Dynamics, %
	Cluster 1	49 806 845	3.65
	Cluster 2	510 249	7.32
	Cluster 3	797 578	4.13
	Cluster 4	669 663	4.74
	Cluster 5	509 320	6.08
	Cluster 6	709 322	4.66
	Cluster 7	649 013	7.12

Source: Made by the authors on the basis of the openly available published materials with the use of the developed text-mining instruments.

regard, several cases that illustrate the development and possible applications of the innovative waste processing technologies in Russia are demonstrated next. For each type of trend (leading, emerging, stable and weak) a case with a particular technology developed by the Russian scientists and related to the improvement of WMS is described in terms of the technology itself, its advancement compared to the other technologies existing, its current stage of development, effects on the economy, society and environment and potential areas of application, including the prospects for future commercialization.

CASE 1: LEADING TREND 'PRODUCTION OF BIOGAS AND BIOFERTILIZERS BY ORGANIC WASTE PROCESSING'

Technology: In 2017 the scientists of the Federal Research Center 'Fundamentals of Biotechnology' of the Russian Academy of Sciences started applied research on the development of integrated resource-saving biotechnology for recycling the organic components of MSW into biogas and biofertilizers with the use of advanced methods (electrophysical effects, anammox and aeration) (Ministry of Science and Higher Education of the Russian Federation, 2017). The creation of biotechnology is planned for the production of biogas with high content of methane and biofertilizers with increased agronomic value by processing organic waste in the anaerobic bioreactors using electrophysical intensifying microbial methanogenesis, nitrification-anammox processes (to clean the liquid fraction from nitrogen) and composting solid waste under reduced aeration.

Advancement: The technological and economic advantages of the developed anaerobic reactors are represented by the simplicity of the reactor design, increased rate of conversion of organic waste into biogas with high methane content and low power consumption of the electrical stimulating device, as well as a significant reduction in capital and operating costs for the construction and running of the reactor. The usage of the advanced methods (electrophysical effects, anammox and aeration) in this technology corresponds to the global level of scientific and technological development.

Effects: The new integrated biotechnology will accelerate the process of the biodegradation of organic waste, which is considered to be a secondary renewable raw material, and reduce the emission of greenhouse gases. In this way, it is anticipated to decrease the anthropogenic pressure on the environment and increase public health. Moreover, this resource-saving eco-biotechnology will allow the production of biogas and biofertilizers with increased energy and agronomic value.

Stages of development: At this stage, an experimental composting plant with optimized aeration was already tested. Also, the process of purification of sludge water from nitrogen using the laboratory bench – a reactor for partial nitrification and anaerobic oxidation of ammonium – was tested. A programme and methods for testing an experimental sample of a composting plant with optimized aeration during the start-up period have been developed. In 2019 the technology was patented as a utility model.

Areas of application: The technology is expected to be widely used in the public utilities sector as well as food and processing industries. Application of the developed technology will ensure import substitution of the reactors and anaerobic fermentation systems for liquid and semi-liquid waste, aerobic fermentation systems for solid organic waste and systems for microbiological purification of the liquid fraction of the fermented mass from nitrogen. For the large-scale implementation of the technology, it is necessary to provide additional economic benefits for the stakeholders (e.g. tax deductions), establish a well-organized system of waste collection from the population and enterprises and increase fees (fines) for waste disposal at unauthorized landfills. Wide application of this technology is projected to reduce the areas of landfills, which could be used for the construction of residential buildings or allocation of industrial enterprises.

CASE 2: EMERGING TREND 'INTELLIGENT SYSTEM FOR AUTOMATED MSW SORTING'

Technology: In 2018 the scientists of the Institute of Thermophysics named after S.S. Kutateladze of the Siberian Branch of the Russian Academy of Sciences started an applied study on the development of an environmentally friendly and cost-effective technology and experimental model of a robotic system for MSW sorting based on artificial neural networks (Ministry of Science and Higher Education of the Russian Federation, 2018).

Advancement: The key advantage of this technology is the usage of artificial intelligence for waste sorting. The novelty of the research is characterized by the machine learning algorithms for the classification of the MSW components with the simultaneous identification of their location. The system has no analogues in Russia.

Effects: The technology is expected to increase the level of automation of waste sorting process and, thus, decrease the share of manual labour in it. Moreover, the technology is anticipated to improve the quality of sorting and raise the share of recovered recyclables. In addition, it is projected to reduce the anthropogenic pressure on the environment by reducing the amount of MSW sent to landfills.

Stages of development: At this stage, a machine learning algorithm for the classification of the MSW components was developed. A library of images of MSW objects, collected on the basis of MSW samples used for secondary processing in the Siberian region of Russia, was created for training and testing this algorithm. The software for its implementation was tested. The approaches to defining the types of plastic in the composition of MSW were also developed on the basis of spectral analysis. The technical documentation for the experimental sample of the waste sorting complex was prepared as well. The final result of the research will be an experimental sample of the robotic sorting system, which will be subsequently brought to the stage of a pilot and serial sample. At the moment, no intellectual property rights are registered for the technology.

Areas of application: The most promising direction for the implementation of this technology is its usage on the manual sorting lines of pre-sorted waste, which should increase the quality of secondary raw materials. The technology is planned to be commercialized as an element of a modular garbage disposal system, developed by a consortium of scientific and industrial enterprises of Novosibirsk. It may be also commercialized as an individual robotic sorting module for other new or existing waste sorting complexes. Nowadays such systems are not represented on the Russian market. Considering the implementation of the National Project 'Ecology' in 2019–2024, the number of waste sorting complexes is expected to increase significantly in the near future (today in Russia there are about 250 waste processing complexes and 50 waste sorting complexes). The estimated gross revenue from the implementation of a series of such systems (at least three modules per year) amounts to 30 million rubles. In case of adaptation of this technology to the other industries, the revenue may be increased by several times.

CASE 3: STABLE TREND 'PLASMA GASIFICATION OF ORGANIC WASTE IN ELECTRIC FURNACE'

Technology: In 2015 the scientists of the Institute of Thermophysics named after S.S. Kutateladze of the Siberian Branch of the Russian Academy of Sciences started an applied study on the development of electric furnaces (electroplasma plant) for the environmentally friendly and waste-free processing of organic waste and production of fuel gas used for combustion in power-generating plants (Ministry of Science and Higher Education of the Russian Federation, 2015). As a result of the research, an energy-efficient and safe electroplasma installation with a capacity of 10–20 kilograms of organic raw materials processed per hour with low operating costs and a high degree of processing was created.

Advancement: The key advantage of this technology is a large concentration of energy in a small volume and high temperatures in the reaction zone, which make it possible to process the organic waste and produce fuel gas efficiently. The technology of plasma gasification of organic waste is non-waste and environmentally friendly in comparison to the other existing thermal methods of waste processing. The plasma technology is based on highly efficient arc plasmatrons, which allow the process to be conducted in an oxidizing environment without adding hydrocarbon fuel, the slag melt to be produced without glass-forming additives and a low degree of entrainment of toxic components with flue gases.

Effects: The technology is expected to reduce the amount of resources, including the energy ones, used for the production processes. Moreover, it is anticipated to decrease the anthropogenic pressure on the environment and improve human health.

Stages of development: The electric furnace, electric arc plasmatron with a capacity of 50 kW and laboratory electroplasma installation with a productivity of 10–20 kilograms per hour were developed and manufactured. Programmes and methods for experimental studies of electroplasma equipment have been developed. The composition of the organic waste was analysed. The consumables for manufacturing of the electroplasma installation were procured. Experimental studies of the designed and manufactured samples of the plasma electric furnace, electric arc plasmatron and a laboratory electroplasma installation were carried out. A business plan for the commercialization of the electric furnace was developed. The technology was patented as a utility model in 2017. The intellectual property rights for it are also protected by production secrets (know-how).

Areas of application: The potential areas for technology application: Electrical technology, industrial production, heat power engineering, ecology, municipalities. The electric furnaces are going to be distributed across the countries of Southeast Asia and municipalities of the Russian Federation. Estimated sales volumes in the domestic and foreign markets: Dozens of electric furnaces of various productivity (from 500 to 2,000 kg of waste per hour). Estimated payback period of one plasma installation – 3–3.5 years.

CASE 4: NICHE TREND 'CLOUD PLATFORM FOR COMPLEX MSW MANAGEMENT'

Technology: In 2016 the scientists of the St. Petersburg National Research University of Information Technologies, Mechanics and Optics (ITMO) started an applied study on the implementation of Internet of Things (IoT) technologies into the urban waste management systems as a part of the smart city concept (Ministry of Science and Higher Education of the Russian Federation, 2016). As a result of the research, the architecture of the cloud platform for the exchange of data between the city services providers (city administration, municipalities, housing and communal services, police) and commercial organizations (MSW carriers and processing plants) is going to be developed in order to support the monitoring and decision-making process in the MSW management field.

Advancement: In Russia the IoT is used for the automation of MSW collection and transportation in smart cities for the first time. The novelty of this system is characterized by the unique Context-as-a-Service (CoaaS) platform developed within the framework of the European bIoTope project (Horizon-2020). Moreover, this system is integrated with the ecosystem of the European bIoTope project (Horizon-2020), which makes it possible to exchange the data on MSW management at the international level.

Effects: The system is expected to increase the comfort of citizens and improve the urban environment. In addition, it is anticipated to ensure timely garbage collection and continuous monitoring of the collection process for the city administration. Finally, it is projected to increase the efficiency of MSW collection and transportation.

Stages of development: To date, the following results have been achieved. The architecture of the cloud-based decision support platform based on IoT technologies was developed. The prototypes of a new generation of a dispatcher console using cloud technology as well as a garbage container, equipped with level sensors, RFID tags and air pollution sensors, were created. Moreover, the web applications for citizens, carriers and government agencies and mobile applications for drivers and housing utilities workers were developed. Furthermore, the decision support system for the optimization of the transportation routes of the MSW carriers was created along with an algorithm for the extraction of the contextual information from the cloud service and its provision to the participants of MSW collection process.

Areas of application: The results of the research will be used to prototype the cloud system for automated control of the MSW collection and transportation process in the smart cities. Also it is planned to spread the developed system across the housing cooperations and private garbage collection and recycling companies.

To sum up, it can be concluded that the Russian scientists are following the global agenda in the waste management field. For example, the leading trends, particularly the production of renewable sources of energy from waste (biomass, biogas, biofuel), are considered by the scientists of Federal Research Center 'Fundamentals of Biotechnology', who have recently started the development of technology for biogas and biofertilizers production by means of organic waste processing. The emerging trends are also considered by the Russian scientists, who have initiated a project on the development of an intelligent system for automated waste sorting using machine learning algorithms at the Institute of Thermophysics in 2018. The stable trend – plasma gasification of organic waste – is investigated at the same research organization, where employees improved the technology of plasma gasification and created and manufactured highly efficient electric furnaces. One of the niche trends – usage of IoT technologies for waste management – is analysed by the ITMO researchers, who are prototyping the architecture of a cloud platform for it.

7.5 CONCLUSION AND DISCUSSION

The problem of waste management is rather acute in Russia: Nowadays the major part of MSW is landfilled, while only a small share of it is recycled or incinerated. At the same time, the amount of waste generated in Russia is annually increasing. One of the reasons for this is the constant growth of packaging materials wasted, another one – the growth of food waste. In response to these issues the Russian Government has implemented a hierarchy of waste management, which prioritizes the first stage of it – the maximum use of raw materials and prevention of waste. Still, this stage is almost ignored by the regional authorities, who plan to address these problems by the construction of a large number of treatment, recycling and disposal facilities and introduction of innovative technologies.

Russian scholars have actively investigated the problems and ways of improvement of the WMSs in foreign countries (e.g. EU, USA), Russia and its regions in recent years. However, the domestic scholars paid much more attention to the management practices, concepts and tools as well as legislative initiatives that addressed these issues rather than novel technologies that can significantly improve the whole WMS. In case they analysed the technological advances in the waste management field (e.g. aerobic and anaerobic treatment, incineration with heat and energy recovery), the researchers ignored the global technological landscape and used such traditional research methods as literature review and statistical analysis.

This study, in turn, was aimed to explore the global agenda in the waste management field with the use of a relatively novel research method – semantic analysis – conducted by means of the intellectual system of big data analysis developed in the National Research University Higher School of Economics. On this basis, a semantic map of the topics related to the issues of waste recycling, disposal and landfilling and mentioned in the research papers published for the last ten years (2009–2019) worldwide was constructed. This map revealed such large clusters of topics as the production of renewable sources of energy from waste (biomass, biogas, biofuel), energy and heat recovery during waste processing, wastewater treatment and recycling. Moreover, a trend map of the same topics analysed for the same period was constructed, which uncovered the dynamics and significance of these topics as well as areas of leading, emerging, stable and niche trends in the global waste management field.

In this regard, a number of Russian cases related to the leading, emerging, stable and niche trends in the global waste management field were analysed. As for the leading trends, the technology for the production of biogas and biofertilizers by means of organic waste processing was described. Regarding the emerging trends, the technology for automated waste sorting based on machine learning algorithms was explained. The technology of plasma gasification of organic waste in electric furnaces, in turn, was described as one of the stable trends. The cloud platform for complex waste management, including its collection and transportation, based on IoT technologies illustrated the niche trend. All of these cases were described in terms of the technologies developed or implemented, their advancements in comparison to the other technologies existing, effects on the economy, society and environment, stages of development, including the intellectual property rights obtained, and potential areas of application, especially prospects for commercialization.

As a result of this analysis, the following management concepts and practices are suggested to be implemented to support the introduction and wide adoption of these innovative technologies. First, the key principles of the concepts of sustainable development and circular economy should be considered by the major stakeholders of the WMS (federal and regional government entities, regional waste management operators, business companies and households) so as to maximize the usage of raw materials and prevent the increase of the amount of waste generated. This issue was already highlighted by the Russian scholars, but ignored by the stakeholders. Second, the practice of separate waste collection should be organized by the federal and regional government entities, implemented by the regional waste management operators and rationally considered by business companies and households in order to provide the recycling plants with secondary raw material of high quality. This issue was also emphasized by the domestic scholars, but still needs more attention from the stakeholders. Third, the concept of smart cities should be adopted by the regional governments so as to automate the processes of waste collection, transportation and sorting and increase their efficiency for the waste management operators and transparency for the citizens. This issue was not mentioned by the Russian researchers, and it should be considered by the stakeholders of WMS in the context of the recent technological advances in the waste management field.

Regarding the legislative initiatives that can contribute to the introduction and wide adoption of the innovative technologies mentioned above, the following aspects should be highlighted. To start with, economic benefits for households that collect waste separately as well as business companies that recycle, but not landfill, waste should be created (e.g. tax deductions). This should stimulate the development of the recycling industry by providing it with high-quality secondary raw materials and increased demand for the waste treatment services. Moreover, the control and supervisory activities over unauthorized landfilling should be strengthened. This, in turn, should eliminate the illegal landfills and also increase the demand for recycling. Furthermore, the state initiative on the development of electronic schemes for waste management should be completed in all regions of Russia. This should make possible the implementation of IoT technologies, particularly the cloud platform, into the national WMS. Finally, the grant programmes that support applied research projects on the development of innovative resource- and energy-saving technologies of waste processing should be extended. These projects should contribute to further technological improvement of the Russian WMS.

ACKNOWLEDGEMENTS

The book chapter was prepared within the framework of the Basic Research Program at the National Research University Higher School of Economics.

REFERENCES

Alekseyev, A. A. (2014). Pererabotka otkhodov—innovatsionnyy segment promyshlennosti. Izvestiya Sankt-Peterburgskogo gosudarstvennogo ekonomicheskogo universiteta, 3(87).

Andreassen, N. (2016). Arctic energy development in Russia—How “sustainability” can fit?. *Energy Research & Social Science*, 16, 78–88.

Compose. (2016). [Electronic resource] URL: https://www.compose.com/articles/how-scoring-works-in-elasticsearch/ (date of access: 08.06.2020).

Council. (2019). Recommendations based on the results of the 67th special (133rd) meeting Council under the President of the Russian Federation on the development of civil society and human rights on the topic “Prevention of waste generation as a necessary condition for ensuring the environmental rights of citizens and the successful implementation of the national project Ecology” [Electronic resource] URL: http://president-sovet.ru/documents/read/648/#_ftn10 (date of access: 08.06.2020).

Directive 2008/98/EC on waste (Waste Framework Directive). [Electronic resource] URL: https://ec.europa.eu/environment/waste/framework/ (date of access: 08.06.2020).

Federal Law of 06.24.1998 N 89-FZ (as amended on December 27, 2019) "On Production and Consumption Wastes". [Electronic resource] URL: http://www.consultant.ru/document/cons_doc_LAW_19109/ (date of access: 08.06.2020).

Fedotkina, O., Gorbashko, E., & Vatolkina, N. (2019). Circular economy in Russia: drivers and barriers for waste management development. *Sustainability*, 11(20), 5837.

Ghosh, S. K., Debnath, B., Baidya, R., De, D., Li, J., Ghosh, S. K., … & Tavares, A. N. (2016). Waste electrical and electronic equipment management and Basel convention compliance in Brazil, Russia, India, China and South Africa (BRICS) nations. *Waste Management & Research*, 34(8), 693–707.

Government of the Russian Federation. (2018). Decree of the Government of the Russian Federation of January 25, 2018 N 84-p "On approval of the Industry Development Strategy for the processing, utilization and neutralization of production and consumption waste for the period until 2030". [Electronic resource] URL: http://static.government.ru/media/files/y8PMkQGZLfbY7jhn6QMruaKoferAowzJ.pdf (date of access: 08.06.2020).

JRC (Joint Research Centre). (2017). Top Marine Beach Litter Items in Europe. [Electronic resource] URL: https://publications.jrc.ec.europa.eu/repository/bitstream/JRC108181/technical_report_top_marine_litter_items_eur_29249_en_pdf.pdf (date of access: 08.06.2020).

Kaazke, J., Meneses, M., Wilke, B. M., & Rotter, V. S. (2013). Environmental evaluation of waste treatment scenarios for the towns Khanty-Mansiysk and Surgut, Russia. *Waste Management & Research*, 31(3), 315–326.

Kirillova, A. N., & Musinova, N. N. (2018). Organizatsionno-ekonomicheskiy mekhanizm upravleniya sistemoy obrashcheniya s tverdymi kommunal'nymi otkhodami. Vestnik universiteta, 10.

Kolotyrin, K. P. (2016). Ekologo-ekonomicheskiye riski v sfere obrashcheniya s otkhodami i puti ikh snizheniya. Gornyy informatsionno-analiticheskiy byulleten' (nauchno-tekhnicheskiy zhurnal), 2.

Kommersant. (2018). V Rossii kazhdyy shestoy otkazalsya ot plastikovykh paketov. [Electronic resource] URL: https://www.kommersant.ru/doc/3793675 (date of access: 08.06.2020).

Kommersant. (2019). Otkhody stanut tsementom. Sborshchiki musora nashli sposob sbyta. [Electronic resource] URL: https://www.kommersant.ru/doc/4165569 (date of access: 08.06.2020).

Kuzminov, I. F., Loginova, I. V., & Lobanova, P. A. (2018). Perspektivy ispol'zovaniya tekhnologiy analiza bol'shikh dannykh dlya strategicheskoy analitiki agropromyshlennogo kompleksa. Nauchno-prakticheskiy zhurnal osnovan v marte 1956 g., 9.

Likhacheva, O. I., & Sovetov, P. M. (2017). Metodologicheskiye aspekty upravleniya sferoy obrashcheniya s tverdymi bytovymi otkhodami. Ekonomicheskiye i sotsial'nyye peremeny: fakty, tendentsii, prognoz, 4(52).

Ministry of Natural Resources and Ecology of the Russian Federation. (2013). Order of the Ministry of Natural Resources and Ecology of the Russian Federation dated August 14, 2013 No. 298 "On approval of a comprehensive strategy for the treatment of solid municipal (household) waste in the Russian Federation". [Electronic resource] URL: http://www.garant.ru/products/ipo/prime/doc/70345114/ (date of access: 08.06.2020).

Ministry of Science and Higher Education of the Russian Federation. (2018). Priority directions of research and development of the Russian scientific and technological complex for 2014–2020. [Electronic resource] URL: http://www.fcpir.ru/participation_in_program/contracts/14.604.21.0203/ (date of access: 08.06.2020).

Ministry of Science and Higher Education of the Russian Federation. (2017). Priority directions of research and development of the Russian scientific and technological complex for 2014–2020. [Electronic resource] URL: http://fcpir.ru/participation_in_program/contracts/14.604.21.0190/#collapse-2 (date of access: 08.06.2020).

Ministry of Science and Higher Education of the Russian Federation. (2016). Priority directions of research and development of the Russian scientific and technological complex for 2014–2020. [Electronic resource] URL: http://fcpir.ru/participation_in_program/contracts/14.587.21.0031/ (date of access: 08.06.2020).

Ministry of Science and Higher Education of the Russian Federation. (2015). Priority directions of research and development of the Russian scientific and technological complex for 2014–2020. [Electronic resource] URL: http://fcpir.ru/participation_in_program/contracts/14.607.21.0118/ (date of access: 08.06.2020).

NRU HSE. (2018). Praktiki produktovykh riteylerov po sokrashcheniyu oborota i otkazu ot odnorazovykh plastikovykh tovarov, tary i upakovki v pol'zu mnogorazovykh al'ternativ. [Electronic resource] URL: http://www.president-sovet.ru/files/d7/16/d7169274e860a822be03574e26fb3a04.pdf (date of access: 08.06.2020).

OECD. (2018). Workshop: "Semantic analysis for innovation policy". [Electronic resource] URL: https://www.innovationpolicyplatform.org/www.innovationpolicyplatform.org/workshop-semantic-analysis-innovation-policy/index.html (date of access: 08.06.2020).

Pakhomova, N. V., Rikhter, K. K., & Vetrova, M. A. (2016). Formirovaniye sovremennoy sistemy obrashcheniya s otkhodami—ot bezopasnogo zakhoroneniya k remanufakturingu. Problemy sovremennoy ekonomiki, 4(60).

Panina, I. A. (2015). Sovremennyye problemy kontrol'no-nadzornoy deyatel'nosti v sfere obrashcheniya s otkhodami v Rossiyskoy Federatsii. Legal Concept, 4.

Plyaskina, N. I., & Kharitonova, V. N. (2016). Upravleniye v sfere obrashcheniya s tverdymi kommunal'nymi otkhodami: sovremennoye sostoyaniye. Vserossiyskiy ekonomicheskiy zhurnal EKO, 12(510).

Prianikova, A. A., Akhmadieva, A. F., & Khrustaleva, A. Y. (2018). Evaluation of the waste processing system in the concept of management of the sustainable development of the Arctic Zone of the Russia. In 2018 IEEE International Conference "Management of Municipal Waste as an Important Factor of Sustainable Urban Development" (WASTE) (pp. 7–9). IEEE.

Prisyach, E. Y., & Shvetsova, O. A. (2018). Elements of Innovative Scenario's Development of Waste Management System in Russia. In 2018 IEEE International Conference "Management of Municipal Waste as an Important Factor of Sustainable Urban Development" (WASTE) (pp. 63–65). IEEE.

Putintseva, N. A., & Chekalin, V. S. (2018). Obzor mer po organizatsii upravleniya otkhodami v Rossii, kak faktora povysheniya yeye energoeffektivnosti. Izvestiya Sankt-Peterburgskogo gosudarstvennogo ekonomicheskogo universiteta, 4(112).

Rodionov, M., & Nakata, T. (2011). Design of an optimal waste utilization system: a case study in St. Petersburg, Russia. *Sustainability*, 3(9), 1486–1509.

Roscongress. (2019). The National Project "Ecology". [Electronic resource] URL: https://roscongress.org/en/knowledge/natsionalnyy-proekt-ekologiya/download/ (date of access: 08.06.2020).

Selezneva, A. (2016). Current environment for development of source separated waste management system in Saint Petersburg, Russia, 2016.

Smetanina, T., Pintassilgo, P., & Matias, A. (2014). Environmental management tools: international practicies for Russia. Вестник Удмуртского университета. Серия «Экономика и право», 1.

Starostina, V., Damgaard, A., Rechberger, H., & Christensen, T. H. (2014). Waste management in the Irkutsk Region, Siberia, Russia: Environmental assessment of current practice focusing on landfilling. *Waste Management & Research*, 32(5), 389–396.

TASS. (2019). Eksperty: v Rossii neobkhodimo sozdat' usloviya dlya pererabotki organicheskogo musora. [Electronic resource] URL: https://tass.ru/nacionalnye-proekty/6310152?fbclid=IwAR0S5tqqmlI0BU3GARR4MJq4MShJv1rHZ0ygI34GRB1BUsJcy1LVXfrIwsw (date of access: 08.06.2020).

Tulokhonova, A., & Ulanova, O. (2013). Assessment of municipal solid waste management scenarios in Irkutsk (Russia) using a life cycle assessment-integrated waste management model. *Waste Management & Research*, 31(5), 475–484.

Verenikina, A. Yu, & Gorokhova, Ye. Yu. (2016). Resursnyy podkhod k upravleniyu otkhodami proizvodstva i potrebleniya. Rossiyskoye predprinimatel'stvo, 17(7).

VTsIOM. (2019). Ekologicheskaya situatsiya v Rossii: monitoring. [Electronic resource] URL: https://wciom.ru/index.php?id=236&uid=9544 (date of access: 08.06.2020).

Zhuchkov, O. A., & Tupikova, O. A. (2017). K reformirovaniyu sistemy obrashcheniya s otkhodami na territorii Rossii. Universum: tekhnicheskiye nauki, 8(41).

8 Different Treatment Technologies Used for Municipal Solid Waste Management in India

Material and Energy Recovery

Karan Chabhadiya, Yash Pujara, Harshit Patel, Janki Govani, Pankaj Pathak and Deepak Mashru

CONTENTS

8.1 INTRODUCTION

For many years, India has been known as a country of traditions and customs that believes in resource conservation and protection of the environment. However, due to rapid urbanization and industrialization, the Government of India is struggling to fulfil all the necessary requirements of ~1.3 billion people (Joshi and Ahmed, 2016; Rehman and Hussain, 2017). Energy and materials are the two important sectors that govern the economy of a developing nation like India (Chabhadiya et al., 2021; Kumar et al., 2018). Nonetheless, the government has to deal with 62 million tons of solid waste every year, and its generation is increasing at a rapid rate (Sharma and Jain, 2018). Moreover, looking at the current scenario, the government estimates that ~50% of the population will be residing in urban areas by 2020 which would result in high waste generation (Kaza et al., 2018). Major sources of waste in India are from domestic, commercial, industrial, agricultural and medical organizations (Desai and Shah, 2018). It includes waste like organic, inorganic, inert, hazardous and infectious waste in different amounts, type, size and density (Joshi and Ahmed, 2016; Pathak and Gupta, 2020). The most common element in the composition of Indian municipal solid waste (MSW) is organic waste (50–60%) that includes fruits and vegetable waste, leftover food, garden waste, etc. (Gupta et al., 2015). Whereas 30–40% is inert waste including dust, stones, pebbles, paper, wood, plastics, rubber, etc., and the rest are recyclables like high-grade plastic, glass, metals, etc. (Joshi and Ahmed, 2016; MSWM Rules, 2000). All kinds of waste have their own potential to be used as a resource in the form of material or energy.

From the total waste generation, 60% of the waste ends up in landfills or water bodies which deteriorates the air, water and soil environment leading to severe human health hazards, and only 12–15% of waste is treated or recycled scientifically (Kumar et al., 2019; Govani et al., 2020). The MSW generated in India majorly consists of organic waste that is high in nutrients which can be utilized in the form of compost via various composting methods (Pujara et al., 2019). Composting serves as one of the best examples of material and nutrient recovery. Organic waste can also be used for the generation of biogas, and the slurry remains can be used as the best quality manure by the farmers of the nation. Also, due to technological advancement and consumption patterns, various types of wastes are being generated. These include synthetic materials like plastics, nylon, rubber, thermocol, sponge, etc., and they are not so environmentally friendly. These wastes can be recycled up to a certain amount and, thereafter, they become trash (Moore, 2008; Govani et al., 2020).

Today, there are various technologies to recover material and energy from waste. Primarily, two treatment technologies are available, namely thermal and mechanical biological treatment. Thermal technologies include energy recovery methods like incineration with energy recovery, gasification, pyrolysis and plasma technology used in waste-to-energy plants (Havukainen et al., 2017; Waste to Energy, 2016). Whereas, biological technologies include composting for nutrient recovery, anaerobic digestion and bioreactor landfilling of organic waste. Energy and material recovery from solid waste generated in India is illustrated in Figure 8.1.

Nevertheless, all the technologies are different from each other; they serve well in material and energy recovery. Thermal processes for waste treatment are generally selected based on the nature and characteristics of the waste. All the technologies have certain benefits along with a few drawbacks. They help in solving the problem of solid waste but result in polluting the air. Thus, an air pollution control device is to be installed along with all of the mentioned thermal treatment technologies. However, in the case of biological treatment, environmental parameters play a very important role. These processes need favourable temperatures, a certain amount of moisture and certain species of microorganisms that can degrade or digest organic waste.

In the present chapter, thermal and biological treatment technologies are elaborated along with a few case studies on material and energy recovery like the consumption of MSW in a cement power plant as an alternate fuel and resource. Another case is described where flower waste from Indian

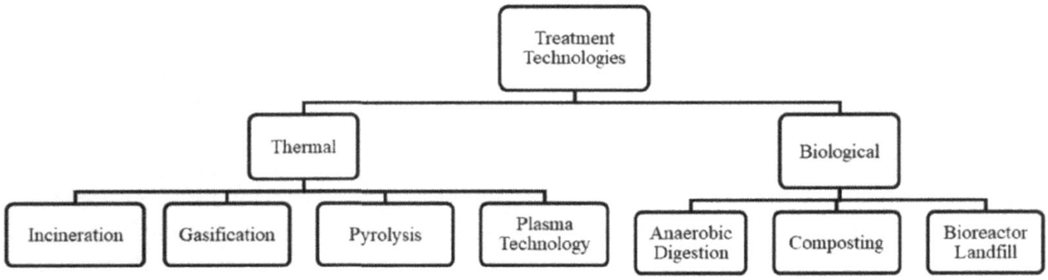

FIGURE 8.1 Various treatment technologies for material and energy recovery from solid waste (Bhatia et al., 2018).

religious places is processed into instance stick as a part of material recovery. Lastly the concept of 'zero waste management' of the organic waste that is consumed by animals, giving by-products like biogas, eggs and high-grade manure, is discussed. Moreover, a common perspective of the Indian solid waste management can be well understood in the chapter.

8.2 TREATMENT TECHNOLOGIES FOR INDIAN MSW

8.2.1 Thermal Treatment

Thermal treatment technologies for waste management aim at the reduction of the waste volume, the conversion of waste into harmless materials and the utilization of the energy of waste as heat, steam, electrical energy or combustible material. They include all processes converting the waste content into gas, liquid and solid products with simultaneous or consequent release of thermal energy.

Waste treatment methods are categorized either as a disposal method or recovery method, and the thermal treatment technologies that are significant for energy recovery are included in the recovery method (Govani et al., 2020). The waste management hierarchy pyramid shows energy recovery is a more favourable option than final disposal; hence many countries have developed and applied waste-to-energy technologies in order to handle the constantly increasing municipal solid waste (Kalyani and Pandey, 2014).

Prominent thermal technologies for solid waste management are incineration, mass burn combustion, pyrolysis, gasification and plasma technology as shown in Figure 8.2. Refuse-derived fuel (RDF) is another upcoming technology, which can be effectively used to produce alternative energy from MSW and industrial waste, reducing the load on landfills (Govani et al., 2019).

8.2.1.1 Incineration

The incineration (combustion) of carbon-based materials in an oxygen-rich environment carried out at temperatures higher than 850°C produces waste gases primarily composed of carbon dioxide (CO_2), water (H_2O), NOx, SOx, etc., which are emitted during operation. The required air emission control devices are shown in Figure 8.3. The inorganic content of the waste is converted to ash. This is the most common and well-proven thermal process using a wide variety of fuels. During the full

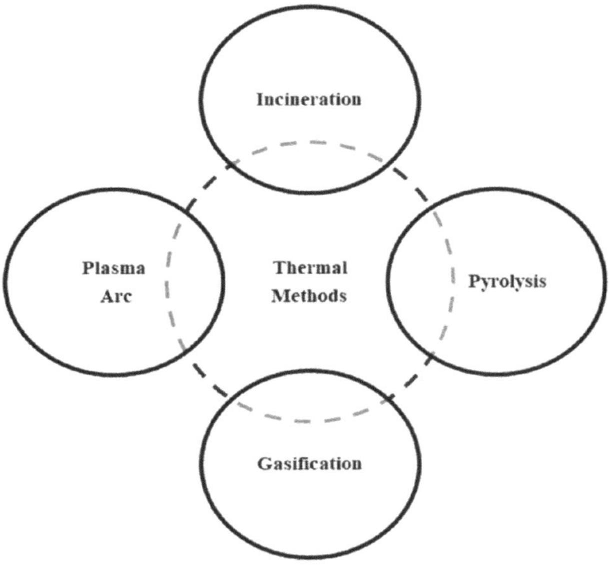

FIGURE 8.2 Thermal treatment technologies of waste (Bhatia et al., 2018).

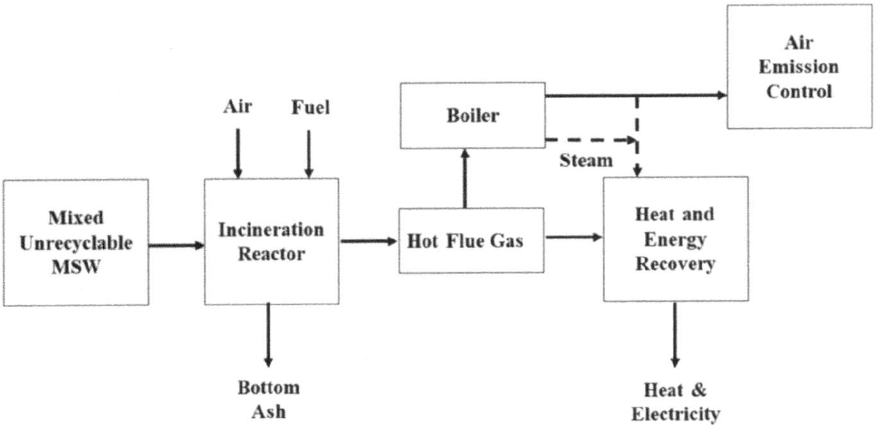

FIGURE 8.3 Schematic flow diagram of MSW incineration (Sharifah et al., 2013).

combustion there is oxygen in excess, no carbon monoxide (CO) is produced and the average gas temperature is 1200°C (Alfons et al., 2016). The reactions are given below:

$C + O_2 \rightarrow CO_2 + 393.77\ J$
$CxHy + (x+ y/4)\ O_2 \rightarrow xCO_2 + y/2\ H_2O$

In the case of lack of oxygen, the reactions are characterized as incomplete combustion where the produced CO_2 reacts with C that has not been consumed and is converted to CO at higher temperatures.

$C + CO_2 + 172.58\ J \rightarrow 2CO$

8.2.1.2 Gasification

Gasification is the thermal process that converts carbon-containing municipal solid waste to syngas which can then be used to produce electric power, valuable products, such as chemicals, fertilizers, substitute natural gas, steam and transportation fuels (Yong-Chil et al., 2018). Gasification is defined as a thermal reaction with insufficient oxygen present for the chemical reaction of hydrocarbons, and it produces a composite gas (syngas) that is comprised primarily of hydrogen (H_2) and carbon monoxide (CO). The gasification process is shown in Figure 8.4 (Konstantinos et al., 2010); and the reactions taking place during gasification are shown in the following:

$C + O_2 \rightarrow CO_2$ Oxidation (exothermic)
$C + H_2O \rightarrow CO + H_2$ Water evaporation reaction (endothermic)
$CO + H_2O \rightarrow CO_2 + H_2$ (exothermic)
$C + CO_2 \rightarrow 2CO$ Boudouard Reaction (endothermic)
$C + 2H_2 \rightarrow CH_4$ Methane formation reaction (exothermic)

8.2.1.3 Pyrolysis

Pyrolysis is the thermal degradation of carbon-based materials using an indirect, external source of heat at a temperature range of 450°C to 750°C. However, pyrolysis occurs in the absence of free oxygen. It drives off the volatile portions of the organic materials, resulting in a syngas that is composed primarily of H_2, CO, CO_2, CH_4 and complex hydrocarbons (Wang et al., 2015). The syngas can be utilized in boilers, gas turbines or internal combustion engines to generate electricity. The balances of the organic materials that are not volatile are left as char material. Inorganic materials of the bottom ash can be used for manufacturing brick as an alternative to disposal (Govani et al., 2019).

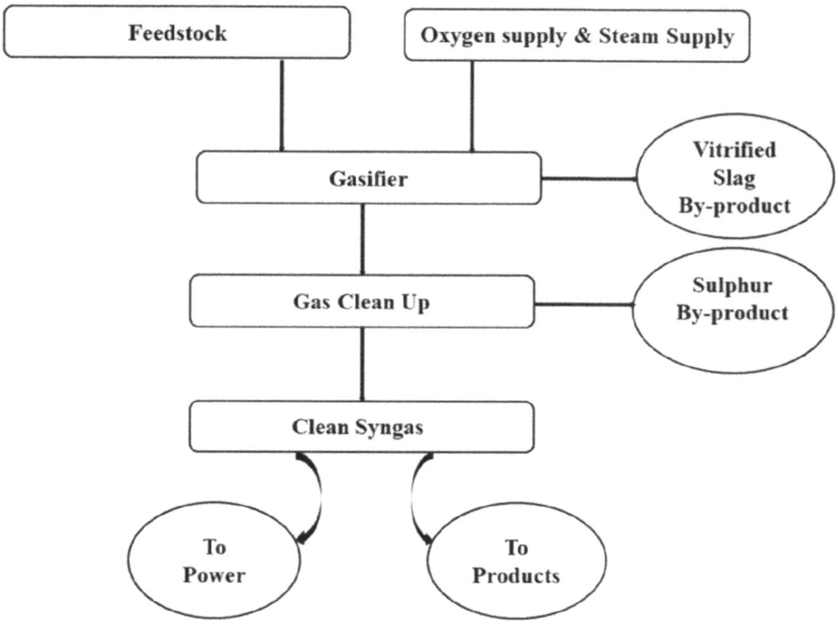

FIGURE 8.4 Schematic flow diagram of gasification process (Konstantinos et al., 2010).

The products produced from pyrolysing materials are the solid residue and a synthetic gas (syngas), while some of the volatile components from tars and oils can be removed and reused. The solid residue (described as a char) is a combination of non-combustible materials and carbon. The syngas is a mixture of gases (combustible constituents include carbon monoxide, hydrogen, methane and a broad range of other VOCs). A proportion of these can be condensed to produce oils, waxes and tars. The syngas typically has high net calorific value and furthermore, a condensable fraction can be collected by cooling the syngas, potentially for use as liquid fuel. The schematic flow of the pyrolysis process is shown in Figure 8.5.

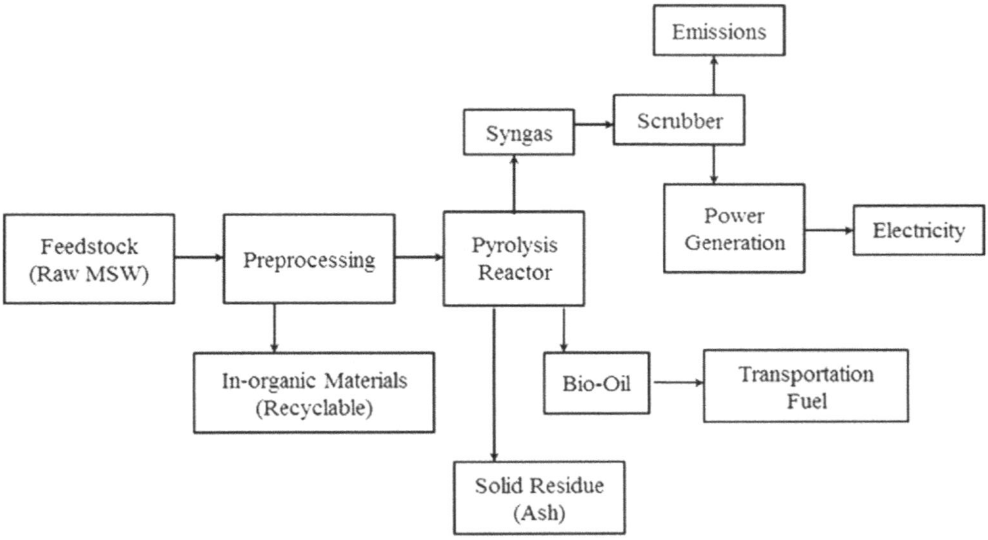

FIGURE 8.5 Schematic process flow diagram of pyrolysis process (Dezhen et al., 2014).

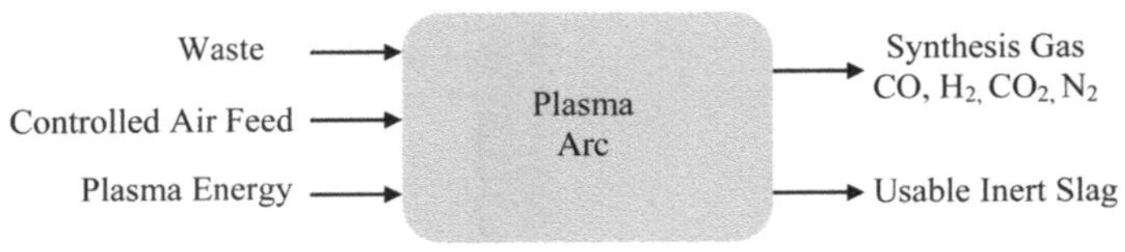

FIGURE 8.6 Inlet and outlet of plasma technology (Gomez et al., 2009).

8.2.1.4 Plasma Technology

Plasma technology partially or totally ionizes the atoms or molecules of gaseous forms. In the plasma technique free electrons occur at reasonably high concentrations and the charges of electrons are balanced by positive ions. Plasma is quasi-neutral and generated from electric discharges, from the passage of current (continuous, alternate or high frequency) through the gas and use of the dissipation of resistive energy in order to make the gas sufficiently hot (Gomez et al., 2009). Plasma is characterized as the fourth state of matter and differs from the ideal gases, because it is characterized by collective phenomena (originates from the wide range of coulomb forces). As a result, the charged particles do not interact only with neighbouring particles through collisions, but they also bear the influence of an average electromagnetic field, which is generated by the electric charges (Blahos, 2000). The inlet and outlet of plasma technology are shown in Figure 8.6.

8.2.2 Mechanical Biological Treatment (MBT)

8.2.2.1 Composting

Composting is a decomposition/degradation process of organic substances through the biological reaction. This process is beneficial to increase the nutrients in the soil as a conditioner, compost manure and addition of humic acid and natural pesticides for earth (Kumar, 2010). Indian municipal solid waste are more than 60% biodegradable matter, and therefore composting is one of the best possible options to treat the MSW (Pujara et al., 2019). It decreases the environmental degradation as compared with incineration and landfilling. For composting, it is mandatory to maintain various stages carefully to avoid contamination from MSW. The segregation of materials at source has the highest potential to recover biodegradable waste and convert it into a resource. Further, it is necessary to determine the organic fraction for effective composting and cellulose degradation activity. Some important parameters such as waste size and proportion, temperature, carbon:nitrogen (C:N) ratio, aeration and moisture content are to be maintained for an effective composting process (Liu et al., 2019). The key components of composting are shown in Figure 8.7. Additionally, different types of composting and its procedures are described in the following.

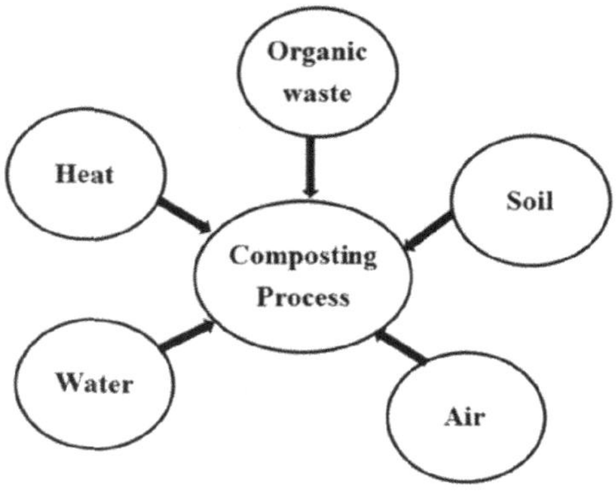

FIGURE 8.7 Vital components for composting process (Kumar, 2010).

Primarily, three major types of composting are categorized for the MSW treatment facilities: (A) windrow composting; (B) in-vessel composting/enclosed composting; (C) aerated static pile composting.

8.2.2.1.1 Windrow Composting

This is a simple composting method where the organic wastes are piled up for the degradation process automatically through the air and heat. Windrow composting is employed when the volume of organic waste is very high and it includes all types of degradable waste together. This method is also useful for producing manure/compost at a very large scale (Kumar, 2010). However, the shape of pile, i.e. triangular, trapezoidal or rectangular, varies based on the turning equipment and composting material. The frequency of turning of the waste pile depends on the aeration requirement, and the turning of windrow can be at a regular interval and availability of aeration machines. Temperature plays a significant role in the turning of composting mass; it varies between 55 and 60°C (Manu et al., 2019). Windrow composting is beneficial when it comes to the cost-effectiveness for an industrial scale, and aeration for windrow can be achieved by mechanical equipment. Periodic turning of a windrow pile maintains the temperature, height and width of the waste material.

8.2.2.1.2 Enclosed Composting/In-Vessel Composting

Enclosed or in-vessel composting provides temperature control, oxygen demand and odour control. The enclosed system is made up of concrete or steel structure material. Thus, the enclosed system has the advantages of both the aerated static pile and windrow pile systems. However, this in-vessel type of composting is rare in India because of the high construction cost of the vessel or enclosed system (Kumar et al, 2018). The in-vessel composting process is either the agitated bed type or silo type. Also, it ensures control of moisture, temperature, odours and amount of oxygen (Aziz et al., 2018).

8.2.2.1.3 Aerated Static Pile Composting

Aeration is provided to degrade the waste during the composting process in either an open pile or closed vessel type pile. High temperatures can be achieved through the aerated closed vessels whereas aeration is provided externally through the mechanical equipment. During the primary composting process, organic waste is degraded without any physical processes. It has been reported that without aeration, the temperature variation in the pile from top to bottom is huge, and this can be moderated by aeration (Kumar et al., 2018).

8.2.2.1.4 Vermicomposting

Vermicomposting is a method where earthworms are used to degrade the semi-composed organic matter of MSW. Vermicompost is highly rich in nutrients compared to the compost produced by windrow composting, in-vessel composting and aerated static pile composting. It also encompasses millions of microorganisms that support the breakdown of complex organic materials and further nutrients spread in the soil. However, in this process, worm castings contain worm mucus that prevents nutrients from washing away with the primary sprinkling and grasps moistness, which is better than the plain soil. Worm compost is nutrient rich but sticky for use as a seed starter. Therefore, it is used as a top dressing or mixed with soil in a ratio of one to four. It has been reported that most of the fruits and seed pits are germinated in vermicomposting effortlessly.

8.2.2.2 Anaerobic Digestion (AD)

Anaerobic digestion is a process for decomposing organic waste under anaerobic conditions or without oxygen. The end product of AD contains biogas where the amount of methane is 60–70%, CO_2 is 30–40% and other impurities. This technology is well established in India to recover energy. It decreases the organic loading from the agriculture waste, food waste and sludge of wastewater.

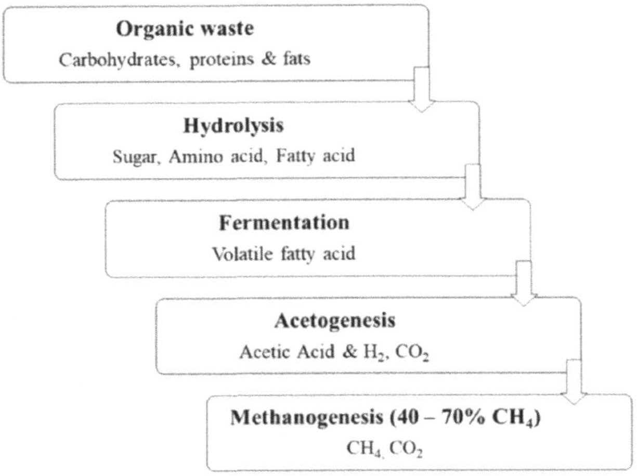

FIGURE 8.8 Mechanism of anaerobic digestion (Bhatia et al., 2018; Siddique et al., 2018).

However, the design of the anaerobic digester and maintaining its operating parameters depend on the AD process such as continuous or batch design, operating temperature psychrophilic, mesophilic and thermophilic, covered lagoons, complete mix and plug flow (Jain et al., 2015; Siddique et al., 2018). There are four prime steps for anaerobic digestion which are discussed below and shown in Figure 8.8.

(a) **Hydrolysis**
This is the very first step of the digestion process; in this step microorganisms break down the organic polymer chain into very small parts and make a simple soluble molecule with the help of enzymes. With this reaction, carbohydrates, lipids and proteins are hydrolysed to sugar, fatty acids and amino acids respectively (Govani et al., 2020).

(b) **Fermentation**
The fermentative microorganisms reduce the organic compounds and convert them into CO_2, H_2, acetic acid and fatty acids (Aryal et al, 2018).

(c) **Acetogenesis**
Furthermore, acetogenic microorganisms help to convert the acetic acid to acetate. The CO_2 and H_2 are directly used for the production of methane (Bhatia et al., 2018).

(d) **Methanogenesis**
This is the final step for the AD method, where methanogenic bacteria help to degrade the acetate, CO_2 and H_2 to produce the methane which can be called biogas (Jain et al., 2015; Bhatia et al., 2018).

8.3 CASE STUDIES

8.3.1 Recovery of Material and Energy from the Cement Industry

Alternate fuel and raw material (AFR) are used at Ambuja Cements Ltd. It is located at Kodinar in Gujarat, India. It is a unit with three plants, viz. Ambuja Cement Ltd. having one kiln, Gajambuja Cement Ltd. having two kilns, and a 90-MW capacity thermal power plant. Initially, a 10% thermal substitution rate (TSR) of AFR was targeted as a pre-processing project in 2014 under Swachh Bharat Mission but only 3.95% was achieved. Subsequently, 5.95%, 4.73% and 6.24% of TSR were achieved for the consecutive years in 2015, 2016 and 2017 respectively (Ambujacement .com). Ambuja Cement is functioning with geo-cycle to expand the waste stream and ensure its availability. Geo-cycle was involved in the co-processing of 1.78 lakh tonnes of AFR in 2014. The waste materials are used in the co-processing for the energy or resource recovery and reduction of conventional fuels or raw materials or both through substitution. The prime objective of the project was to handle the waste in a safe and sustainable manner through segregation, shredding and homogenizing of waste material for co-processing as AFR in the cement kilns and to reduce the cost of thermal energy. The generation of hazardous waste (HW) has been significantly decreased from 2012 to 2014, from 653 to 315 tonnes. The high temperature (1,200–1,700°C) inside the kiln and controlled conditions ensure that no harmful gases are emitted and no residue is left, but the appropriate size and low sulphur and chloride content have to be ensured. Thus, the progression of AFR to reduce the consumption of coal and lower emissions of CO_2 is shown in Figure 8.9. A leading global certification body has awarded prominent endorsement on sustainable product labelling 'PRO-SUSTAIN' to Ambuja Cement. This certification indicates that the company promotes the implementation of accountable and cost-effective actions for integrating sustainability into product design, development, production and supply chain management. Through its societal and sustainable works, Ambuja Cement has been certified as the first cement company in the world to be 'Water Positive' (surplus water by four times) and the Ambujanagar unit is declared as 9.6 times water positive.

8.3.2 Recovery of Materials from Temple Waste

The constructive and divine vibrations from temples, churches, gurudwaras and mosques always charge the atmosphere with positive vibes. People offer flowers, garlands, coconut, milk, etc., in such religious places as a symbol of devotion. However, every year, 8 million tonnes of temple waste are dumped into rivers or dumpsites, disturbing the aesthetic of the places and aquatic systems. As reported by WHO in 2005, 36% of the chemicals used in the plantation of flowers are exceptionally toxic and have a hostile effect on health, in particular respiratory problems, skin rashes and eye problems. Several social entrepreneurs are recycling the flowers offered in the temple for the natural dying of clothes (Environment-health-safety, 2019). The process involves washing and drying flowers in the sun. Then, they are powdered and mixed with natural resins resulting in a dough which leads to the generation of handmade incense sticks. The temple waste can be utilized to manufacture several products like natural dye, incense sticks, coconut shell ash as cement, handmade paper, sugar syrup and floral oils (Ask-ehs, 2019). Burning coconut shells in the open for three hours produces coconut shell ash which can partially replace cement. Additionally, marigold flower extract has a place in industries and veterinary feeds in the preparation of sugar syrup. The oils from flowers also have medicinal properties, viz. lily cures jaundice, madhuca treats bronchitis, etc. This novel approach to reusing and recovering temple waste can additionally generate revenue and empower underprivileged people, considerably promoting bio-economy.

8.3.3 Zero Waste Management: Vellore Model

Zero waste management (ZWM) is where animals, viz., cows, hens, frogs, fish, ducks and earthworms, are used to consume wet waste, and it is also known as the Vellore Model. This concept of escalating garbage to resource earned many accolades in the field of solid waste management (Zerowastemanagement.org.in). There are nine different stages involved in attaining zero waste management (Zerowastemanagement.org.in). The input is alike (i.e. wet waste) for the whole system but the outputs are useful distinctive products, viz. biogas, compost, chicken eggs, and duck eggs. The individual unit process is in the circle form as presented in Figure 8.10. The colour coding indicates the desirability of the waste to be consumed by animals. The red colour represents most unfavourable while dark green represents most favourable. The lighter shades represent intermediate products. Further, each unit has its own sub-processes and functions on its own time schedule. In this model, waste is collected every day but compost is created aerobically after 35 days from organic matter; thus partial fresh waste can be added to the system. Similarly in vermicomposting,

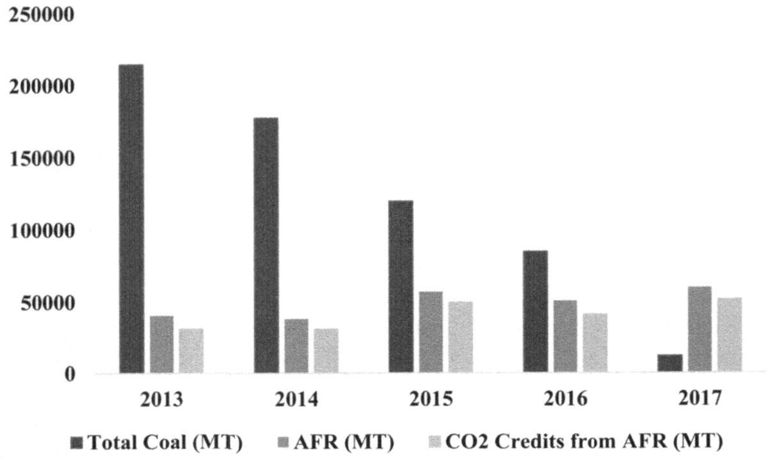

FIGURE 8.9 Progression of AFR to reduction of col and CO_2 emission (Ambuja Cement.com).

earthworms can be introduced to fresh vermicompost to continue the process. The process is environmentally friendly, locally designed and not energy intensive. It can be observed from Figure 8.10 that each unit is interlinked with the others. To increase the desirability of waste at each stage, the process is systematically designed to utilize the waste at every step. The Vellore Model is recommended as a 'centralized project with decentralized processes'.

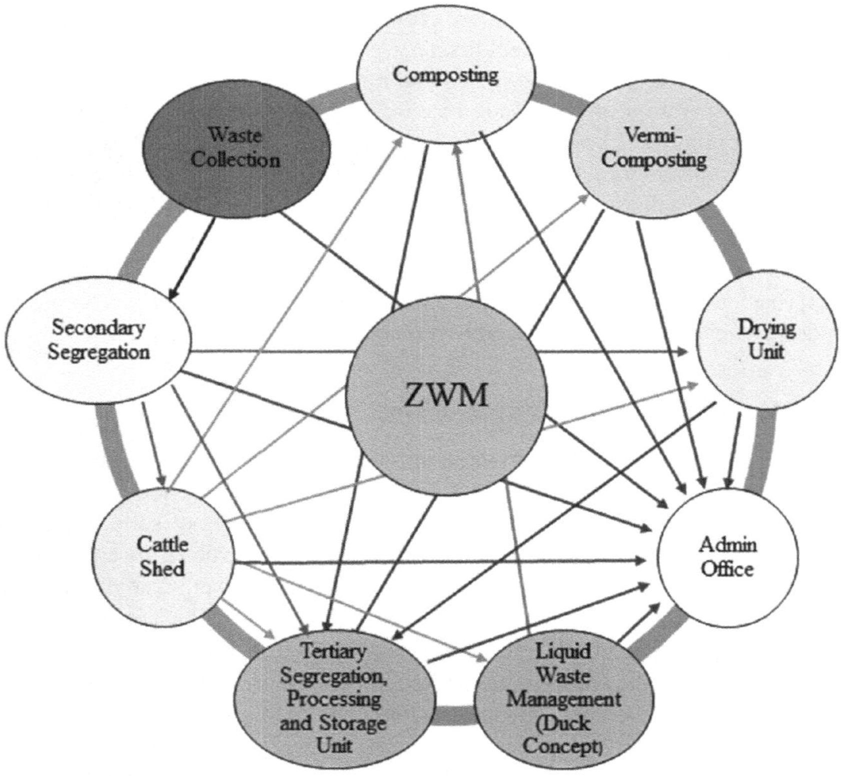

FIGURE 8.10 Vellore Model of zero waste management (Zerowastemanagement.org.in).

8.4 SOLID WASTE ISSUE OR SOLUTION: COMMON PERSPECTIVE

India, being a developing country, has unlimited issues and limited solutions to address the solid waste management problems. The management of municipal solid waste which is generated from residential, commercial, industrial, institutional, construction and demolition and municipal services is one of the most challenging issues (MSWM rules, 2000). Most of the municipalities face major environmental challenges concomitant with waste generation and inadequate waste collection, transport, treatment and disposal. The challenges include all the administrative, financial, legal and planning functions as well as the physical aspects of solid waste handling.

There have been many questions raised about human and environment health issues in MSW management as a large number of people who are downtrodden and marginalized have been exposed to occupational biohazards and safety risks due to their involvement in the collection of recyclable materials like plastic, glass and paper. These rag pickers are at greater risk of adverse health issues. Moreover, when these infected people move around in the city they tend to spread diseases. On the other hand, when waste is not segregated at source, it becomes next to impossible to segregate later and use the same for the purpose of recycling. Hence, such waste is dumped and the society faces consequences in multiple ways like the disposal of organic and recyclable waste in the landfills. The products cannot be recycled or reused, which creates scarcity in the market, resulting in higher prices, leading to a financial burden on the people. As per the revised rules of Solid Waste Management in 2016 by the Ministry of Environment, policy-level interventions must be implemented. There should be promotion of waste-to-energy (WTE) plants and provision for local bodies to levy waste collection fees on waste generators (Waste to Energy, 2016). There are three major stakeholders in the waste management value chain: (i) the waste generator, (ii) the waste collection agent and other middlemen in charge of collection and transportation, and (iii) the final contractor assigned to dump the collected waste. The waste collected is often found unsegregated from this chain; hence, a well-structured reward system if introduced can bring hope to the people. As a waste generator, when one does not have any financial and/or non-financial gain and sources readily available, people will not be keen to keep the waste segregated. In such a case, if the municipal authority offers incentives/discounts in property taxes for the people who support the municipality in segregating waste at source, the major task of solid waste management can be addressed. Such government schemes and offers can create an opportunity for public participation and awareness programmes can be aired all over the nation. Thus, segregating solid waste at source saves energy and resources, and valuable products can be shaped.

8.5 CONCLUSIONS

Management of solid waste in a scientific manner is one of the major issues faced by India. Different technologies are available to treat the MSW based on its composition. However, it is extremely heterogeneous in nature, and segregation of MSW is one of the biggest tasks in reutilizing the MSW effectively. Different technologies based on thermal and biological treatment are incorporated to treat the MSW. The dry organic waste preferably undergoes thermal treatment where incineration, gasification and pyrolysis are the common technologies. These technologies convert waste into energy; however, waste-to-energy conversion causes environmental impact, therefore, it is recommended to install an appropriate air pollution control device in the plants. Further, wet organic waste undergoes biological treatment such as composting and anaerobic digestion, which convert wet waste into a resource. Material and energy both can be recovered through biological treatment. In this context, several organizations and NGOs are employing these treatment technologies to manage the MSW and convert it into a resource. Therefore, three different case studies on waste-to-resource technologies entail the significance of the technologies if they are implemented successfully. Consequently, public awareness of the need to segregate the MSW is mandatory, and for effective segregation government should make a plan to provide incentives and rewards to those who are doing their job effectively.

REFERENCES

Alfons, B., Cuicui, D., Jie, Y., Jianhua, Y., Longjie, J., Shengyong, L., Zhiliang, C. (2016). Municipal solid waste incineration in China and the issue of acidification: a review. *Waste and Resource Management* 34(4), 280–297.

Ambujacement.com. (2020). https://www.ambujacement.com/html/sustainability-report-2014/alternate-fuel-and-raw-material-2.html (Accessed 25 January 2020).

Aryal, N., Kvist, T., Ammam, F., Pant, D., Ottosen, L.D.M. (2018). An overview of microbial biogas enrichment. *Bioresource Technology* 264, 359–369.

Ask-ehs. (2019). https://www.ask-ehs.com/blog/environment/recycling-temple-waste-preserving/ (Accessed 24 June 2020).

Aziz, S.Q., Omar, I.A., Mustafa, J.S. (2018). Design and study for composting process site. *International Journal of Engineering Inventions* 7(9), 9–18.

Bhatia, S.K., Joo, H.S., Yang, Y.H. (2018). Biowaste-to-bioenergy using biological methods—a mini-review. *Energy Conservation Management* 177, 640–660.

Blahos, L. (2000). Plasma physics, the fourth state of matter. In: Giolas (ed.) *Introduction Terms*, Salonica, 1–12.

Chabhadiya, K., Srivastava, R.R., Pathak, P. (2021). Growth projections against set-target of renewable energy and resultant impact on emissions reduction in India. *Environmental Engineering Research* 26(2), 200083.

Desai, S.N., Shah, M.A. (2018). Decentralized solid waste management in urban areas: a review. *International Journal of Current Engineering and Technology* 8, 21–23.

Dezhen, C., Lijie, Y., Huan, W., Pinjing, H. (2014). Pyrolysis technologies for municipal solid waste: a review. *Waste Management* 34, 2466–2486.

Environment-health-safety. (2019). https://www.ask-ehs.com/blog/environment/recycling-temple-waste-preserving/ (Accessed 25 January 2020).

Gomez, E., AmuthaRani, D., Cheeseman, C., Deegan, D., Wise, M., Boccaccini, A. (2009). Thermal plasma technology for the treatment of wastes: a critical review. *Journal of Hazardous Materials* 161, 614–626.

Govani, J., Patel, H., Pujara, Y., Chabhadiya, K., Pathak, P. (2020) Technological upgradation and advancement for treatment of plastic waste in developing countries. In: Hussain, C. (eds) *Handbook of Environmental Materials Management*. Springer, Cham, ISBN 978-3-319-58538-3.

Govani, J., Patel, H.T., Chabhadiya, K., Jadeja, U., Pathak, P. (2019). Transformation of industrial waste into alternate resource: a critical review. National Environmental Conference (NEC-2019), IIT Bombay.

Gupta, N., Yadav, K.K., Kumar, V. (2015). A review on current status of municipal solid waste management in India. *Environmental Research* 37, 206–217.

Havukainen, J., Zhan, M., Dong, J., Liikanen, M., Deviatkin, I., Li, X., Horttanainen, M. (2017). Environmental impact assessment of municipal solid waste management incorporating mechanical treatment of waste and incineration in Hangzhou, China. *Journal of Cleaner Production* 141, 453–461.

Jain, S., Jain, S., Wolf, I.T., Lee, J., Tong, Y.W. (2015). A comprehensive review on operating parameters and different pretreatment methodologies for anaerobic digestion of municipal solid waste. *Renewable and Sustainable Energy Reviews* 52, 142–154.

Joshi, R., Ahmed, S. (2016). Status and challenges of municipal solid waste management in India: a review. *Cogent Environmental Science* 2(1), 1139434.

Kalyani, K.A., Pandey, K.K. (2014). Waste to energy status in India: a short review. *Renewable and Sustainable Energy Reviews* 31, 113–120.

Kaza, S., Yao, L., Bhada-Tata, P., Van Woerden, F. (2018). What a waste 2.0: a global snapshot of solid waste management to 2050. World Bank. http://datatopics.worldbank.org/what-a-waste/ (Accessed 15 December 2019).

Konstantinos, M., Maria, L. (2010). Solid waste management through the application of thermal methods. *Waste Management*. ISBN: 978-953-7619-84-8. 255–285.

Kumar, S. (2010). Composting of municipal solid waste. *Critical Reviews in Biotechnology* 31(2), 112–136.

Kumar, S., Dhar, H., Nair, V.V., Rena, Govani, J., Arya, S., Bhattacharya, J.K., Vaidya, A.N., Akolkar, A.B. (2019). Environmental quality monitoring and impact assessment of solid waste dumpsites in high altitude sub-tropical regions. *Journal of Environmental Management* 252, 109681.

Kumar, S., Negi, S., Mandpe, A., Singh, R.V., Hussain, A. (2018). Rapid composting techniques in Indian context and utilization of black soldier fly for enhanced decomposition of biodegradable wastes—a comprehensive review. *Journal of Environmental Management* 227, 189–199.

Liu, T., Awasthi, M.K., Chen, H., Duan, Y., Awasthi, S.K., Zhang, Z. (2019). Performance of black soldier fly larvae (*Diptera: Stratiomyidae*) for manure composting and production of cleaner compost. *Journal of Environmental Management* 251, 109593.

Manu, M.K., Kumar, R., Garg, A. (2019). Decentralized composting of household wet biodegradable waste in plastic drums: effect of waste turning, microbial inoculum and bulking agent on product quality. *Journal of Cleaner Production* 226, 233–241.

Moore, C.J. (2008). Synthetic polymers in the marine environment: a rapidly increasing, long-term threat. *Environmental Research* 108(2), 131–139.

MSWM Rules, Municipal Solid Wastes Management (2000). http://www.moef.nic.in/legis/hsm/mswmhr.html (Accessed 04 January 2020).

Pathak, P., Gupta, D.K. (2020). Strontium contamination in the environment, volume series 88. In: *Handbook of Environmental Chemistry*, Springer Nature, Switzerland, ISBN 978-3-030-15314-4.

Pujara, Y., Pathak, P., Sharma, A., Govani, J. (2019). Review on Indian Municipal Solid Waste Management practices for reduction of environmental impacts to achieve sustainable development goals. *Journal of Environmental Management* 248, 109238.

Rehman, S., Hussain, Z. (2017). Renewable energy governance in India: challenges and prospects for achieving the 2022 energy goals. *Journal of Resources, Energy and Development* 14(1), 13–22.

Sharifah, A., Chun, Y., Muhamad, R., XiChen, M. (2013). Incineration of municipal solid waste in Malaysia: salient issues, policies and waste-to-energy initiatives. *Renewable & Sustainable Energy Reviews* 24, 181–186.

Sharma, K.D., Jain, S. (2018). Overview of municipal solid waste generation, composition, and management in India. *Journal of Environmental Engineering* 145(3), 04018143.

Siddique, M.N.I., Wahid, Z.A. (2018). Achievements and perspectives of anaerobic co-digestion: a review. *Journal of Cleaner Production* 194, 359–371.

Wang, H., Wang, L., Shahbazi, A. (2015). Life cycle assessment of fast pyrolysis of municipal solid waste in North Carolina of USA. *Journal of Cleaner Production* 87, 511–519.

Waste to Energy. (2016). World energy council (Accessed 19 January 2020). https://www.worldenergy.org/wp content/uploads/2017/03/WEResources_Waste_to_Energy_2016.pdf.

Yong-Chil, S., Md Tanvir, A., Won-Seok, Y. (2018). Chapter 7. Gasification of municipal solid waste. In: *Gasification for Low-Grade Feedstock*. 115–141.

Zerowastemanagement.org.in. (2020). http://www.zerowastemanagement.org.in/readmore.php (Accessed 30 January 2020).

9 Technological Innovations for Waste Management in China

Qian Li, Zhiyang Xu and Yunlong Zhou

CONTENTS

9.1 INTRODUCTION

As the world moves towards sustainable development, cities in China integrate cyberinfrastructure to foster all-around economic growth, including better quality of life and more efficient management of resources (Albino et al., 2015). The increasing quantity of waste and its management are major concerns to many cities in China, due to the ineffective and inefficient operations of waste collection and management (Jacobsen et al., 2018). At present, cities are increasingly focusing upon developing solutions to waste collection by using smart enabling technologies, such as the Internet, electronic metrology, global positioning systems (GPS), geographic information systems (GIS) and so on. Municipal solid waste (MSW) is a main component of solid waste, occupying the largest volume of residues produced worldwide. With the development of urbanization during the past 30 years in China, increasing burdens in MSW management are becoming a huge problem to the government. Thus, this chapter mainly pays attention to introducing the present situations of several treatment techniques applied in China.

Landfill and incineration are two methods applied widely in MSW treatment in China. As a traditional method, landfill is always the main choice for MSW, which is related to its relatively low cost both in facilities construction and in process run and the simple operation. However, this extensive disposal mode can't meet the needs of urbanization development and economic growth, which bring a series of problems such as environmental pollution of groundwater and the atmosphere

(Kalyani & Pandey, 2014). Several improvements to landfill are introduced in this chapter, mainly focused on promoting disposal efficiency, resolving environmental pollution and increasing stability. Compared with landfill, incineration is a treatment with higher disposal efficiency, which is reflected in the high removal rate and low production of residues. Moreover, the heat generated in the whole combustion process can be reused. Two main techniques of incineration are compared in this chapter. And the disposal of fly ash is listed, which has an advantageous impact on the atmospheric environment.

As a developing country, food waste (FW) dominates the composition of municipal solid waste (MSW), reaching up to more than 55% of the total amount. On one hand, this substrate with high moisture content is unsuitable for traditional disposal ways, which may bring about many problems like pollution of groundwater and the atmosphere in landfill and low fuel value in incineration. On the other hand, the excellent biodegradability of FW related to its high organic content results in huge potentialities on investment of economic value. Based on the FW features in China, the main techniques of the whole FW treatment process in China are introduced (e.g. pretreatment, anaerobic digestion, use of digestion residues and intermediate product). And the present situations of different techniques are analysed. This chapter tries to present the situation of solid waste treatment in China through introducing various techniques of different waste objects. We hope that this work can provide some references on the developing direction of solid waste treatment techniques in China.

9.2 NEW DIRECTION AND CONCEPT OF WASTE MANAGEMENT IN CHINA

9.2.1 Explore the Construction of a 'Zero-Waste City'

At present, China is the country with the largest population and solid waste production in the world. According to the data released by the Ministry of Ecology and Environment, 1.55 billion tons of solid wastes were generated in 202 large and medium-sized cities in China in 2017. In terms of waste management in China, the reduction and recycling of waste started early. In 2009, the Circular Economy Promotion Law of the People's Republic of China was issued. And in 2018, it was revised according to the social and economic development of China. In February 2018, Li Ganjie, the former Minister of Environmental Protection, said that the construction of a 'zero-waste city' would be speeded immediately at the National Environmental Protection Conference. In December 2018, Experimental work-plan of "zero-waste city" was printed and distributed by the General Office of the State Council of the People's Republic of China. On 29 April 2019, '11 + 5' pilot projects for the "zero-waste city" (Baotou in Inner Mongolia, Tongling in Anhui, Weihai in Shandong, Shengzhen in Guangdong, main district of Chongqing, Shaoxing in Zhejiang, Sanya in Henan, Xuchang in Henan, Xuzhou in Jiangsu, Panjing in Liaoning, Xining in Qinghai and Xiongan New Area in Hebei, Beijing Economic and Technology Development Area, Sino-Singapore Tianjin Eco-city, Guangze in Fujian, Ruijin in Jiangxi) were announced by the Ministry of Ecology and Environment. In terms of management scope, China's 'zero-waste city' construction plan is more comprehensive and includes both social and industrial source wastes. In terms of construction goals, China's 'zero-waste city' construction plan not only promotes the source reduction of solid waste, resource and harmless utilization of solid waste, but also establishes a 'zero waste' waste management quantitative index system to form a replicable and extendable construction model. Unlike developed countries or regions with relatively mature and complete waste management systems, China's waste management system still leaves room for practical progress. In terms of overall development goals, China needs to take the sustainable development of the economy and society into account. This is also the first consideration of other developing countries when implementing more stringent waste management strategies. China's 'zero-waste city' construction can provide other countries with valuable waste management transformation experience (Zaman & Lehmann, 2011).

9.2.2 Modern Information Management of Domestic Waste

In recent years, China's Internet technology, artificial intelligence and modern logistics technology have reached the world's most advanced level. Related technologies have been applied in the management of domestic waste in many cities in China, which have improved the efficiency of sanitation management. At present, the information management of the whole domestic waste collection process including transportation, disposal and metering and charging and the integrated application of the above-mentioned multiple technologies are gradually being explored. Taking Shanghai as an example, the domestic waste has a corresponding technical basis in the process of collection, transportation and treatment including the application of metrology technology to achieve the measurement of domestic garbage, the use of vehicle supervision technology to achieve the supervision and management of the logistics process and so on to realize information management of domestic waste. On this basis, China is further studying the technological integration of various M2M (Machine to Machine) such as RFID readers, GPS positioning, satellite positioning, vehicle measurement, pressure weighing and other sensor measurement. And information collection and technology management are conducted from the source of domestic waste collection and transportation. Collection data of waste classification such as transportation vehicles, garbage stations, transfer stations and various sorting and disposal sites is integrated systematically. Based on GPRS and 3G wireless data communication networks, large domestic garbage logistics management platforms are researched and built. In this way, each step of the whole process of domestic garbage collection and transportation is supervised. At the same time, the content, management process and system of waste classification are further studied. The construction of domestic waste source documentation, the quantitative assessment, the monitoring of the whole process of collection and transportation and so on all start from the management of the domestic waste source, which can provide data support for assisted decision-making on household waste (Zhang, 2015).

9.3 SITUATIONS OF SOLID WASTE MANAGEMENT IN CHINA

In 2016, the Ministry of Science and Technology, the Ministry of Finance and the State Administration of Taxation of the People's Republic of China revised the administrative measures for the identification of new and high-technology enterprises. In these new measures, a number of advanced and new technologies supported by the state were introduced, among which the new technologies related to solid waste treatment in the field of environmental protection are as follows.

As one of the four main solid wastes, municipal solid waste (MSW) is the largest volume of residues produced worldwide (Potdar et al., 2016). And the development of urbanization during the past 30 years has been putting increasing burdens on MSW management in China (Zhang et al., 2010)

The choice of treatment technologies in MSWs management is dependent on compositions of MSW and significantly. Differences in economic level and living customs among countries and regions influence the composition of MSWs considerably. In general, lower income countries have a higher percentage of organic waste than higher income countries. In China, MSWs are dominated by 55.86% food waste, which is a larger portion of organic content than the high-income countries (Hoornweg & Perinaz, 2012). The high percentage of organic waste leads to a negative effect on incineration energy recovery and proper landfilling, which is caused by the high moisture content in food waste, thus, demands high classification on MSW in China.

9.3.1 Improvement of Landfill

In China, landfill has been widely used as it involves relatively low construction and operating costs (Robinson A, 2003). Moreover, the quantity of landfill has been increasing continuously with the improving of quality over the period of time. From 2004 to 2014, the numbers of landfills increased by 36%, which provided nearly 56% growth of disposal by landfill. However, a series of problems are unavoidably caused by extensive disposal, such as groundwater pollution by landfill leachate and air pollution by greenhouse gasses. Combined with landfill disposal, different technologies have been applied in the treatment of these products.

The degradation of landfill substrates produces plenty of pollutants such as acidic and alkali (Kusterer et al., 1998) and other pathogens; the acidic condition enhances heavy metal leaching from the landfill site. Furthermore, improper leachate treatment from non-engineered landfill generated by biodegradation or rainfall causes serious surface- and groundwater contamination. It was reported that nearly 50% of landfills in China didn't treat leachate before discharge, which included 10% of landfill sites discharging leachate into the sewage system (Xu WL, 2006). Most other cities just treated leachate partially (Hui et al., 2006). Biological treatment is an effective way to resolve problems from landfill leachate, which can degrade organic compounds efficiently. However, a non-negligible fault of biological treatment is the negative separation between treated water and sludge in the final clarifier, which cannot satisfy the regulatory discharge standards. A membrane bioreactor connected to the water treatment system is a feasible way to solve this problem. Several projects in major cities of eastern China have implemented this technology in landfill leachate treatment, such as the waste leachate project of Wuxi Huilian waste incineration plant, the landfill leachate treatment project of Poly GCL (Xuzhou) Renewable Energy Power Generation Co., Ltd, the landfill leachate treatment project of Tianjin Dagang waste incineration power plant and the leachate treatment system of Shanghai Jinshan permanent domestic waste comprehensive treatment plant. An anaerobic membrane bioreactor (AnMBR) is a technology combining anaerobic digestion with membrane separation technology, which is going to be in the field of the treatment of landfill leachate in China as its various advantages (Ng et al., 2017). (i) The quantity of excess sludge is minimized due to the lower growth rate of microbes in anaerobic digestion than in aerobic processes, ii) the rate of organic matter removal and quality of effluent are promoted efficiently, which is related to the stability of microorganism concentration in anaerobic reactors intercepted by membrane, (iii) the replacement of the aerobic process with anaerobic digestion reduced treatment cost through the generation of energy from combustible gas and energy savings from aeration. The landfill leachate treatment project of Xishan waste transfer station in Wuxi Jiangsu province has implemented this technology.

Control of the gas recovery system is another improvement on MSW landfill disposal, which can generate energy and solve exhaust emission. Shuige landfill is the largest landfill of Nanjing, which has a GEF-funded landfill methane power plant with a capacity of 1,200 kW. In China, landfill gas-to-electricity projects have been operated in major cities (e.g. Hangzhou, Guangzhou, Nanjing, Xian, Beijing, Changsha, Wuxi and Jinan). In 2008, there were nearly 20 plants of landfill gas utilization all over the Chinese mainland, which achieved electrical power generation using landfill gas, with a total capacity of 40 MW. The electricity from landfill gas will be input into the electricity grid preferentially as a 0.25 Yuan RMB/kW, which is higher than the price of electricity from coal (Habil P, 2007).

9.3.2 Incineration in China

Along with the rapid urbanization, municipal solid waste (MSW) generation is increasing at annual rates of 8–10%, with more than 1.506108 t of MSW generated each year (Nie, 2008). A growing number of cities in China are facing the problem of MSW disposal. China will not be able to further dump raw MSW because of the lack of space for new landfills. Some municipal authorities already have great difficulty in finding appropriate landfill sites for MSW. Thus, an increasing number of cities have constructed or are planning to construct incineration plants, in spite of the significant capital and operating costs involved. MSW incineration is playing an increasingly important role in municipal solid waste management in China. Among MSW incineration plants in China, there are four major types of MSW incineration technologies, i.e. stoker, fluidized bed, rotary kiln and paralysis. Depending on actual conditions and specific demand, all those incineration systems have been constructed and operated in different areas of China. The stoker incineration system, with an average capacity of 500 t/d or high capacity of 1,000 t/d, needs a minimum calorific value of 6,000 to 6,500 kJ/kg, which is suitable for large and coastal areas with enough high calorific value MSW. The relatively lower capacity than stoker and lower demand for calorific value make fluidized bed systems adaptable to mid-sized cities. It is widely known that the thermal process in MSW incineration has the potential of emitting diverse types of air pollutants to the environment. Progress has been made on emission control though adding a secondary high-temperature afterburner, which can reduce the dioxin levels of the incinerator stack below an emission limit of 0.1 ng TEQ/Nm^3. Fly ash is another problem during incineration. Sintering technology has been developed by Tsinghua University to convert the MSWI fly ash into non-toxic material and make the decontaminated products safely reusable in construction materials (Nie, 2008). The results obtained from a pilot plant with a treatment capacity of 5 t/d of fly ash show that the decomposition rate of dioxin-like compounds during the sintering process was about 99.3%. The remained rate of the heavy metal in the sintered product depended on pretreatment of the fly ash. Water extraction pretreatment could remove at least 90% of chlorides from the MSWI fly ash, which are considered as the main factors controlling vaporization during the heating process. Another beneficial result of water-extraction is the significant increase in the content of SiO_2 in pretreated fly ash, contributing to strength forming in the sintering process.

9.3.3 Food Waste Treatment Technology

As a developing country, nearly 56% of China's MSW is food waste (Hoornweg & Perinaz, 2012), which means food waste management must be treated individually. This strategy of individual management of food waste deriving from other MSW can cut off more than 50% of MSW treatment using landfill and incineration. Moreover, the high moisture content in MSW caused by food waste can be resolved, which promotes the efficiency of landfill and incineration substantially.

Several modes of food waste treatment are widely used in China. Direct utilization as animal feed is a common food waste disposal method because of the high energy content generated from the oxidation of organic compounds in the feed (e.g. carbohydrates, lipids and proteins) (Lin et al., 2011). There are several methods for converting FW to an animal feed, such as boiling to produce a feed; producing a dry feed by dehydration, drying, disinfection and crushing in sequence; and biological treatment (Chen et al., 2014). The fact of huge breeding scale in agricultural country and the demand of a population of 1.4 billion for food result in a huge need for animal feed, which could be an effective reuse method for FW through its production. However, the complex sources and composition of FW may introduce potential biological security hazards from raw feed materials, such as homologous problems, pathogens, heavy metals and organic pollutants (Westendorf, 2000). Meanwhile, the differences of FW components caused by a diversity of dietary structures in China make it difficult to stabilize the quality of animal feeds, which limits the increase in animal feed production.

Composting is another suitable method of FW treatment, because of the high organic content of FW and minimal concern about heavy metals and pathogens in fertilizer production. Moreover, the huge amount of carbon in FW is immobilized in the soil though composting and fertilizing, which reduces GHG emissions (Cerda et al., 2018). Compared with the limitations placed on animal feeds by the Chinese government, FW composting has been attracting considerable attention in recent years in China. In addition, the high inorganic content in raw FW and poor quality of separation make it hard to produce animal feeds meeting requirements in China, which guide an attention transition to composting. At present, three composting routes are mainly applied in China, including solid phase FW composting with liquid phase after AD, aerobic composting and rapid aerobic composting. Similar with producing animal feed via FW, FW composting processes face several issues that influence the process efficiency and product quality. (i) A long duration is required to complete the composting process (Awasthi et al., 2014), which limits the disposal rate and production scale. (ii) The specific characteristics of FW (e.g. high moisture, acidity, high oil content and high salinity) (Chan et al., 2016; Sundberg, 2005) lead to frequent acidification during the whole process (Wang et al., 2018), which impacts on microbiological activity significantly (Sundberg & Jönsson, 2005). (iii) Although less common than with direct utilization as animal feed, heavy metal contamination still exists, which may eventually accumulate in the human body via the food chain (Siti Aminah A.M. et al., 2016). (iv) Similar to applications in animal food, the huge differences in FW components from different areas in China frequently change the composting conditions (Chang and Hsu, 2008), which makes process control difficult and leads to instability in fertilizer quality.

9.3.3.1 Pretreatments of FW

There are two main characteristics of food waste in China. (i) Relatively high content of inorganic substances, which is attributed to the low separation level of food waste. This problem will negatively impact further treatment both on bioprocess efficiency and stability of facility operations. (ii) Nonnegligible proportion of oil in food waste, which is related to Chinese dietary structures. On one hand, it is a disadvantage to bio-contact between microbes and substrates; on the other hand, the considerable content of oil in food waste is worth recycling. As above, before further processing, a serial process of pretreatment of food waste is necessary in China.

9.3.3.1.1 Mechanical Separation

Although source separation of MSW has been introduced in some prosperous cities in China, such as Beijing Shanghai and Hangzhou, both classification of the waste and compliance are of low quality and this activity is not well-established in China (Zhang et al., 2010). Mechanical separation is applied at the beginning of whole food waste treatment process, which is aimed to remove the inorganic compounds in food waste (e.g. bones, plastics and metals).

9.3.3.1.2 Thermal Pretreatment

Thermal pretreatment is usually applied after mechanical separation to achieve the effects of sterilization (Chen et al., 2015), homogenization and solubilization of organic matters so that subsequent biological processes can be enhanced (Jin et al., 2016). Meanwhile, both separation and extraction can be enhanced (Li et al., 2016). Most thermal pretreatment conditions are controlled between a temperature range of 95 to 120°C with a duration time more than 20 min. A wider range of treatment temperature was set-up with a wider range of duration time to investigate the effects of organic solubilization of FW and subsequent AD efficiency. It is reported that 120°C is the upper limit to ensure disposal effects. Compared with biological methods and mechanical pretreatment, thermal pretreatment at low temperatures (less than 110°C) is proved as the most sustainable method for enhancing AD processes (Ariunbaatar et al., 2014). In consideration of the factor of running cost, finding a low pretreatment temperature is necessary. Actually, many FW treatment plants have been trying to lower pretreatment temperatures to reduce running cost (Shanghai, Xuzhou, Tianjin, Zhengzhou and Chongqing), which obtains the ideal effect on sterilization, homogenization, and solubilization of organic matters. Thermal pretreatment may be more effective than other pretreatment methods for enhancing the efficiency of AD, oil-liquid separation efficiency and the sterilization rate (Jin et al., 2016; Li et al., 2016).

Three-phase separation is an essential process for facilitating the use of split waste phases, such as a liquid phase for AD, a solid phase for fertilizer and an oil phase for biodiesel. The oil phase separated from FW has been widely used for the production of biodiesel. Due to China having limited crop land and the world's largest population, it is important for the country to develop non-food-based feedstocks for biodiesel production instead of only oil produced from crops for biodiesel production (Li et al., 2011). Moreover, oil phase from FW was recycled into kitchens, which threatened people's health significantly. Utilization for biodiesel is an effective solution to this problem (Figure 9.1).

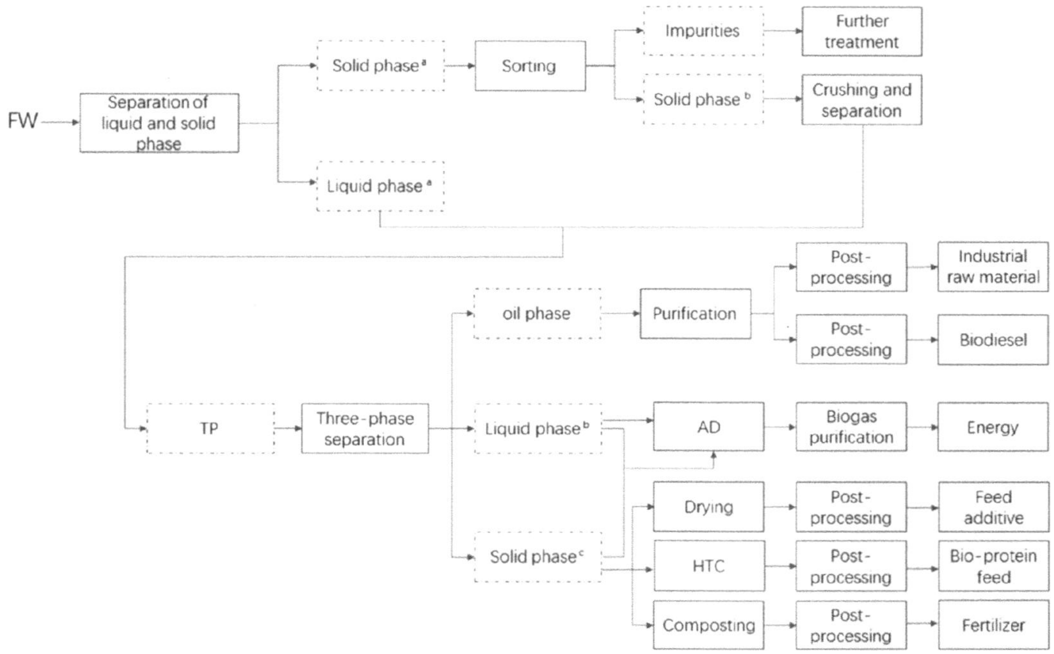

FIGURE 9.1 Main technical routes of FW treatment in China (Li, Y., et al., 2019).

9.3.3.2 Anaerobic Digestion (AD)

Anaerobic digestion (AD) has been widely applied in China, and is the most commonly used approach in terms of number of facilities and the overall capacity of FW treatment facilities (Li et al., 2019), accounting for 76.1% of all treatment capacities (as of September 2015) (Wei Xiaofeng, 2016). According to the difference of substrate, AD treatment of FW is classified as whole material digestion (WMD) and liquid digestion (LD). WMD means that all of the materials after thermal pretreatment (both solid and liquid) participate in anaerobic digestion. And LD means that only the liquid part of material after thermal pretreatment participates in anaerobic digestion, and the solid part enters further processing to other products (e.g. animal feed and fertilizer). Relative to LD, the effect of limited and inhibited in WMD process is significant caused by the solid of FW, which dominates the organic portion. (i) Studies on digestion temperature demonstrate a series of advantages associated with thermophilic digestion, such as faster reaction rates (Sánchez et al., 2001), more biogas production (El-Mashad et al., 2004), less inhibition by ammonia accumulation (Gallert et al., 1998), higher rates of pathogen destruction (Kim et al., 2006) and improved solid separation (Kaparaju & Angelidaki, 2008) in comparison to mesophilic digestion. However, the sensitivity to changes in environmental conditions of the thermophilic process (De la Rubia et al., 2006; Kim et al., 2002) can lead to instability of the process. Therefore, a common utilization of mesophilic digestion due to a higher overall solubilization rate (Komemoto et al., 2009). (ii) Additionally, low methane yields and process instability are two of the main problems under the WMD process, which are caused by inhibition from excessive accumulation of free ammonia (NH_3) (Yenigün & Demirel, 2013) and long-chain fatty acids (LCFA) (Palatsi et al., 2010; Rinzema et al., 1989). These two problems can crush the whole process under running conditions of high organic loading rates. Different strategies were implemented, running with low organic loading rates or co-digestion of waste with low nitrogen and lipid content (e.g. sewage sludge or MSW) (Pham et al., 2015; Xu et al., 2018; Zhang et al., 2014), to overcome these problems are hard to improve rate of FW treatment. The utilization of two-stage digestion in FW treatment proved the effectivity of high organic loading rates with higher methane generation than single-stage digestion processes (Ren et al., 2018; Uçkun Kiran et al., 2014), which increased investment and running costs apparently. With

WMD it seems hard to achieve both effective disposal of FW and low cost at the same time. LD is another attempt at AD in FW disposal in which the solid part is transported to further processing without participating in AD. Most organics in FW are used in other ways, which can lighten the burden on the AD process caused by high organic loading rates. Currently, single-stage anaerobic digesters with continuously stirred tank reactors (CSTR) have mainly been applied in Chinese FW treatment plants with LD. It should be noted that the solids of FW can be utilized partially in LD, which aims to adjust methane yield. This adjustment may result in a long hydraulic retention time, generally in the range of 20–40 days (Pham et al., 2015). In addition, there may be some variation in methane yield with the same capacity and technical routes due to differences in the organic composition of the FW, ranging from 119 to 780 mL/g VS with a VS reduction of 80–90% as reported in previous studies (De Clercq et al., 2016; Kim et al., 2006; Shahriari et al., 2013; Zhang et al., 2013).

9.3.3.3 Disposal of Digestion Residues

Digestion residues include solid phase and liquid phase, which consist of microbial biomass and semi-degraded organics and inorganics. Nutrients content in the digestion residues are abundant such as nitrogen, phosphorus, potassium, calcium, and magnesium for macro nutrients; and boron, zinc, iron, manganese, molybdenum, nickel and boron for micro nutrients or trace elements. This feature makes digestion residues a raw material of valuable solid amendment or organic fertilizer (Nkoa, 2014), which has wider utilization prospects than chemical fertilizers due to its environmental and economic benefits (Lukehurst, 2010; Owamah et al., 2014). Effective degradation of FW during anaerobic digestion increases absorption and decreases odours and percolates, which reduces possible risks of air pollution and leachate pollution. The utilization of digestion residues as organic fertilizer in China is suitable for the WMD process via solid phase and liquid phase, which can provide assurance of the quality and quantity of fertilizer. The digestion mode that takes the liquid phase from the FW as the only back flow of AD can provide liquid digestate as liquid fertilizer. Traditionally treatment of the liquid digestate was focused on pollutant degradation as wastewater treatment and discharging after meeting the standard (Integrated Wastewater Discharge Standard (GB 8978-1996)). Apparently, liquid digestate used as fertilizer is more economically beneficial than waste treatment. Agricultural benefits and environmental risks are two of the main considerations regarding the reuse of digestion residues (either solid or liquid). Some FW treatment facilities use both the liquid phase and minor amounts of solid phase FW as feedstock for AD; in this case, digestion residues were separated into solid phase and liquid phase (e.g. dehydration) used as solid fertilizer and liquid fertilizer respectively. Although lacking further research on effective use of digestion residues from AD of FW as an organic amendment and fertilizer (Möller & Müller, 2012; Nkoa, 2014; Teglia et al., 2011), plenty of technologies aimed at digestate processing and nutrient recovery have been attempted, such as phosphorus precipitation and crystallization, ammonia stripping and absorption, acidic air scrubbing, biomass production and harvesting (Vaneeckhaute et al., 2017). These attempts provide cheerful prospects of widespread application in China. Stability of the quantity and quality of digestion residues has been increasingly emphasized, which is strongly related to their wide utilization as fertilizer. For the reason of content fluctuations between batches from the same digester and even the same batch, keeping the levels of heavy metal content low is also important to ensure the quality of digestion residues as fertilizer. Quality management is a trend of ensuring the safety and beneficial value of digestion residues when used as fertilizers, soil conditioners or as a growing medium, which involves permits and quality control standards. An increasing number of AD facilities have been paying attention of the storage and application of the digestion residues, that must comply with codes of good agricultural practice and follow national guidelines and legislation.

9.3.3.4 Resource Utilization of Solid Phase Separated from FW

The solid phase separating from FW dominant proportion of all nutrients. For the whole FW management, the effective utilization of solid phase from three-phase separation is key for economic benefit increase.

Animal feed is a feasible method for solid phase utilization in China. Compared with direct use of FW, the solid phase is more suitable for animal feed, which provides quality and security through a series of treatments, involving classification (to remove inorganics), thermal pretreatment (to kill microbes) and separation (to reduce moisture content). Some other ways utilizing solid phase as animal feedstuff have been spread recently. Different insects are fed, which can use organic waste for growth (Diener et al., 2011a, b; Newton, 2005; Zheng et al., 2012).

In addition, the utilization of solid phase using insects can be harvested for biodiesel production. It was reported that the fuel properties of biodiesel derived from black soldier flies, such as the density, viscosity and cetane number, are close to those of biodiesel from rapeseed (Li et al., 2011). Many FW treatment plants with small capacities in China have been focused on this treatment, which has shown effective reduction of solid phase. However, several factors prevent the efficiency of reduction ratio, such as low egg hatching and high larval mortality related to heavy metals, moisture tolerance and process optimization (Diener et al., 2011b).

9.3.4 Other Waste Treatments

9.3.4.1 Hazardous Solid Waste Disposal Technology

This includes efficient incineration of hazardous solid wastes; disposal technology of incineration slag, fly ash, sintering ash and calcination ash; transportation and safe landfill disposal of hazardous solid wastes; curing technology of hazardous wastes; collection, transportation and disposal technology of clinical wastes; comprehensive utilization technology of toxic and harmful solid wastes; disposal technology of radioactive solid wastes and so on.

9.3.4.2 Industrial Solid Waste Comprehensive Utilization Technology

This includes harmlessness, reduction, resource and comprehensive utilization technology of industrial solid waste.

9.3.4.3 Construction Waste Disposal and Recycling Technology

This includes construction waste classification and recycling technology; recycling construction waste key technology; new application technology of recycled building materials; key technologies for the preparation of recycled concrete and its products; key construction technology of recycled concrete and its products; application technology of recycled inorganic materials in road engineering and so on.

9.3.4.4 Disposal and Recycling Technology of Solid Waste from Social Sources

This includes waste battery, waste electrical and electronic equipment, waste plastic, and other social sources of solid waste harmless disposal and recycling technology, etc.

9.4 PERSPECTIVES

Solid waste management is a systematic project. A complete solid waste collection and classification system is the basis of technique utilization for solid waste treatment. During solid collection (especially for the FW collection in MSW), the level of waste classification affects treatment techniques and disposition effect directly. Thus, actions need to be taken in China to enhance the quantity and quality of solid waste, including constructing different collection systems for different waste groups to prevent mixing. Furthermore, people's awareness of waste classification should be advanced through a serious of policies and measures. Solid waste source separation has different levels of difficulty, while suitable treatment methods should be variable. Both diversification of technical processes for the same waste treatment and synergism among different waste treatments are necessary, which can achieve wider adaptability and better treatment effects under different conditions. During waste treatment processes, some countermeasures need to be taken to advance innovation of waste treatment techniques, such as legislated pretreatment methods, diversified treatment capacities and technical routes, and the advanced use of end products and so on.

REFERENCES

Albino, V., Berardi, U., Dangelico, R. 2015. Smart cities: definitions, dimensions, performance, and initiatives. *Journal of Urban Technology*, 22, 2015.

Ariunbaatar, J., Panico, A., Esposito, G., Pirozzi, F., Lens, P.N.L. 2014. Pretreatment methods to enhance anaerobic digestion of organic solid waste. *Applied Energy*, 123, 143–156.

Awasthi, M.K., Pandey, A.K., Khan, J., Bundela, P.S., Wong, J.W.C., Selvam, A. 2014. Evaluation of thermophilic fungal consortium for organic municipal solid waste composting. *Bioresource Technology*, 168, 214–221.

Cerda, A., Artola, A., Font, X., Barrena, R., Gea, T., Sánchez, A. 2018. Composting of food wastes: status and challenges. *Bioresource Technology*, 248, 57–67.

Chan, M.T., Selvam, A., Wong, J.W.C. 2016. Reducing nitrogen loss and salinity during 'struvite' food waste composting by zeolite amendment. *Bioresource Technology*, 200, 838–844.

Chang, J.I., Hsu, T. 2008. Effects of compositions on food waste composting. *Bioresource Technology*, 99, 8068–8074.

Chen, T., Jin, Y., Qiu, X., Chen, X. 2014. A hybrid fuzzy evaluation method for safety assessment of food-waste feed based on entropy and the analytic hierarchy process methods. *Expert Systems with Applications*, 41, 7328–7337.

Chen, T., Jin, Y., Shen, D. 2015. A safety analysis of food waste-derived animal feeds from three typical conversion techniques in China. *Waste Management*, 45, 42–50.

De Clercq, D., Wen, Z., Fan, F., Caicedo, L. 2016. Biomethane production potential from restaurant food waste in megacities and project level-bottlenecks: a case study in Beijing. *Renewable and Sustainable Energy Reviews*, 59, 1676–1685.

De la Rubia, M., Romero Garcia, L., Sales, D., Perez, M. 2006. Pilot-scale anaerobic thermophilic digester treating municipal sludge. *AIChE Journal*, 52, 402–407.

Diener, S., Studt Solano, N.M., Roa Gutiérrez, F., Zurbrügg, C., Tockner, K. 2011a. Biological treatment of municipal organic waste using black soldier fly larvae. *Waste and Biomass Valorization*, 2(4), 357–363.

Diener, S., Zurbrügg, C., Roa-Gutiérrez, F., Hong Dang, N., Morel, A., Koottatep, T., Tockner, K. 2011b. Black soldier fly larvae for organic waste treatment - prospects and constraints. In: *Proceedings of the waste safe — 2nd international conference on solid waste management in the developing countries*. Khulna, Bangladesh.

El-Mashad, H.M., Zeeman, G., van Loon, W.K.P., Bot, G.P.A., Lettinga, G. 2004. Effect of temperature and temperature fluctuation on thermophilic anaerobic digestion of cattle manure. *Bioresource Technology*, 95(2), 191–201.

Gallert, C., Bauer, S., Winter, J. 1998. Effect of ammonia on the anaerobic degradation of protein by a mesophilic and thermophilic biowaste population. *Applied Microbiology and Biotechnology*, 50(4), 495–501.

Habil, P.R.B. 2007. Renewable energy for rural areas in China. *GTZ Training*, ICEEE, Shenyang.

Hoornweg, D., Perinaz, B.T. 2012. What a waste: a global review of solid waste management. *Urban Dev Ser Knowl Pap*, 15, 87–88.

Hui, Y., Li'ao, W., Fenwei, S., Gang, H. 2006. Urban solid waste management in Chongqing: challenges and opportunities. *Waste management (New York, N.Y.)*, 26, 1052–1062.

Jacobsen, R., Willeghems, G., Gellynck, X., Buysse, J. 2018. Increasing the quantity of separated post-consumer plastics for reducing combustible household waste: the case of rigid plastics in flanders. *Waste Management*, 78, 708–716.

Jin, Y., Li, Y., Li, J. 2016. Influence of thermal pretreatment on physical and chemical properties of kitchen waste and the efficiency of anaerobic digestion. *Journal of Environmental Management*, 180, 291–300.

Kalyani, K.A., Pandey, K.K. 2014. Waste to energy status in India: a short review. *Renewable and Sustainable Energy Reviews*, 31, 113–120.

Kaparaju, P., Angelidaki, I. 2008. Effect of temperature and active biogas process on passive separation of digested manure. *Bioresource Technology*, 99(5), 1345–1352.

Kim, J.K., Oh, B.R., Chun, Y.N., Kim, S.W. 2006. Effects of temperature and hydraulic retention time on anaerobic digestion of food waste. *Journal of Bioscience and Bioengineering*, 102(4), 328–332.

Kim, M., Ahn, Y.-H., Speece, R.E. 2002. Comparative process stability and efficiency of anaerobic digestion; mesophilic vs. thermophilic. *Water Research*, 36(17), 4369–4385.

Komemoto, K., Lim, Y.G., Nagao, N., Onoue, Y., Niwa, C., Toda, T. 2009. Effect of temperature on VFA's and biogas production in anaerobic solubilization of food waste. *Waste Management*, 29(12), 2950–2955.

Kusterer, T., Willson, R., Bruce, S., Tissue, E., Lau, P. 1998. Municipal landfill Leachate management. *Environmental Progress*, 17, 278–284.

Li, Q., Zheng, L., Cai, H., Garza, E., Yu, Z., Zhou, S. 2011. From organic waste to biodiesel: black soldier fly, Hermetia illucens, makes it feasible. *Fuel*, 90(4), 1545–1548.

Li, Y., Jin, Y., Borrion, A., Li, H. 2019. Current status of food waste generation and management in China. *Bioresource Technology*, 273, 654–665.

Li, Y., Jin, Y., Li, J. 2016. Enhanced split-phase resource utilization of kitchen waste by thermal pre-treatment. *Energy*, 98, 155–167.

Lin, C., Wu, E., Lee, C.-N., Kuo, S. 2011. Multivariate statistical factor and cluster analyses for selecting food waste optimal recycling methods. *Environmental Engineering Science*, 28, 349–356.

Lukehurst, C.T., Frost, P., Al Seadi, T. 2010. Utilisation of digestate from biogas plants as biofertilise. *IEA bioenergy*, 1–36.

Möller, K., Müller, T. 2012. Effects of anaerobic digestion on digestate nutrient availability and crop growth: a review. *Engineering in Life Sciences*, 12, 242–257.

Newton, L., Sheppard, C., Watson, D.W., Burtle, G., Dove, R. 2005. Using the black soldier fly, Hermetia illucens, as a value-added tool for the management of swine manure. In: *Report for mike williams, director of the animal and poultry waste management center.* North Carolina State University.

Ng, C.A., Wong, L.Y., Yee, C.H., Bashir, M.J.K., Ho, C.-D., Nisar, H., Lo, P.K. 2017. Investigation on the performance of hybrid anaerobic membrane bioreactors for fouling control and biogas production in palm oil mill effluent treatment. *Water Science & Technology*, 76, 1389.

Nie, Y. 2008. Development and prospects of municipal solid waste (MSW) incineration in China. *Frontiers of Environmental Science & Engineering in China*, 2, 1–7.

Nkoa, R. 2014. Agricultural benefits and environmental risks of soil fertilization with anaerobic digestates: a review. *Agronomy for Sustainable Development*, 34(2), 473–492.

Owamah, H.I., Dahunsi, S.O., Oranusi, U.S., Alfa, M.I. 2014. Fertilizer and sanitary quality of digestate biofertilizer from the co-digestion of food waste and human excreta. *Waste Management*, 34(4), 747–752.

Palatsi, J., Illa, J., Prenafeta-Boldú, F.X., Laureni, M., Fernandez, B., Angelidaki, I., Flotats, X. 2010. Long-chain fatty acids inhibition and adaptation process in anaerobic thermophilic digestion: batch tests, microbial community structure and mathematical modelling. *Bioresource Technology*, 101(7), 2243–2251.

Pham, T.P.T., Kaushik, R., Parshetti, G.K., Mahmood, R., Balasubramanian, R. 2015. Food waste-to-energy conversion technologies: current status and future directions. *Waste Management*, 38, 399–408.

Potdar, A., Singh, A., Unnikrishnan, S., Naik, N., Naik, M., Nimkar, I., Patil, V. 2016. Innovation in solid waste management through clean development mechanism in developing countries. *Procedia Environmental Sciences*, 35, 193–200.

Ren, Y., Yu, M., Wu, C., Wang, Q., Gao, M., Huang, Q., Liu, Y. 2018. A comprehensive review on food waste anaerobic digestion: research updates and tendencies. *Bioresource Technology*, 247, 1069–1076.

Rinzema, A., Alphenaar, A., Lettinga, G. 1989. The effect of lauric acid shock loads on the biological and physical performance of granular sludge in UASB reactors digesting acetate. *Journal of Chemical Technology & Biotechnology*, 46, 257–266.

Robinson, A., Wu, S., Damodaran, N., Kalas Adams, N. 2003. *Landfill data from China: addressing information needs for methane recovery.* Environmental Protection Agency.

Sánchez, E., Borja, R., Weiland, P., Travieso, L., Martín, A. 2001. Effect of substrate concentration and temperature on the anaerobic digestion of piggery waste in a tropical climate. *Process Biochemistry*, 37(5), 483–489.

Shahriari, H., Warith, M., Hamoda, M., Kennedy, K. 2013. Evaluation of single vs. staged mesophilic anaerobic digestion of kitchen waste with and without microwave pretreatment. *Journal of Environmental Management*, 125, 74–84.

Siti Aminah, A.M., Sharifah, N.S.I., Sarva, M.P. 2016. Application of effective microorganism (EM) in food waste composting: a review. *Asia Pacific Environmental and Occupational Health Journal*, 2, 37–47.

Sundberg, C. 2005. *Improving compost process efficiency by controlling aeration, temperature and pH.* Swedish University of Agricultural Sciences.

Sundberg, C., Jönsson, H. 2005. Process inhibition due to organic acids in fed-batch composting of food waste – influence of starting culture. *Biodegradation*, 16(3), 205–213.

Teglia, C., Tremier, A., Martel, J.L. 2011. Characterization of solid digestates: part 1, review of existing indicators to assess solid digestates agricultural use. *Waste and Biomass Valorization*, 2(1), 43–58.

Uçkun Kiran, E., Trzcinski, A.P., Ng, W.J., Liu, Y. 2014. Bioconversion of food waste to energy: a review. *Fuel*, 134, 389–399.

Vaneeckhaute, C., Lebuf, V., Michels, E., Belia, E., Vanrolleghem, P.A., Tack, F.M.G., Meers, E. 2017. Nutrient recovery from digestate: systematic technology review and product classification. *Waste and Biomass Valorization*, 8(1), 21–40.

Wang, X., Selvam, A., Lau, S.S.S., Wong, J.W.C. 2018. Influence of lime and struvite on microbial community succession and odour emission during food waste composting. *Bioresource Technology*, 247, 652–659.

Wei Xiaofeng, S.W., Guanping, W., Wenpeng, P., Xuesen, J., Qin, K. 2016. Situation of kitchen waste treatment market in China. *Environmental Sanitation Engineering*, 24, 28–30.

Westendorf, M. 2000. *Food waste to animal feed*. John Wiley & Sons.

Xu, F., Li, Y., Ge, X., Yang, L., Li, Y. 2018. Anaerobic digestion of food waste-Challenges and opportunities. *Bioresource Technology*, 247, 1047–1058.

Xu, W.L., Higuchi, L.J.S. 2006. Review and forecast on MSW treatment and landfill technology in China. http://www.cis.fukuokau.ac.jp/*higuchis/research/h18/H18.happyou%201.pdf. Accessed 8 May 2011.

Yenigün, O., Demirel, B. 2013. Ammonia inhibition in anaerobic digestion: a review. *Process Biochemistry*, 48(5–6), 901–911.

Zaman, A.U., Lehmann, S. 2011. Urban growth and waste management optimization towards 'zero waste city'. *City, Culture and Society*, 2(4), 177–187.

Zhang, C., Su, H., Baeyens, J., Tan, T. 2014. Reviewing the anaerobic digestion of food waste for biogas production. *Renewable and Sustainable Energy Reviews*, 38, 383–392.

Zhang, C., Su, H., Tan, T. 2013. Batch and semi-continuous anaerobic digestion of food waste in a dual solid–liquid system. *Bioresource Technology*, 145, 10–16.

Zhang, D.Q., Tan, S.K., Gersberg, R.M. 2010. Municipal solid waste management in China: status, problems and challenges. *Journal of Environmental Management*, 91(8), 1623–1633.

Zhang, J. 2015. Management technology of modern domestic waste logistics information. *Environmental Sanitation Engineering*, 4, 43–47.

Zheng, L., Li, Q., Zhang, J., Yu, Z. 2012. Double the biodiesel yield: rearing black soldier fly larvae, Hermetia illucens, on solid residual fraction of restaurant waste after grease extraction for biodiesel production. *Renewable Energy*, 41, 75–79.

10 Solid Waste Management in Brazil

Policies, Strategies, Treatment Technologies and Disposal Methods

Luíza Santana Franca, Ingrid R.F.S. Alves, Isabelli Dias Bassin, Erimar Pontes Santiago and João Paulo Bassin

CONTENTS

LIST OF SYMBOLS AND ACRONYMS

ABNT	Brazilian Association of Technical Standards
Ancat	Association of Waste Pickers and Recyclers
CEMPRE	Business Commitment to Recycling
CH_4	methane
CO_2	carbon dioxide
CSTR	continuous stirred tank reactors
H_2S	hydrogen sulphide
GHG	greenhouse gases
LHV	lower heating value
MBT	mechanical biological treatment
MSW	municipal solid waste
NGOs	non-governmental organizations
NO_x	oxidized nitrogen species
NSWP	National Solid Waste Policy
OECD	Organization for Economic Co-operation and Development
PET	polyethylene terephthalate

PP	polypropylene
PROBIOGÁS	Brazil-Germany Project for the Promotion of Energy Use of Biogas in Brazil
PVC	polyvinyl chloride
RDC	residue-derived compound
SCP	sorting and composting plants
Sisnama	National Environment System
SO_4^{2-}	sulphate
SNVS	National Health Surveillance System
TS	total solids
UCY	urban composting yards
VFA	volatile fatty acids
WDP	waste delivery points
WTE	waste to energy
WWTP	wastewater treatment plants

10.1 BRAZILIAN NATIONAL SOLID WASTE POLICY, TECHNICAL STANDARDS AND NEW BASIC SANITATION FRAMEWORK

The purpose of solid waste management is to systematize all stages from the generation to the final destination of the waste, considered as the end of the chain. According to the Brazilian Law 12,305, which instituted the National Solid Waste Policy (NSWP) in Brazil in 2010 (BRASIL, 2010), solid waste management is defined as a set of actions taken, directly or indirectly, in the stages of collection and transport, transfer, treatment and environmentally appropriate final disposal of solid waste, in accordance with the municipal plan for integrated management of solid waste. In this sense, the NSWP has established a series of instruments for better waste management, working to face the main environmental, social and economic problems arising from the inadequate management of solid waste.

In this context, the NSWP determines guidelines for the prevention and reduction of waste generation through conscious consumption, promoting recycling, reuse and the environmentally appropriate disposal of waste. Among the main mechanisms of the NSPW are shared responsibility, solid waste management plan, reverse logistics and selective collection. All these are detailed as follows:

- Shared responsibility: Attributes to each member of the production chain and stakeholders of waste management share the responsibility for the environmentally appropriate waste management activities. The waste generator, the transportation and final destination managers, therefore, are co-responsible and may suffer sanctions from the municipal government if they do not guarantee the destination to suitable locations
- Solid waste management plan: Document created based on social participation, which contains goals and strategies for waste management. The federal government, states, municipalities and companies must have plans that work with integrated waste management
- Reverse logistics: Correspond to a set of actions that aim to enable the collection and refund of certain waste from the business sector. Thus, companies that produce packaging, medicines, batteries, electrical and electronic products, lamps, lubricating oils and tires must collect their waste and be responsible for the correct final destination
- Selective collection: Depending on the possibility of recycling or reuse, waste must be segregated by the generator itself, thus enabling greater awareness among citizens when generating waste

The municipality must promote the contracting of services for the collection, storage, transportation, transfer, treatment and final disposal of solid waste. However, this does not exempt the liability of individuals or legal entities for eventual improper disposal.

The Brazilian Association of Technical Standards (ABNT) has 23 Standards (NBR) related to the generation, classification, management, treatment and final destination of waste from different sources, which include urban, industrial, civil construction, electronics and health services solid waste. Although the current legal provisions contemplate aspects that largely involve waste management there is still a high number of Brazilians without access to basic sanitation services, such as waste (51.2 million) and sewage (99.7 million) collection and drinking water supply (34.6 million).

Established in 2020, the new Regulatory Framework for Basic Sanitation in Brazil (Brazilian Law No. 14.026) (BRASIL, 2020) provides for the universalization of water and sewage services by 2033 and makes it possible to inject more private investments in sanitation services. This law extends the deadline for the end of the dumping sites, until 2024 for municipalities that developed municipal solid waste management plans, and until the end of 2020 for municipalities that did not develop plans. Moreover, the law also facilitates the privatization of state-owned companies in the sector and extinguishes the current contract model between municipalities and state water and sewage companies. The new regulatory framework transforms the current contracts into concessions with the private company that takes over the state-owned company. The text also makes it mandatory to open a bidding process for sanitation services, involving public and private companies.

10.2 SOLID WASTE MANAGEMENT STRATEGIES: FOCUS ON THE BRAZILIAN CONTEXT

According to the European Directive 2008/98/EC, currently amended to Directive (EU) 2018/851, and which was adapted to Brazilian legislation by the NSWP, the management of solid waste must be planned and organized to comply with the entire hierarchy management, making the involved actors responsible throughout the stages. Figure 10.1 presents all management steps, which will be detailed next.

Even before its generation, it is possible to assess whether there is a need to generate a certain waste. From there, it is evaluated whether the amount of waste generated can be minimized and then assess whether this material can be reused, considering the same function as a product, or receive a new function as another type of product. Subsequently, it is evaluated whether the materials can be sent to economically viable and environmentally appropriate treatment processes so that the waste is revalued and may become an input or serve as an energy source for a process.

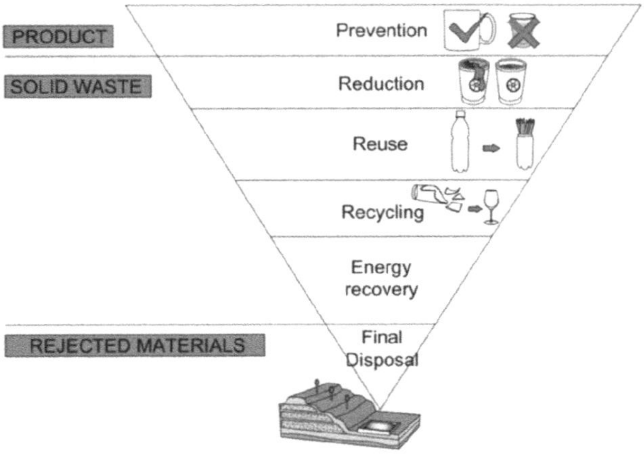

FIGURE 10.1 Hierarchy of waste management in European legislation. Source: Based on World Bank (2012).

FIGURE 10.2 Waste management steps.

The last option of the solid waste management stage is its final destination, and only what is considered to be a rejected material should be sent to be disposed of, that is, only the solid waste that, after all the treatment and recovery by available and economically viable technological processes, exhibits no other possibility than environmentally appropriate final disposal (Brazilian Law 12,305/2010).

Considering this hierarchy, the solid waste passes through the following stages: generation, collection and transportation, sorting, treatment, transfer and final disposal (Figure 10.2). Therefore, solid waste management characterizes, systematizes and plans all actions related to each stage of this waste chain, starting with the generation stage. This first stage is the moment when the generation of waste is evaluated, which is the demand for the service of waste management. For this, it is possible to quantify and qualify the heterogeneous mass of solid waste generated and characterize it according to its physical, chemical and biological characteristics (IBAM, 2001). The quantification of the waste generated will depend on each type of generator.

As for the quality of the waste generated, the physical characteristics are related to the type of material present, which can be determined by gravimetric characterization (Silva et al., 2008), and to the structure and physical behaviour of the particles of the waste aggregate mass, such as particle size, compressibility and shear stress between them. The chemical composition, on the other hand, is associated with the chemical compounds present in the heterogeneous mass of residues or individually present in each fraction of material, being related to chemical properties such as combustibility and corrosion. Finally, biological characteristics refer to the presence of an existing microbial community, and its characterization will be based on the quantification and identification of microorganisms that can be made at different taxonomic levels (from domain to species).

Considering the generation of waste in an urban environment, according to the Organization for Economic Co-operation and Development (OECD) classification, municipal solid waste (MSW) is that which is collected and treated under the responsibility of the municipalities (World Bank, 2012). This group includes residues generated by residences, commercial establishments, gardening and tree pruning and public streets cleaning, as well as industrial and construction and demolition residues. In addition to municipal waste, the OECD classification also distinguishes other groups of solid waste, such as industrial, medical and agricultural waste (World Bank, 2012).

According to Brazilian legislation (BRASIL, 2010 and ABNT, 2004a), solid waste can be classified in two ways: According to the origin and according to its hazard potential, which is related to its physical and chemical characteristics.

- **As to origin:**

a) household waste: those originating from domestic activities in urban residences;
b) urban cleaning residues: those originating from sweeping, cleaning of streets and public roads and other urban cleaning services;
c) solid urban waste: those included in items ‘a’ and ‘b’;
d) wastes from commercial establishments and service providers: those generated in these activities, except those referred to in items ‘b’, ‘e’, ‘g’, ‘h’ and ‘j’;
e) wastes from public basic sanitation services: those generated in these activities, except those referred to in item ‘c’;
f) industrial wastes: those generated in production processes and industrial installations;
g) waste from health services: those generated in health services, as defined in regulations or standards established by the National Environment System (Sisnama) and the National Health Surveillance System (SNVS);

h) civil construction residues: those generated in the construction, renovation, repair, and demolition of civil construction works, including those resulting from the preparation and excavation of land for civil works;
i) agriculture and cattle residues: those generated in agricultural and cattle activities, including those related to inputs used in these activities;
j) waste from transport services: those originating from ports, airports, terminals, roadways and railways, and border crossings;
k) mining residues: those generated in the activity of exploration, extraction, or processing of ores.

- **As for hazardousness characteristics:**

a) Class I waste – Hazardous: those that, due to their characteristics of flammability, corrosivity, reactivity, toxicity, pathogenicity, carcinogenicity, teratogenicity, and mutagenicity, present a significant risk to public health or environmental quality, according to the law, regulation or technical standard;
b) Class II waste – Non-hazardous: non-inert waste (class II A) and inert waste (class II B).

To be classified, solid waste must be sampled in accordance with NBR 10.007 – Solid (ABNT, 2004d) and subjected to leaching and dissolution tests, according to NBR 10.005 – Procedures for Obtaining Solid Waste Leached Extract (ABNT, 2004b) and NBR 10.006 – Procedures for Obtaining Solubilized Extract from Solid Waste (ABNT, 2004c). The results of the concentrations of a series of chemical species in the leachate and solubilized extracts are compared to the maximum limits stipulated in NBR 10.004:

- Values higher than the limits in the leachate extract indicate that the residue is class I
- Values above the limits in the dissolved extract indicate residues of class IIA
- All limit values classify the waste as class IIB

Note that class II waste classification considers the concept of inert and non-inert waste, therefore it is necessary to clarify what exactly this characteristic means. According to Von Sperling (2018), inert compounds are all those that are not biodegradable, that is, they are all those compounds that are very difficult to be consumed by living organisms and are not transformed by a digestion process. On the other hand, non-inert compounds are all those that are biodegradable, being easily assimilated by those organisms.

As presented before, according to the NSWP (BRASIL, 2010), the concept of MSW can be understood as the waste originating from domestic activities, in urban residences, originating from sweeping, cleaning of streets and public roads and other services of urban cleaning. However, for the Brazilian reality, waste from commercial establishments and service providers characterized as non-hazardous can, due to its nature, composition or volume, also be treated as household waste by the municipal government (BRASIL, 2010). Therefore, MSW can encompass all residues from domestic activities, urban cleaning and commercial establishments classified as domestic.

Upon generation, solid waste should proceed to the next stage of the management chain, the collection (Figure 10.2). This operation comprises the entire region of waste generation and is responsible for sending the waste directly to the treatment facilities or the final disposal site. The collection operation can also be completed at intermediate receivers in the solid waste chain, which are the transfer or sorting stations.

At this stage, the collection logistics can be strategically planned in two ways: The process can take place differently for each type of waste, or in an aggregate way. When the collection operation is differentiated or selective, inert and non-inert (organic) waste (class II) is collected separately by different types of vehicles and transported to different treatment and final disposal sites, depending on their composition. Therefore, there is no mixing of these two types of materials in the same transport vehicle. On the other hand, for the aggregate collection approach, all waste is collected together by a single vehicle. It is worth mentioning that, according to NSWP (BRASIL, 2010), hazardous waste (class I) requires differentiated management, being transported and treated individually, so that it can receive the appropriate treatment and final disposal.

In Brazil, inert waste is normally collected in vehicles that allow the integrity of the material to be kept to the maximum, which is the case, for example, with box trucks (IBAM, 2001). Organic waste, on the other hand, can be collected in compactor trucks, since they allow the addition of a larger amount of material by compressing its particles, releasing the mass of water present in food in the form of fresh leachate, which will be stored in a basin coupled to the truck compaction system.

The selective collection of inert waste in Brazil can be carried out by local public agencies, by private companies, by waste pickers cooperatives and even by companies who make the intermediation of their sale to recycling companies. Companies work for profit with the resale of these materials, unlike cooperatives, which receive donations. The selective collection can be carried out door to door, in order to cover the entire road network for the collection of waste generated in each location, or it can be carried out by waste delivery points (WDP), in which the generator itself forwards the waste to a specific point accessed by the collection truck (Ferri et al., 2015). It is also possible to observe that, in some cases, the WDP can work as a sorting station; manual or mechanized segregation processes of each type of inert material, such as metal, plastic, paper and glass, may occur.

As the recycling market receives few incentives in Brazil, the recycling rate is still very small due to the low demand for this process. As a result, a small mass of waste is traded at low prices, requiring the sorting and selling of a large amount of material for these facilities to become economically sustainable. However, the reality shows that the majority of the sorting companies and cooperatives within the country are small, with little space and deficient infrastructure, implying the need for several units along the chain in order to sort and aggregate an increasing mass of a certain type of material. Therefore, the chains have about three to four intermediate points, before reaching the recycling unit (Ferri et al., 2015).

10.3 WASTE SORTING AND TRANSFER

This stage aims to meet a logistical strategy for transporting waste to the treatment stage so that a greater volume of waste can be transported to its treatment and final disposal site. Generally, waste transfer stations and sorting sites are closer to urban centres than treatment and final disposal sites, thus being an interesting alternative for logistics planning.

Here it is also possible to have a differentiation in the collection strategy, segregating or not the type of waste. In the case of selective collection, inert waste is collected separately and transported to a sorting station, in which the different types of materials that can be sold later to the recycling industries will be sorted. Then, in a sorting station, there is a manual and electromechanical separation of several types of materials, such as plastic, cardboard, paper, glass, ferrous metals and non-ferrous metals, among others.

According to IBAM (2001), a sorting unit has three stages of operation: Reception, feeding and sorting. In the reception stage, the weight or volume is checked using a weighing scale or is estimated by theoretical calculation, so that the residues are stored in appropriate silos or deposits to be processed in a day. In the feeding stage, the loading is carried out in the processing line, employing machines, such as loaders, cranes, polyps and hydraulic arm. It is also possible to adopt devices that allow the discharge of waste from trucks directly into the processing lines. Finally, in the sorting stage, the waste stream is dosed in the sorting lines so that the processes for separating the types of materials that can be recycled take place. The most used flow metering equipment consists of metal conveyor belts and rotary drums. The drums are more suitable for small units with a line capacity of up to 10 ton/h. The sorting conveyor belts must have a speed between 10 and 12 m/min, to allow good performance of the workers who do the manual picking. The pickers must be positioned along the picking belt, next to containers, separating the bulkiest materials such as paper, cardboard and plastic film at the beginning of the conveyor belt so that smaller materials can be viewed and separated by the pickers at the end of the line. Usually, the first position is occupied by a 'bag-breaker', who also has the task of spreading the waste on the conveyor belt to facilitate the work of the other pickers. After the segregation of each type of material, they go through a press and packaging process so that they are prepared for sale in the form of bales. It is noteworthy that waste cooperatives are responsible for carrying out exactly this type of operation, segregating each type of material for later sale.

Concerning the differentiated collection of organic waste, the sorting usually takes place in the treatment facility for this type of waste, and any type of inappropriate and undesirable material for the process is removed in a preliminary treatment step. Thus, the type of treatment and its operational needs must determine how a particular waste is sorted and transported for treatment. Figure 10.3 illustrates the inert and organic waste management chain when a segregation strategy at the generating source and individualized selective collection are implemented.

On the other hand, in an aggregated collection process, all waste can be disposed of in a transfer station, which will allow the aggregation of a larger amount of material in an undifferentiated way and this can be transported later for its final disposal, without prior treatment. Usually, transfer stations are used when the distance between the centre of waste generation and the landfill is greater than 25 km (IBAM, 2001). So, the transportation to the landfill of the waste discharged at the transfer stations is done by vehicles of greater capacity and of lower unit transport cost compared with collection vehicles. According to IBAM (2001), transfer stations in Brazil can be of the direct waste transhipment or with storage. In the case of direct transhipment, the collection trucks unload the waste directly into the transfer trucks through a pit in which they are disposed of. In stations with storage, they may or may not have compaction mechanisms, in order to store a greater amount of waste. Thus, in the latter type, the waste is stored for a certain time at the station, before it is disposed of in transfer trucks through wheel loaders. From there, the transfer trucks then proceed to transport the aggregate mass of waste from the transfer station to the landfill, which is the final disposal of the waste. Therefore, for this management strategy, the waste is collected and handled

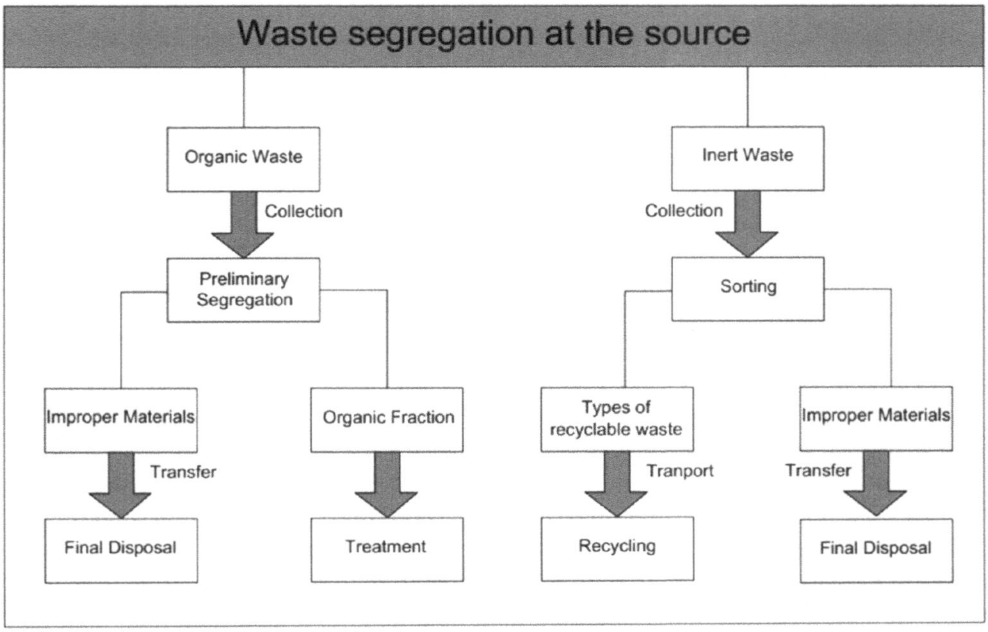

FIGURE 10.3 Management chain of the waste segregated at the generating source.

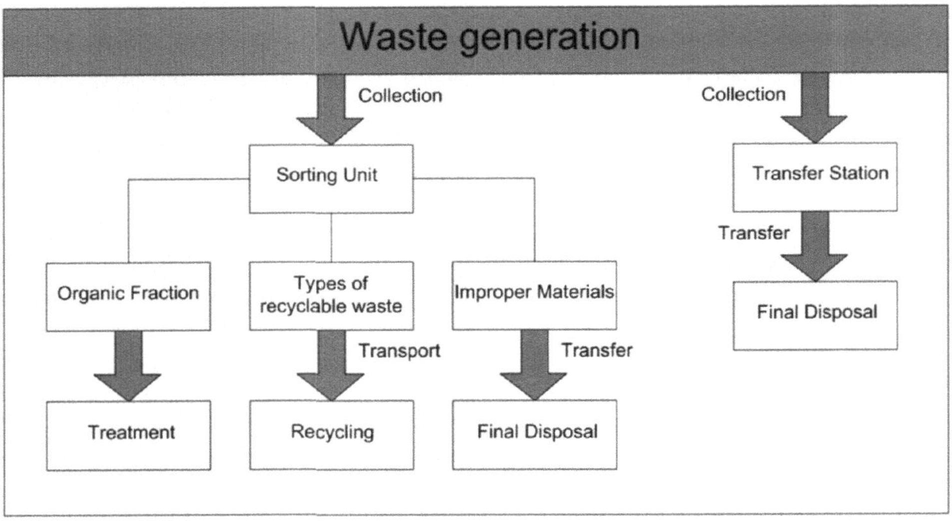

FIGURE 10.4 Waste management chain with aggregate collection.

as an undifferentiated mass, being transported for disposal in landfills, with no type of processing along the chain.

Another option that is adopted in Brazil, for the case of the aggregate strategy, is the use of sorting stations as the destination of the collection routes. In these places, the collection trucks discharge inert and organic solid waste mixed in receiving areas, and, from then on, an operation to segregate inert materials from organic materials begins. Generally, these stations have screens and conveyor belts for prior separation of the organic from the inert waste fractions. Thereafter, inert waste undergoes a process similar to that adopted by cooperatives, with materials segregated by each type, while organic waste is sent to a treatment unit (Angelo, 2014). In such cases, it is common for the treatment unit itself to be located in the same facilities as the sorting station, with no transport necessary until this stage. Thus, as in these stations both inert and organic waste are processed using mechanical equipment and biological processes, they are also called mechanical biological treatment (MBT) stations, which are widely adopted in European countries. This concept was brought to Brazil and started to be widely used in the early 1990s as a waste management strategy for large cities, such as Rio de Janeiro and São Paulo. Figure 10.4 illustrates the waste management based on the aggregate collection strategy.

After collection and sorting of the solid waste generated, each type of waste is sent to a treatment system where it is transformed into another material by promoting changes in its physical, chemical and biological composition.

10.4 RECYCLING OF INERT WASTE AND CONSTRUCTION WASTE

Recyclable urban inert waste can be segregated in sorting units, as described in Section 10.3, so that there is separation by each type of inert material that is sent to the recycling plant to serve as an input in the manufacturing process of a given product. Thus, recycling plants generally process only a certain type of material, such as plastic, paper or glass, transforming the initial composition of the waste so that it has a final characteristic similar to the raw material used as a process input.

For example, for the plastics industry, sorting cooperatives separate plastic into several types according to its constitution and density, including polyvinyl chloride (PVC), polypropylene (PP), polyethylene terephthalate (PET) and polyethylene of high and low density (IBAM, 2001). For each of these types of plastic, there is a recycling unit that will be responsible for transforming the raw material that was benefited by the sorting unit and that arrives packaged in bales, into a new material that will be commercialized for the secondary and tertiary production industries. Thus, a recycling unit is usually present in the primary stage of the production chain for a given product, being responsible for providing one of the inputs for its production. In a conventional production chain, these inputs would be produced from raw materials extracted from natural systems and which are processed to become inputs. Therefore, a recycling unit aims to transform the raw inert waste, which is nothing more than a finished and post-used final product, into a productive input, being able to partially or fully replace the mass of raw material needed for its production.

In Brazil, the state and national institutions are responsible for disseminating information regarding the recycling market, compiling data from the units of sorting and absorption of the mass of waste benefited by the productive market. The coalition of industry and commerce organizations, coordinated by the Business Commitment to Recycling (CEMPRE), was organized after the creation of the NSWP and in 2015 prepared a study to evaluate the sector. According to this study (CEMPRE, 2015), 371 municipalities engaged with the establishment and expansion of selective collection operation and sorting stations after the creation of the NSWP. The number of business investment actions in waste sorting warehouses, training, improving the structure and purchasing vehicles and equipment, recorded by CEMPRE (2015), increased from 80 to 956 between 2012 and 2014. Besides, the report described the installation of 1,646 VDP by companies between 2012 and 2014, a value 255% higher than the target set for 2015.

The Recycling Yearbook, created by the Brazilian National Association of Waste Pickers and Recyclers (Ancat), gathered the results of a representative sample of 260 waste pickers' organizations in 2018, which compose a database of 1,710 entities. In 2018 alone, 116,610 tons were recycled, 4.4% more than in 2017 – that is, there was a recovery of 22% of the post-consumer packaging placed on the market by the companies participating in the programme, moving around R$62 million (around $11 million) (ABRELPE, 2019).

Because it is also inert waste, construction waste also undergoes a similar process to that of inert urban waste. Also called construction and demolition waste, debris or rubble, these residues are composed of materials generated by the civil construction sector such as bricks, concrete, mortar, wood, steel, tiles, lime and plaster, among others. The most widespread treatment of construction waste is segregation, followed by crushing, so that the resulting material is reused by the construction industry, as a base and sub-base for highways, coarse aggregate in the execution of construction structures of buildings, reinforced concrete artwork and precast pieces.

According to IBAM (2001), three factors must be considered when evaluating the implementation of a construction waste recycling process in a given region: Population density, the supply of natural aggregates for civil construction and the level of industrialization of the place. The population density directly influences the amount of waste that is generated. The ease or difficulty of obtaining natural aggregates may or may not favour the development of the market for recycling these wastes, as it directly affects the amount of material mass demanded by the sector. Finally, the level of industrialization directly affects a society's need for and awareness of recycling.

In a civil construction waste sorting and processing unit, the operation goes through a sequence of the following steps (IBAM, 2001): Unloading the material in the receiving yard, checking the material's characteristics to assess whether they are compatible with the beneficiation process, manual separation of waste materials, removal of large materials that are not compatible with the equipment, adjustment of humidity through a water sprinkler system and material processing.

There are two ways of processing civil construction waste: Automatic and semi-automatic approaches. The fully automatic form consists of robust, high-powered equipment, capable of receiving and crushing the waste without prior separation of the hardware that is retained in the concrete blocks. Subsequently, the crushed material passes through a magnetic separator that removes the ferrous material, leaving only the crushed inert material. The ferrous material goes to a press, and the bales are subsequently sold, while the crushed material falls into a rotating sieve that performs the segregation of the material in its various granulometric portions. In the semi-automatic mode, the most used in Brazil, the material to be processed must undergo prior segregation of the hardware, and it is not recommended to grind the materials together.

From the crushing machine, the material goes to a small conveyor belt equipped with a magnetic segregator, where the separation of iron residues that escaped the sorting and were introduced into the crushing device is carried out. After this initial segregation, the material is sent to the vibrating screen, which separates the material in selected particle sizes. From the sieve, each of the fractions is transported to its respective storage yard using a conventional, constant-speed conveyor belt. Finally, the segregated materials are classified as marketable (in the case of ferrous scrap, for example) and unserviceable (remaining material), being deposited in separate places for storage and future disposal.

10.5 BIOLOGICAL WASTE TREATMENT TECHNIQUES

The biological treatment of organic solid waste is nothing more than a degradation process that transforms the fresh organic mass into a stabilized organic material. This process is carried out using microorganisms that break down complex organic compounds and transform them into organic compounds of smaller and simpler chains so that they can obtain energy and nutrients for cellular metabolism. Although the degradation of these compounds can be done by complex and relatively large living beings, it is often mediated by microorganisms, such as protozoa, fungi, bacteria, some algae and even archaea.

The metabolism of these living beings is composed of two types of biochemical processes: Catabolism, from which they obtain energy through chemical reactions involved in the degradation of a certain organic substrate; and anabolism, which uses the nutrients obtained by the degradation of the complex organic substrate and the energy produced by the catabolism to form cellular material (Von Sperling, 2018).

In the anabolism process, cell division occurs and cells can multiply. In the case of prokaryotic organisms, such as bacteria, which do not have a cell membrane, the nucleus of the cells is dispersed in the protoplasm (Von Sperling, 2018) and they only absorb and assimilate dissolved compounds. This is one of the reasons why water should be present in the material that will be degraded. Besides, the water content inside the cells of bacteria is 70 to 85%, the rest of its composition being characterized by the presence of carbon and other compounds (Fricke and Bidlingmeier, 2016).

The microorganisms are distinguished and classified according to the carbon sources, used in anabolism, and energy sources, used in catabolism. Regarding the anabolism, the types of microorganisms can be classified as follows (Von Sperling and Chernicharo, 2005):

- Heterotrophic organisms: Those which obtain carbon from organic substrates, playing a central role in the decomposition process of these compounds
- Autotrophic organisms: Those whose carbon source for cellular synthesis is carbon dioxide (CO_2); therefore they consume a substantial amount of energy for this purpose and grow more slowly compared to heterotrophic organisms

As for the catabolism, living organisms can be differentiated by their source of energy, which can be obtained from solar radiation (phototrophic organisms) or chemical reactions (chemotrophic organisms). For the case of chemotrophic organisms, catabolism can be differentiated in two ways (Von Sperling and Chernicharo, 2005):

- Oxidative catabolism: This process occurs through a series of oxidation-reduction (redox) reactions between electron donor compounds, which serve as an energy source, and electron acceptor compounds, with the energy being obtained from the transfer of electrons through the action of enzymes. Thus, the compound that receives electrons is called an electron acceptor or oxidizing agent (inorganic compounds such as oxygen, nitrate or sulphate), having a reduced oxidation state at the end of the reaction. The compound that donates electrons, on the other hand, is called an electron donor or reducing agent (e.g. the organic substrate to be digested), presenting, at the end of the reaction, a higher oxidation state

- Fermentative catabolism: In this process, energy is obtained by rearranging electrons between an organic compound resulting from oxidative catabolism, which donates electrons, and a reduced organic compound that receives electrons. This rearrangement will occur until the organic compounds stabilize, that is, until the oxidized and reduced compounds reach their maximum and minimum oxidation stage, respectively

Given the electron transfers in the redox reactions and the electron rearrangements in the fermentation processes, the amount of energy obtained in the catabolism is directly related to the oxidation and reduction potential of organic and inorganic compounds. Therefore, both in anabolism and catabolism, microorganisms need organic and inorganic compounds for their development. Organic compounds, based on carbon, nitrogen, sulphur and phosphorus, can be present in the mass of organic residues in more complex forms such as proteins, lipids and carbohydrates. On the other hand, the inorganic compounds can be, for example, in the form of calcium bicarbonate, present in residues such as eggshells, and phosphate ions, present in bone residues of animals.

In a given treatment process, there will be a rich microbial community composed of different species, each of which will carry out its processes of anabolism and catabolism. Thus, it will be the external conditions imposed by the system that will guarantee its harmony or competition for access to the electron donor compounds present in the environment. In general, the conditions can be aerobic (with the presence of oxygen), anoxic (with the absence of oxygen, but with the presence of oxidized nitrogen species (NO_x), such as nitrate and nitrite) or anaerobic (with the absence of oxygen and NO_x and the presence of sulphate (SO_4^{2-}) and CO_2) (Von Sperling, 2018).

In this context, microorganisms can be differentiated in terms of the use of free oxygen as the only oxidizing agent for the oxidation reaction of organic compounds (strict aerobic organisms), they may have a preference for free oxygen but also use nitrate and nitrite as an electron acceptor (facultative aerobic organisms) or only perform catabolism in the absence of free oxygen, thus using other compounds as electron acceptors, such as CO_2 (strict anaerobic organisms).

For the treatment of organic waste, the most commonly used aerobic biological process in Brazil is composting, which is very similar to the natural degradation process of organic compounds. Through microbial activity, complex organic molecules are broken down into simpler organic molecules that are oxidized by a rich and diverse microbial community, comprised of bacteria, fungi and other microorganisms. The biological activity existing in a waste stream prepared for the treatment of composting is complex and depends on environmental factors such as the availability of energy and nutrients, pH, humidity, oxygen and temperature, which will directly influence the dynamics of the ecological relationships existing between species (Inácio and Miller, 2009).

According to Fricke and Bidlingmeier (2016), the biochemical processes that occur in a waste stream prepared for composting treatment can be summarized by the following steps:

1) Release of extracellular enzymes by microorganisms to come into contact with complex organic compounds
2) Hydrolysis of complex organic compounds
3) Diffusion of the simplest organic compounds, which are dissolved, into the cells of the microorganisms
4) Transport of oxygen molecules through the pores of the composting windrow
5) Transfer of oxygen in the gaseous state to the liquid phase (leachate)
6) Diffusion of oxygen through the liquid phase
7) Diffusion of oxygen into the cell
8) Aerobic degradation of simple organic compounds within the microorganism cell

According to Siqueira and Assad (2019), from the Brazilian experiences with composting, strategies for treating organic waste were developed in centralized and decentralized manners. For the centralized form, there are two modalities: Sorting and composting plants (SCP) and urban composting yards (UCY). For the decentralized strategy, institutional, home and community composting are distinguished.

The SCP are private or public enterprises characterized by receiving the mass of solid urban waste in an aggregate form, that is, the collection trucks unload the organic waste mixed with the inert waste in the reception area. A unit of this type may have, in addition to the sorting belt, different equipment for separating rejected material, a composting yard, a landfill for rejected waste and a leachate treatment system. The sorting belt separates waste into its various components, and the degree of separation depends on the number and efficiency of the employees involved and the speed of the belt which, the smaller it is, the more efficient the sorting will be. The SCP is considered a conventional composting method in Brazil and has the advantage of treating all the waste from one or more municipalities in a single place. On the other hand, they have the disadvantage of generating a low-quality final compost, resulting from the processing of waste collected in an undifferentiated way. In addition to insufficient nutritional characteristics, the presence of heavy metals and inert residues discourages the demand for the compound produced, which often remains stored in large quantities in these units. Technical, economic and managerial infeasibility are the main reasons why most SCP have been disabled in Brazil since the 1980s.

Finally, UCY are spaces located in an urban environment that process organic waste from different origins. The previously described modalities were considered as *in situ* composting because they process the waste generated *in loco*. Urban composting yards are considered *ex situ*, because the waste is transported and treated elsewhere and not within the generating community, and the compost may or may not return to the taxpayers of the system. As for its structure, the UCY can adopt manual or mechanized management, with or without waterproofing the soil and cover. These units generally receive large volumes of organic fraction of solid waste, prioritizing the quality of the final product. Thus, these units are generally sized to serve an entire municipality and are restricted to receiving the organic waste segregated at the source and which is collected in a differentiated way.

The decentralized and *in situ* composting strategy, developed by most Brazilian cases, uses the adoption of these systems for the individualized treatment of organic waste generated by public agencies and education institutions (Siqueira and Assad, 2019). Of the treatment systems implemented, the methods adopted range from natural composting and vermicomposting to the most modern technologies, such as bio-accelerators and electric composters. In this respect, the method employed has a crucial weight in the cost of the operation, because the faster and more compact the system, the more costly it will be. On the other hand, natural composting systems that have employees and use large facilities can present lower costs.

Household composting, on the other hand, corresponds to the modality developed inside homes using the waste generated by the residents themselves. The compost produced is generally used locally in household gardens. Community composting presents initiatives developed in neighbourhoods that require the local treatment of kitchen and/or pruning waste. Composting windrows are worked within the community and can be located in community areas, public spaces, vacant lots or even within institutions. A key feature of this modality is the participation of residents in the process. This involvement can occur at several levels, from the correct segregation of organic waste in homes and voluntary delivery of waste at collection points to participation in joint efforts, system planning and search for solutions. The final compost is always donated to residents and local institutions. In this case, neighbourhood associations, community agents, social entrepreneurs, non-governmental organizations (NGOs), university extension groups and private companies that carry out educational activities can act efficiently by establishing a close interaction of monitoring and exchange with the community.

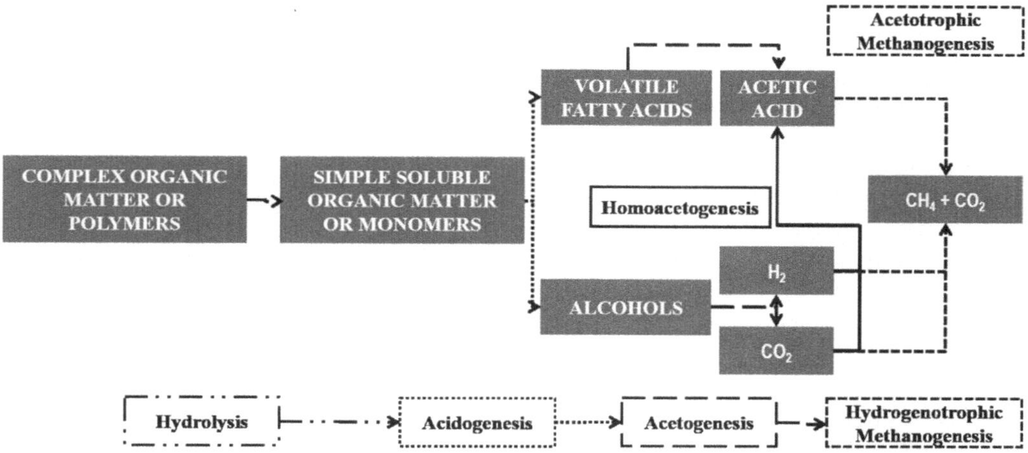

FIGURE 10.5 Stages of the anaerobic digestion process. Source: Based on Fagbohungbe et al. (2016), Jha and Schmidt (2016), Zhang et al. (2014), Jain et al. (2015), Chatterjee and Mazumder (2016), Henze et al. (2017) and Von Sperling and Chernicharo (2005).

Another process for the treatment of organic waste that, although adopted on a smaller scale in Brazil, has been constantly expanding in recent years, is anaerobic digestion. Anaerobic processes aim to degrade organic substrates in controlled environments and the absence of oxygen, to promote the development and proliferation of microorganisms that will be responsible for anaerobic digestion. This consists of a biochemical process that takes place through four steps, outlined in Figure 10.5.

In the first stage of the anaerobic digestion process, hydrolysis occurs, which is responsible for breaking the intermolecular bonds of complex organic compounds, carried out by fermentative bacteria, which excrete extracellular enzymes so that the long-chain organic compounds are transformed into simpler compounds (Von Sperling and Chernicharo, 2005). Therefore, compounds like carbohydrates will be broken down into sugars, proteins will be broken down into amino acid molecules, lipids into fatty acids and so on. The simplest substrates are characterized by being soluble, associated with the water molecule. This first stage is extremely important for the occurrence of the other stages since it allows the compounds to be prepared for their degradation by increasing their surface area and thus becoming available for assimilation by fermentative bacteria.

The second stage of anaerobic digestion is acidogenesis, in which a diverse community of fermentative bacteria can metabolize the organic compounds generated in the hydrolysis stage, using intracellular enzymes. First, the soluble compounds are assimilated, passing through the cell membrane, and oxidative catabolism reactions occur inside the cell, releasing usable energy for anabolism (Von Sperling and Chernicharo, 2005). Thus, compounds resulting from hydrolysis, such as amino acids and sugars, fatty acids and monosaccharides, will be transformed into volatile fatty acids (VFA) and alcoholic compounds. These generated compounds are expelled out of the cell. In the specific case of VFA, due to their low molecular weight, they volatilize under atmospheric pressure.

The third stage is acetogenesis, in which acetogenic microorganisms perform the chemical oxidation of compounds resulting from acidogenesis, transforming them into simpler chains. Then, volatile fatty acids, such as propionic and butyric acids, are transformed into acetic acid, dissociated to acetate present in the liquid state of the material, and also into hydrogen and carbon dioxide, present in the gaseous state. Alcoholic compounds are transformed into hydrogen and carbon dioxide also during acidogenesis. H_2 and CO_2 may be converted to acetate; this process is also known as homoacetogenesis, but it is not the most direct way to obtain this compound. As the stages of acidogenesis and acetogenesis are interconnected, both processes are also known as fermentation.

The fourth stage of anaerobic digestion is methanogenesis, which can be considered the most important stage in the entire process. It is responsible for the generation of methane, which is present in biogas and gives it a high energy potential due to its calorific value. Thus, methanogenic microorganisms (archaea) can generate methane through the acetotrophic pathway, characterized by the decarboxylation of acetic acid, or by the hydrogenotrophic pathway, with the reduction of carbon dioxide (electron acceptor) by hydrogen (electron donor). Then, in the absence or limitation of hydrogen, the acetotrophic pathway is activated so that the methyl group of acetic acid is reduced to methane and the carboxylic group is oxidized to carbon dioxide (Von Sperling and Chernicharo, 2005). According to Henze et al. (2017), 70% of the methane is produced by this route.

It is also worth mentioning that, from the by-products of acetogenesis, a last stage called sulphetogenesis may take place, in which the previously generated sulphates undergo a reduction or desulphation reaction, being transformed into hydrogen sulphide (H_2S), present in the gas phase (Von Sperling and Chernicharo, 2005).

Considering the stages of anaerobic digestion of organic waste, three main by-products are generated: The liquid resulting from anaerobic digestion (digestate), containing all compounds that have not been fully converted to CH_4; the biosolid, a digested solid material that has already been biochemically stabilized and can be used as an organic compound (as a soil conditioner, for example); and biogas, which is the gas formed by all gaseous compounds released during the process. According to Von Sperling and Chernicharo (2005), about 70 to 90% of biodegradable organic matter is converted to biogas, with only a small part of 5 to 15% being converted into microbial biomass.

The components present in greater quantity in the biogas are methane (CH_4) and carbon dioxide (CO_2), as can be seen in Table 10.1. Thus, the higher the concentration of methane in biogas, the greater its calorific value and, consequently, the better its energy use.

For the treatment of organic waste by anaerobic digestion, there are different and established technologies worldwide and they differ by the conditions considered in the treatment stages. Therefore, the technologies can be classified in terms of the concentration of total solids (TS), which is related to the humidity of the organic waste to be treated, the process temperature, the type of operating system and the division of the stages of anaerobic digestion (André et al., 2018).

According to the TS concentration of organic residues, wet systems are those in which the organic material has a maximum concentration of 15% TS, and is most commonly treated at a concentration between 3 and 7% TS. For dry systems, the material is treated at a concentration above 15%, reaching up to 35% TS. Finally, in extra dry (also referred to as high solid and solid state) systems, the TS

TABLE 10.1
Composition of Biogas

Component	Concentration (%)
CH_4	40–75
CO_2	25–40
H	1.0–3.0
N	0.5–2.5
H_2S	0.4–0.5
O_2	0.1–1.0
NH_3	0.1–0.5

Reference: Brazilian Ministry of Cities (2015).

content varies from 25 to 50%, with no need to add water for the digestion to occur, a practice often used in wet systems. The term anaerobic digestion in the solid state is therefore usually adopted (André et al., 2018) to indicate dry and high solid systems, with TS content above 15%.

Regarding temperature, the process can be classified as psychrophilic, performed in the range between 4 and 15°C; mesophilic, between 37 and 42°C; or thermophilic, between 50 and 60°C. It is common for the mesophilic temperature to be reached naturally by the biological process in tropical regions, however, the thermophilic range is usually obtained through an external heat supply system.

Regarding the operation of the system with respect to the feed flow, it can follow a continuous, semi-continuous or batch flow regime. In the case of continuous flow, the entrance of the organic substrate to be treated is done continuously and without any interruption. In the semi-continuous flow, the input is constant and periodic, with a frequency of insertion of the organic substrate in the system. Concerning batch systems, the input of organic substrate occurs at a certain time and the treatment process is therefore intermittent. For the insertion of fresh material, all material that has already been digested must be removed from inside the reactor.

Finally, in relation to the development of the stages of anaerobic digestion, treatment systems can be classified into single or multiple stages. For single-stage systems, all stages of anaerobic digestion occur in aggregate in a single reactor. For the multistage system, it is possible to divide each stage of anaerobic digestion into a different reactor, so that each stage can be optimized, that is, the most appropriate environmental conditions for each stage are adjusted so that the process develops as efficiently as possible.

Organic waste treatment systems are often employed in mechanical-biological treatment (MBT) plants, which receive municipal solid waste in an aggregate manner (Gomes et al., 2012). On the other hand, it was possible to observe the adaptation of wastewater treatment systems for the introduction of waste, which caused a series of constraints, mainly due to the particularities of these residues, such as heterogeneity, the high content of total solids and of materials unsuitable for biological treatment, among others, which led to the development of specific technologies for these materials.

It is worth mentioning that the biggest difference between the treatment systems for anaerobic digestion is in the content of TS used. Following the trend of wastewater treatment systems, the wet route was initially the most used for the treatment of organic waste. In this type of system, water, or another fluid organic substrate, is added to the process so that the desired total solids concentration is achieved. As a result, the treatment facilities make use of continuous stirred tank reactors (CSTR), with internal agitators. Among the advantages, scientific studies point to the possibility of adding one or more organic substrate to be digested together with organic waste, a process also known as co-digestion, with the best balance of nutrients for the biological process (Chiu and Lo, 2016).

In Brazil, over the years 2013 to 2017, the Brazil-Germany Project for the Promotion of Energy Use of Biogas in Brazil (PROBIOGÁS) came into force, as a result of technical cooperation between the Brazilian Government, through the Ministry of Environmental Sanitation at the Ministry of Cities, and the German Government, through Deutsche Gesellschaft für Internationale Zusammenarbeit GmbH (GIZ). The project aimed to develop and disseminate a collection of technical publications in order to contribute to the expansion of the efficient energy use of biogas and, therefore, to the reduction of greenhouse gas emissions.

Therefore, according to PROBIOGÁS (Brazilian Ministry of Cities, 2015), in CSTR-type reactors, it is common to have the formation of a floating layer, also called foam, composed of plastics and other low-density materials, which accumulate on the fluid surface to be digested. The problem is that this foam may hinder the diffusion of the formed biogas and its escape into the free space above the material (commonly referred to as headspace). There are also records of damage to mechanical agitators, due to the presence of inappropriate materials that are not digested by microorganisms, in addition to the sedimentation of denser inert materials, accumulating at the bottom of the reactor. Finally, there are other obstacles concerning the addition of organic residues in co-digestion with another wastewater in wet processes, which is the possibility of the accumulation of toxic compounds to biological treatment, such as ammonia, functioning as an inhibitor of the process at high concentrations. It is worth mentioning that, to prevent or remediate these problems, the installations may have high energy consumption in the preliminary treatment processes, before the material enters the reactor, or even in post-treatment systems so that the material is suitable for disposal in water bodies.

Given these obstacles, anaerobic digestion systems in solid state (dry and high solid modes) began to develop worldwide. In this type of process, there is no introduction of water to dilute the organic waste to be digested, and it is common to use a unit for conditioning the waste so that it receives a material rich in microorganisms responsible for anaerobic digestion, also known as an inoculum, which usually consists of digested sludge. Thus, the material is homogenized and fluidized before its introduction into the reactor, which can be of the complete mixture or plug-flow type (Ren et al., 2018). The advantages that can be observed compared to wet processes are the possibility of applying a higher organic load, the need for a smaller reactor volume and the lower generation of liquid effluents at the process outlet. However, this type of system also faces some difficulties similar to the wet process, such as sedimentation of inert materials at the bottom of the reactor and difficulty in homogenizing the material (André et al., 2018).

In the anaerobic digestion processes in the solid state, there is no stage of homogenization of the material before it enters the reactor. The reactors developed for this type of digestion have a discontinuous operation, in which the material is introduced at one time and there is no introduction of new material while it is subjected to the digestion process. For the new material to be introduced, the material already digested must be completely removed from the reactor. There are several types of reactors, with different configurations and specificities.

According to a report by PROBIOGÁS (Brazilian Ministry of Cities, 2015), the following solid state anaerobic digestion treatment technologies with lower energy consumption and less generation of liquid effluents stand out: Dranco (Organic Waste Systems, Belgium), Kompogas (Axpo, Switzerland), Valorga (Steinmuller Valorga Sarl, France), Laran (Linde BRV Biowaste Technologies, Austria), Kompofern Plus (Eggersmann Anlagenbau GmbH & Co. KG, Germany), Bekon Dry Cycle Fermenter (Bekon Energy Technologies GmbH & Co. KG, Germany) and BIOferm System (Viessmann Werke GmbH & Co. KG, Germany).

According to Cibiogás (2020), in a technical note that presents the panorama of the biogas market in Brazil in the period 2018–2019, an increase of 36% was observed in the volume of biogas produced and 31% in the number of plants in operation with energy application, which shows an early stage of development. The current scenario of 1.3 billion Nm^3/year of biogas production is equivalent to 1.5% of the national potential, showing that the biogas chain has an opportunity to expand 98.5% in Brazil. The type of anaerobic digestion technology used in solid waste treatment systems is not openly disclosed, with 416 plants for the treatment of agricultural waste in 2019, 62 dedicated to industrial waste and 43 units that receive municipal solid waste and domestic wastewater (Cibiogás, 2020).

It is worth mentioning here the implementation of the first system in Latin America for the treatment of the organic fraction of solid urban waste through anaerobic digestion in the solid state: The TMethar system, located in the city of Rio de Janeiro. Designed for a treatment capacity of 23.5 ton/day of waste, the system comprises solid-state anaerobic reactors, sequenced and operated in batches, and with the production of biogas aimed at the generation of electric and thermal energy from an engine-generator. The system is illustrated in Figure 10.6.

Overall, there is great potential for biogas production to be explored in Brazil, taking into account the need to ensure adequate treatment of solid waste and wastewater and better efficiency and quality in the sanitation sector, while guaranteeing the business sustainability through the country's energy self-sufficiency.

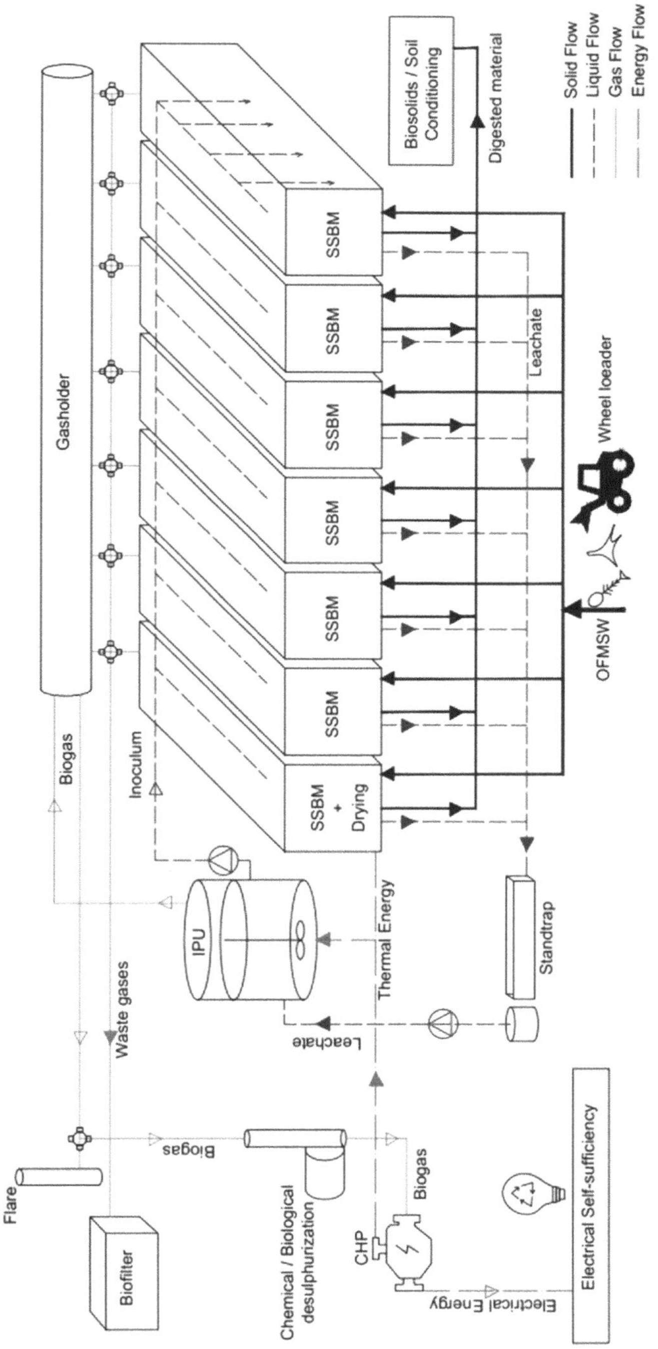

FIGURE 10.6 Solid-state anaerobic digestion plant in Rio de Janeiro, Brazil.Source: Based on Ornelas-Ferreira et al. (2020).

10.6 THERMAL WASTE TREATMENT TECHNIQUES

The thermal treatment of waste is an essential part of an integrated waste management system, which, as previously reported, presents a hierarchy composed of steps of prevention of waste generation, reuse and resource recovery (materials and energy) and final disposal of the rejected waste that can no longer be used.

In this context, it is clear that the use of landfills should be dedicated to pre-treated waste, which is not biologically active and does not contain hazardous substances. Reuse and recycling aim to recover as many materials as possible that can be used in a given production chain. For the waste whose material recovery is not effectively applied, energy recovery is an interesting option to minimize its associated environmental impact. This can be done through the application of biological processes, described in Section 10.5, or thermal processes, a topic that is discussed as follows.

Thermal treatment involves thermochemical processes that occur at relatively high temperatures, causing changes in the chemical structure of the waste that is being processed. In general, there are three main processes available for the thermochemical conversion of solid waste: Combustion, gasification and pyrolysis (Lombardi et al., 2014).

Combustion processes, generally referred to as incineration, are the most widely used and well-established processes worldwide for different types of waste, including MSW, exhausted solid fuels, industrial waste, hazardous or not. Waste incineration is generally associated with energy recovery, either in the form of electricity and/or in the production of heat. In some cases, the incineration of hazardous industrial waste is not accompanied by energy use, since the fuel gas generated has many pollutants that hinder this operation (Lombardi et al., 2014). On the other hand, gasification and pyrolysis arise as an alternative to producing fuels such as coal, coke and urban gas.

Gasification is a process of thermochemical conversion of materials that aims to produce gaseous compounds. It occurs in reducing conditions, achieved by restricting the oxygen supply, normally supplied through air injection. Although there are several gasification processes, these occur at temperatures that can vary between 750 and 900°C. The resulting gas is typically a mixture of carbon monoxide, carbon dioxide, hydrogen, water vapor, methane, nitrogen and small amounts of heavier hydrocarbons. This gas can have several applications, such as being used as fuel in boilers, internal combustion engines, gas turbines and fuel cells, and organic synthesis for the production of various chemicals by Fischer–Tropsch synthesis or synthesis of methanol (Paradela, 2007).

Pyrolysis is the thermal decomposition of materials in the absence of oxygen to obtain liquid by-products. Pyrolysis applied to organic matter (i.e. mainly composed of hydrocarbons) promotes the 'cracking' of the heavier molecules and their conversion into others of lower molecular weight. Moderate temperatures and pressures not much higher than atmospheric pressure are normally used. Currently, the main objective of pyrolysis is to obtain liquid hydrocarbons that can be used as fuels or as raw material for certain industries (Paradela, 2007).

The main interest in the thermal treatment of waste is generally the ability to significantly reduce the solid mass of waste (around 70–80%) and volume (80–90%), reducing the space required in the landfill, at the same time as eliminating the putrefaction of waste (especially the organic fraction), avoiding sanitary problems (Gohlke and Martin, 2007). Due to new technological developments in recent years, another attractiveness related to thermal treatment has become energy use, which, for the case of MSW, has significantly increased as compared to that obtained in the past. This scenario is also related to the change in habits of consumers and increased prior selective collection. Many incinerated MSW have had their lower heating value (LHV), expressed in GJ/Mg, increased in recent decades.

According to estimates made by the Brazilian Energy Research Company (EPE, 2014) on electricity generation in Brazil, from the MSW thermal treatment (incineration, gasification or pyrolysis), it is possible to meet 5.4% of the national energy demand. Considering only the metropolitan regions, the estimated potential of MSW waste-to-energy (WTE) thermal treatment plants is 2.4 GW (1.85% of the national matrix), with an annual generation of 14,400 GWh (2.74% of total generation), with a capacity factor above 90% and with energy being injected close to consumers (distributed generation) (Rizzo, 2019).

Brazil has several biogas capture plants in landfills for electricity or biomethane generation, but it does not have any large WTE mass burning plants in operation, which are the plants used in more than 90% of the occasions to treat mixed MSW. Currently, the country has ten smaller WTE plants, with installed power from 1 to 5 MW, employing national and imported gasification and pyrolysis technologies to treat MSW and hazardous industrial waste. National technologies include those developed by WEG (gasifier) and ZEG (pyrolysis) for the treatment of MSW and other wastes (ANEEL, 2019).

Regarding industrial waste, there are 38 co-processing plants in operation that produce the residue-derived compound (RDC) to be used as fuel in the cement industry, and represent 70% of Brazil's production capacity (ABCP, 2019). RDC is used to power industrial furnaces, generating calorific power for the operation and manufacture of products that require high temperatures. There is no generation of environmental liabilities, and natural resources such as coking coal are preserved. According to the Cement Technological Roadmap (Visedo and Pecchio, 2019) regarding the reduction of GHG emissions, co-processing is one of the most important tools and expects to reach, in 2035, the percentage of 35%, and in 2050, 55%. It is estimated that the cement industry will go from the current 1,000,000 ton/year RDC to 4,000,000 tons/year in 2035, and around 7,500,000 tons/year in 2050. Most of this increase comes from the significant growth of biomass, RDC of MSW and class II waste.

In Brazil, there is also a compact industrial pyrolysis unit in operation since February 2015, where dozens of types of waste have already been tested: MSW, health service waste, various industrial waste, sludge from wastewater treatment plants (WWTP), industrial sludge, biomass in general (coconut husk and fibre, wood chips, organics, leather scraps), scrap tires and electronic waste, among others. The technology is 100% national and developed by RTB Holding Energy in a technical and commercial partnership with INDDRA Energy and Waste. It involves slow pyrolysis in a rotating, humid and catalysed drum, where 90% of synthesis gas is produced for the generation of electricity and/or steam and 10% of biochar intended for agriculture or recovery of degraded soils (ANEEL, 2019).

10.7 FINAL DISPOSITION OF WASTE IN BRAZIL: DUMPS, CONTROLLED LANDFILLS AND SANITARY LANDFILLS (CLASS I AND CLASS II)

Currently, to comply with the NSWP, recycling, treatment and reuse of waste alternatives should be prioritized, and only rejected waste should be sent to the landfill. However, the dumping sites, which consist of areas without any measures to protect the environment or public health, are still a Brazilian reality. Another form of inadequate disposal is controlled landfills, where the waste is covered with soil and eventually compacted, however, without a base waterproofing layer and, in many cases, without a drainage system and treatment of leachate and gases. Therefore, controlled landfills are a strategy to recover an existing dumping site to minimize their environmental impacts. Figure 10.7 illustrates the controlled landfill applied in Brazil, in comparison to the sanitary landfill.

Sanitary landfills have been the most used form by Brazilian municipalities for the destination of generated waste, receiving 59.5% of the waste collected in the country, which corresponds to 43.3 million tons (ABRELPE, 2019). This practice, which is already prohibited in some countries, is only permitted by the NSWP through the use of appropriate containment techniques with engineering criteria and specific standards for hazardous (class I) and non-hazardous (class II) waste, which has its final disposal in class I and class II landfills, respectively. Figures 10.8 illustrates an example of a sanitary landfill, showing the main components: (i) waterproofed soil; (ii) compacted waste; (iii) land layer covering compacted waste; (iv) channels for draining the leachate; (v) leachate treatment plant; (vi) chimneys with filters to release gases; (vii) possible use of the area after deactivating the landfill.

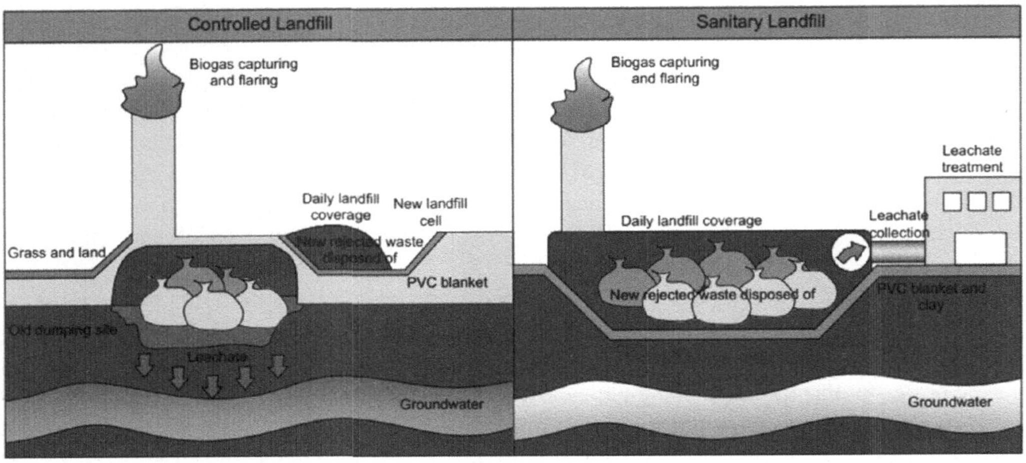

FIGURE 10.7 Controlled landfill system.

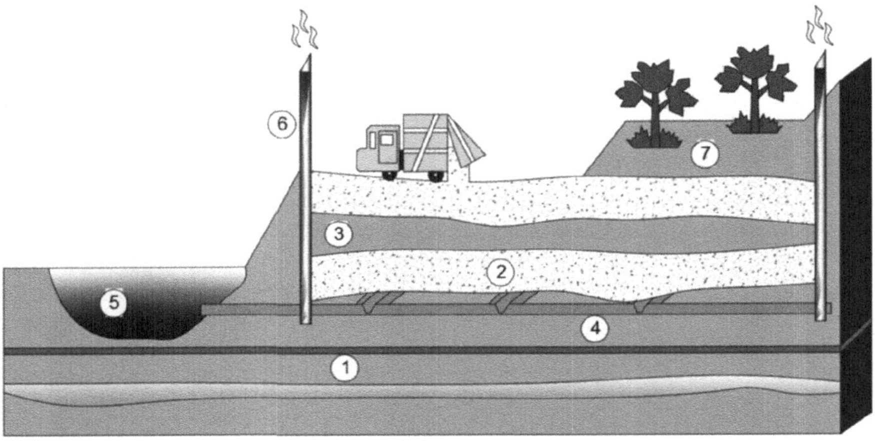

FIGURE 10.8 Schematic representation of a sanitary landfill with the main components necessary for its proper functioning.

According to the Brazilian Standard NBR 10.004 (ABNT, 2004a), hazardous waste, due to its characteristics, needs differentiated attention regarding treatment and final disposal. The waste sent to class I landfills must have a maximum of 70% humidity and be stable, poorly soluble and non-volatile. Acids, strong bases, flammable materials and explosives waste cannot be disposed of in industrial landfills unless special pre-treatment techniques are employed, such as stabilization, encapsulation, solidification and vitrification. NBR 10.157 (ABNT, 1987) establishes the technical criteria for the construction of a landfill for hazardous waste. It is important to note that, in landfills, there is no elimination of liabilities but only their control, so that the useful lifetime of a landfill must be extended through solutions that minimize and prevent the waste mass generation (IPEA, 2012).

According to Jucá et al. (2014), the basic infrastructure of class II landfills is composed of several units (e.g. shed for equipment, guardhouse, weighing scale, office), while urbanization and landscaping services should also be considered for all sizes of landfills. As the geological characteristics of the region affect the costs of the landfill installation, the size of the areas depends on their conformation, the height that can be reached at the landfill (number of layers) and the useful life considered for the operation. Other factors that influence costs are the type of waterproofing adopted, the drainage system and the percolate treatment processes.

Since the NSWP creation, the greatest challenge of MSW management is the closure of dumping sites and the construction of sanitary landfills. However, more than 90% of Brazilian municipalities have a population smaller than 100,000 inhabitants (IBGE, 2020) which show characteristics of small urban areas and a lack of economic and technical resources of municipal management. Therefore, the municipalities have little technical knowledge and money for the construction and operation of sanitary landfills. So, as a consequence of NSWP, the creation of consortiums of municipalities for MSW management and final destination has become a strategy guided by the Brazilian Environmental Ministry (MMA, 2020), which allows the construction of these sites and the closing of dumps little by little.

The size of the municipality appears as a determining factor for the design of MSW landfills in Brazil. In small municipalities (population of up to 20,000 inhabitants or that generate up to 20 ton/day), small (manual operation) or conventional (mechanized) landfills can be implemented. Conventional or mechanized landfills serve all municipalities with a generation of waste that economically justifies the use of machinery for excavation operations, preparing the land, covering material, moving, spreading, compacting and covering the garbage. These types of landfills are generally used by municipalities with a population greater than 20,000 inhabitants or consortium members (Jucá et al., 2014).

In larger municipalities, it should also be considered that landfill systems for the disposal of MSW have the potential to implement biogas recovery projects, aiming at their energy use. It is important to note that this type of technology is only possible in landfills that have a minimum capacity of 300 tons/day, in which it is possible to generate 0.1 to 0.2 MWh per ton of MSW (Tomalsquin, 2003). Regarding the type of technology to be used for energy conversion, Abreu et al. (2009) affirm that it is possible to obtain higher thermal efficiency of modern internal combustion engines in plants with power in the range of 10 to 50 MW in comparison to the combined cycles and turbines, which demand a high investment and are more suitable for projects with more than 50 MW.

According to the Biogas Panorama in Brazil (Cibiogas, 2019), there are in the country about 43 biogas production plants for energy purposes using substrates from landfills (MSW) and sewage treatment plants. These plants correspond to 8% of the total 521 biogas-producing units in operation, being responsible for the production of 1,026,527,529 Nm^3/year, which represents 76% of the total biogas produced in the country.

10.8 CONCLUDING REMARKS

The National Solid Waste Policy (NSWP) was established in Brazil in 2010 by Law 12,305. This policy defined all steps, instruments and guidelines for solid waste management, from the prevention of waste generation to environmentally appropriate final disposal of what is considered reject material. Along this chain, the stages of reduction, reuse, recycling and resource recovery from waste are considered.

Although it represents a milestone in the waste management sector in the country, there is still much to evolve in this regard. There are many people without access to basic sanitation services, such as waste and sewage collection and drinking water supply. To overcome this scenario, the new Regulatory Framework for Basic Sanitation in Brazil (Brazilian Law No. 14.026), established in 2020, provides for the universalization of water and sewage services by 2033 and makes it possible to inject more private investments into sanitation services. The end of the dumping sites was also extended, depending on whether the municipalities developed or not the solid waste management plan.

The recycling market is one that should be carefully considered in the current days in Brazil, as the recycling rate is substantially low, partially because of the low prices at which waste is traded. More incentives must be given to recycling so that companies and cooperatives can improve their infrastructure, offering more job opportunities in this important economic sector.

Resource recovery from waste should also be encouraged throughout the country. There are many options for waste treatment in this respect. One of them is the biological composting process, by which the complex organic matter present in the waste is transformed into simple organic compounds by a rich and diverse microbial community composed of bacteria, fungi and other microorganisms.

Another biological process in constant evolution in Brazil is anaerobic digestion, in which organic substrates are degraded and converted, in the absence of oxygen, in a mixture of gases (referred to as biogas), mainly composed of methane and carbon dioxide. Besides biogas production, a liquid product (digestate) rich in nutrients and a solid fraction (biosolids) which can be used as a soil conditioner are also obtained at the end of the anaerobic digestion process. Different anaerobic digestion technologies and processes are available for full-scale plants, and the most suitable option should be defined for each particular case, depending on local conditions and waste characteristics. The biogas market in Brazil is growing every year, but the national potential for biogas production is far from being reached.

Thermal treatment processes also emerge as an interesting option to minimize the environmental impact of waste. By changing the chemical structure of the waste at high temperatures, thermochemical processes (e.g. combustion/incineration, gasification and pyrolysis) may allow energy recovery (in the form of electricity or heat production) or fuel production such as coal, coke and urban gas. In this sense, resource recovery is accompanied by the reduction of the solid waste mass, minimizing the space required in landfilling practices. As pointed out by the Brazilian Energy Research Company, it is possible to meet 5.4% of the national energy demand by thermal treatment of municipal solid waste. The current scenario involves the use of both national and imported thermal technologies to treat MSW and hazardous industrial waste, but the sector has to take giant steps to contribute to the supply of energy.

When it comes to final disposal of waste, dumps, controlled landfills and sanitary landfills are the main adopted strategies in Brazil. Without any measures to protect the environment or public health, dumping sites are still a reality in the country. Controlled landfills are another inadequate disposal site for rejected materials, where they are covered with soil and eventually compacted, but there is no base waterproofing layer neither drainage system and treatment of leachate and gases. For appropriate disposal of waste with engineering design and specific standards for different types of waste (hazardous and non-hazardous), sanitary landfills are considered the best alternative in Brazil, although this practice has been prohibited in some developed countries.

The implementation of sanitary landfills is more challenging in municipalities with less than 100,000 inhabitants, which corresponds to more than 90% of Brazilian cities, where technical knowledge and financial resources for construction and operation are not scarce. In this respect, the creation of consortiums of municipalities for MSW management is seen as an alternative to overcome such constraints and allow for the closure of dumping sites. For larger municipalities, disposal of MSW should also consider energy recovery projects in the form of biogas, which can also be conducted in association with the waste (sludge) generated at municipal wastewater treatment plants.

REFERENCES

ABCP - Associação Brasileira de Cimento Portland. (2019). *Panorama do Coprocessamento.* São Paulo: ABPC, 19p.

ABNT NBR 10157, de 30 de dezembro de 1987. (1987). *Norma brasileira: fixa as condições mínimas exigíveis para projeto e operação de aterros de resíduos perigosos.* Brasil.

ABNT NBR 10004, de 31 de maio de 2004. (2004a). *Norma brasileira: Resíduos Sólidos – Classificação.* Brasil.

ABNT NBR 10005, de 31 de maio de 2004. (2004b). *Norma brasileira: fixa os requisitos exigíveis para a obtenção de extrato lixiviado de resíduos sólidos.* Brasil.

ABNT NBR 10006, de 31 de maio de 2004. (2004c). *Norma brasileira: fixa os requisitos exigíveis para obtenção de extrato solubilizado de resíduos sólidos.* Brasil.

ABNT NBR 10007, de 31 de maio de 2004. (2004d). *Norma brasileira: fixa os requisitos exigíveis para amostragem de resíduos sólidos*. Brasil.

ABRELPE – Associação Brasileira de Empresas de Limpeza Pública e Resíduos Especiais. (2019). *Panorama dos Resíduos Sólidos no Brasil 2018/2019*. São Paulo.

Abreu, F. V., Costa Filho, M. A. F., Souza, M. C. L. (2009). "Technical and economical feasibility analysis of energy generation though the biogas from garbage in landfill – an alternative of renewable energy generation". *20th International Congress of Mechanical Engineering*, Gramado – RS.

André, L., Pauss, A., Ribeiro, T. (2018). "Solid anaerobic digestion: State-of-art, scientific and technological hurdles". *Bioresource Technology*, 247, 1027–1037.

ANEEL – Agência Nacional de Energia Elétrica. (2019). *Consulta Pública nº 25, de 18 de dez. 2019, publicado no DOU de 17 de ago. 2019, seção 3*. 116p. https://www.aneel.gov.br/consultas-publicas?p_p_id=participacaopublica_WAR_participacaopublicaportlet&p_p_lifecycle=2&p_p_state=normal&p_p_mode=view&p_p_cacheability=cacheLevelPage&p_p_col_id=column-2&p_p_col_pos=1&p_p_col_count=2&_participacaopublica_WAR_participacaopublicaportlet_ideDocumento=39240&_participacaopublica_WAR_participacaopublicaportlet_tipoFaseReuniao=fase&_participacaopublica_WAR_participacaopublicaportlet_jspPage=%2Fhtml%2Fpp%2Fvisualizar.jsp. Access: 25/08/2020.

Angelo, A. C. M. (2014). *Contribuições para o Inventário do Ciclo de Vida dos Resíduos Orgânicos provenientes da Coleta Domiciliar na Cidade do Rio de Janeiro*. Master's thesis, Universidade Federal do Rio de Janeiro (UFRJ/COPPE).

BRASIL. LEI Nº 12.305/2010. (2010). *Institui a Política Nacional de Resíduos Sólidos*. Presidência da República do Brasil.

BRASIL. LEI Nº 14.026/2020. (2020). *Atualiza o marco legal do saneamento básico*. Presidência da República do Brasil.

Brazilian Ministry of Cities. (2015). *Estado da Arte da Tecnologia de Metanização Seca*. Coletânea de publicações do PROBIOGÁS Série Aproveitamento Energético de Biogás de Resíduos Sólidos Urbanos, 1ª Edição. Brasília, BR.

CEMPRE – Compromisso Empresarial para Reciclagem. (2015). *CEMPRE Review 2015*. São Paulo, BR.

Chatterjee, B., Mazumder, D. (2016). "Anaerobic digestion for the stabilization of the organic fraction of municipal solid waste: A review". *Environmental Review*.

Chiu, S. L. H., Lo, I. M. C. (2016). "Reviewing the anaerobic digestion and co-digestion process of food waste from the perspectives on biogas production performance and environmental impacts". *Environmental Science and Pollution Research*, 23, 24435–24450.

CIBIOGÁS (2020). *Nota Técnica: 002/2010 – Panorama do Biogás no Brasil em 2019*. Foz do Iguaçu, Abril de.

DIRETIVA (UE) 2018/851, de 30 de maio de 2018. (2018). *Altera a Diretiva 2008/98/CE relativa aos resíduos*. Jornal Oficial da União Europeia.

EPE – Empresa de Pesquisa Energética. (2014). *Inventário Energético dos Resíduos Sólidos Urbanos. Nota Técnica DEA 18/14*. Rio de Janeiro: EPE. 49p.

Fagbohungbe, M. O., Herbert, B. M. G., Hurst, L., Ibeto, C. N., Li, H., Usmani, S. Q., Semple, K. T. (2016). "The challenges of anaerobic digestion and the role of biochar in optimizing anaerobic digestion". *Waste Management*.

Ferri, G. L., Chaves, G. L. D., Ribeiro, G. M. (2015). "Reverse logistics network for municipal solid waste management: The inclusion of waste pickers as a Brazilian legal requirement". *Waste Management*, 40, 173–191.

Fricke, K., Bidlingmeier, W. (2016). "Biological fundamentals and processing. Teach4waste - international E-Learning platform waste and resource management". *Alemanha*.

Gohlke, O., Martin, J. (2007). "Drivers for innovation in waste-to-energy technology". *Waste Management Research*, 25, 214–219.

Gomes, F. C. S. P, Aquino, S. F., Colturato, L. F. D. B. (2012). "Biometanização seca de resíduos sólidos urbanos: Estado da arte e análise crítica das principais tecnologias". *Engenharia Sanitaria e Ambiental*, 17(3), 295–304.

Henze, M., Loosdrecht, M. C. M, Ekama, G. A., Brdjanovic, D. (2017). *Tratamiento biológico de aguas residuales - Principios, modelación y diseño*. Reino Unido: IWA Publishing.

IBAM (2001). "Manual Gerenciamento Integrado de Resíduos Sólidos". *Instituto Brasileiro de Administração Municipal (IBAM) e Secretaria Especial de Desenvolvimento Urbano da Presidência da República (SEDU/PR)*. Rio de Janeiro, Brasil.

IBGE (2020). *Database of Population from Brazilian Municipalities*. https://www.ibge.gov.br/estatisticas/sociais/populacao/9103-estimativas-de-populacao.html?=&t=resultados>. Access: 27/08/2020.

Inácio, C. T., Miller, P. R. M. (2009). "Compostagem – Ciência e prática para a gestão de resíduos orgânicos". *Embrapa Solos e Universidade Federal de Santa Catarina.* Rio de Janeiro, BR.

IPEA – Instituto de Pesquisa Economica Aplicada. (2012). *Diagnóstico dos Resíduos Sólidos Industriais.* Brasília: IPEA, 67p.

Jain, S., Jain, S., Wolf, I. T., Lee, J., Tong, Y. W. (2015). "A comprehensive review on operating parameters and different pre-treatment methodologies for anaerobic digestion of municipal solid waste". *Renewable and Sustainable Energy Reviews*, 52, 142–154.

Jha, P., Schmidt, S. (2016). "Reappraisal of chemical interference in anaerobic digestion processes". *Renewable and Sustainable Energy Reviews.*

Jucá, J. F. T, Lima, J. D., Mariano, M. O. H., Firmo, A. L. B., Lima, D. G. A, Lucena, L. F. L., Castilho Júnior, A. B., Cavalcanti, R. M. S., Sá, E. V. F. L., Melo, R., Mendonça, M. Z., Galvão, D. (2014). *Análise das Diversas Tecnologias de Tratamento e Disposição Final de Resíduos Sólidos Urbanos no Brasil, Europa, Estados Unidos e Japão.* Jaboatão dos Guararapes, PE: Grupo de Resíduos Sólidos da UFPE e Banco Nacional do Desenvolvimento Econômico e Social - BNDES, 186p.

Lombardi, L., Carnevale, E., Corti, A. (2014). "A review of technologies and performances of thermal treatment systems for energy recovery from waste". *Waste Management*, 37, 26–44.

MMA – Brazilian Environment Ministry. (2020). *Information about Municipal Consortiums for Solid Waste Management.* https://mma.gov.br/cidades-sustentaveis/residuos-solidos/item/10333-consorcios.html. Access: 27/08/2020

Ornelas-Ferreira, B., Lobato, L. C. S., Colturato, L. F. D., Torres, E. O., Pombo, L. M., Pujattia, F. J. P., Araújo, J. C., Chernicharo, C. A. L. (2020). "Strategies for energy recovery and gains associated with the implementation of a solid state batch methanization system for treating organic waste from the city of Rio de Janeiro – Brazil". *Renewable Energy*, 146, 1976–1983.

Paradela, F. M. R. (2007). *Estudo da pirólise de misturas de resíduos plásticos e de biomassa.* Dissertação apresentada na Faculdade de Ciências e Tecnologia da Universidade Nova de Lisboa.

Ren, Y., Yu, M., Wu, C., Wang, Q., Gao, M., Huang, Q., Liu, Y. (2018). "A comprehensive review on food waste anaerobic digestion: Research updates and tendencies". *Bioresource Technology*, 247, 1069–1076.

Rizzo, F. (2019). *Waste-to-Energy Development: Perspective and Lesson Learned. In: Workshop de Fontes Energéticas no Âmbito do Planejamento de Longo Prazo.* Brasília, BR: Recuperação Energética Waste-to-Energy. Ministério de Minas e Energia.

Silva, U. A., Teixeira, A. F. M., Quintaes, B. R., Silva, C. A. M. C., Campos, J. C. (2008). *Caracterização Gravimétrica dos Resíduos Sólidos Domiciliares do Município do Rio de Janeiro –2008.* 25° Congresso Brasileiro de Engenharia Sanitária e Ambiental ABES – Associação Brasileira de Engenharia Sanitária e Ambiental.

Siqueira, T. M. O., Assad, M. L. R. C. L. (2019). "Compostagem de Resíduos Sólidos Urbanos no estado de São Paulo (Brasil)". *Ambiente & Sociedade*, XVIII(4), 243–264. São Paulo, BR.

Tolmasquim, M. T. (2003). *Fontes renováveis de energia no Brasil.* 1a. Edição. Rio de Janeiro: Interciência, 516p.

Visedo, G., Pecchio, M. (2019). *Roadmap tecnológico do cimento: Potencial de redução das emissõesde carbono da indústria do cimento brasileira até 2050.* Rio de Janeiro: SNIC, 64p.

Von Sperling, M. (2018). *Princípios básicos do tratamento de esgotos. Coleção Princípios do Tratamento Biológico de Águas Residuárias*, 2, 2ª ed. Brasil, BR: Universidade Federal de Minas Gerais.

Von Sperling, M., Chernicharo, C. A. L. (2005). *Biological Wastewater Treatment in Warm Climate Regions.* Volume one. India: IWA Publishing.

World Bank. (2012). *What a Waste - A Global Review of Solid Waste Management.* Urban Development Series Knowledge Papers. EUA.

Zhang, C., Su, H., Baeyens, J., Tan, T. (2014). "Reviewing the anaerobic digestion of food waste for biogás production". *Renewable and Sustainable Energy Reviews*, 38, 383–392.

11 Potential Innovative Technique for the Management of Polystyrene Waste in South Africa

Olusola Olaitan Ayeleru, Sisanda Dlova, Ojo Jeremiah Akinribide, Oluwasayo F Olorundare, Freeman Ntuli, Williams Kehinde Kupolati and Peter Apata Olubambi

CONTENTS

11.1 INTRODUCTION

There are four groups of materials all over the world. These materials can be classified as metals, ceramics, polymers and composites (Kam & Kueh, 2015). Polymer plastics (PPs) have been employed in diverse applications, which has led to their exponential increase in the recent years. Polymers themselves are materials comprising of structural units that occur repeatedly; each of the units is obtained from a compound known as a monomer (Tur & Trivedi, 2014, 2015). Polymers can be categorized into two types, the thermoplastic and thermoset, and once additives are combined with them, plastics are formed (Kam & Kueh, 2015). Plastics are materials made up of high molecular weight polymers which are comprised of carbon and hydrogen formed from petroleum and natural gas (Plastic Waste Management Institute, 2009). Plastic can be grouped into seven classes (Figure 11.1). Plastics are non-biodegradable material, with artificial synthetic polymers; their structure does not occur naturally (Kržan, 2012). Global plastic production is in the tens of millions of tons annually. This growth has called for a holistic approach for efficient, effective and sustainable technology for the recycling of the waste polymer plastics (WPPs) generated after their first application or usage (Geyer et al., 2017). Presently, incineration and landfilling are methods extensively used for the disposal of WPPs; only a fraction of waste generated from PPs has been recycled by material recycling, thermal energy recovery or hydrothermal/solvothermal process (Antelava et al., 2019; Awasthi et al., 2017). Therefore, new technologies for WPPs recycling are expected to develop since the conventional methods are having issues with regards to sustainability.

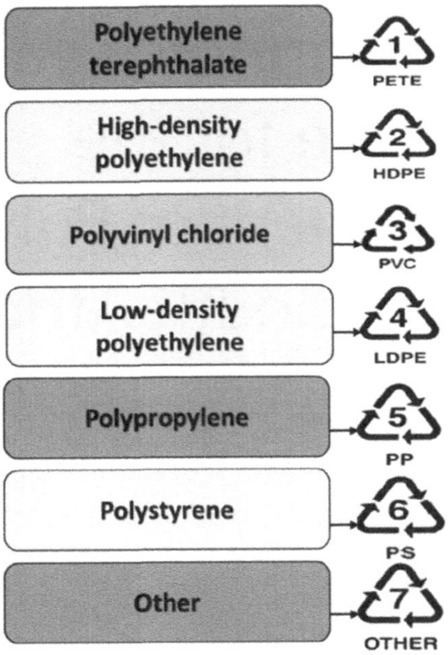

FIGURE 11.1 Various kinds of plastic recycling codes.

11.2 OVERVIEW OF POLYSTYRENE

Polystyrene (PS) has been in use globally as far back as 1839 and is the oldest and most utilized aromatic thermoplastic derived from styrene, a hydrocarbon family (Hansen et al., 2015; Lynwood, 2014). PS is a material that is thermally cracked into aromatic hydrocarbons because of its essential properties such as electrical insulator, optical clarity etc. (Hussain et al., 2012). It is a multipurpose plastic material employed in many applications such as packaging of household items, electronics packages, packaging of food items, laboratory wares and in many other purposes (Al-Maaded et al., 2012; American Chemistry Council, 2014; Samper et al., 2010). A recent study has also shown that PS has been utilized in packaging materials since the early 1950s (Block et al., 2017). PS is the type 6 plastic (appears a 6 in the arrows) (Figure 11.2) which can either be rigid or foamed, and has potential applications in many areas (Cleanaway, 2018). The rigid PS is employed in containers for yoghurt drinks, plastic spoons, etc. The foam aspect of PS is the expanded polystyrene (EPS)

FIGURE 11.2 (a) Polystyrene containers; (b) polystyrene packaging.

which is utilized as cups for hot teas or coffees, packaging for refrigerators or electronic packaging. It is made up of approximately 96% air and 4% of plastic (Cleanaway, 2018; UNEP, 2018). The global growth rate of PS production has been steadily increasing with about a 1.4% rise annually since 2000, and its market demand reached ~15 million tons in 2010. The market demand for PS has been projected to hit ~24 million tons by 2020 (Hansen et al., 2015; Intelligence Global Business, 2012). PS is mostly categorized as crystal PS (CPS) or general-purpose polystyrene (GPPS) because of its sparkling appearance, or as high-impact polystyrene (HIPS) because of its brittleness (Amaniampong, 2015; Hansen et al., 2015). The GPPS is used for making video cassette cases and other materials (Cleanaway, 2018). PS is usually formed through suspension or solution polymerization of styrene monomers obtained from petroleum (Abota, 2012; Amaniampong, 2015; Ampofo, 2015; Chemicals and Petrochemicals Manufacturers' Association, 2012).

However, the advent of PS on the global market and its application in a wide range of product packaging, including the widespread production of products of low-cost, high-performance, energy saving and sustainable development, have led to the acute growth rate of PS (Vilaplana & Karlsson, 2008). The combination of PS with additives and colourants allows the production of several products like toys, automobile parts, electronics, etc. (American Chemistry Council, 2014). The bulk of this PS material has a single-use application after which it eventually becomes waste otherwise known polystyrene waste (PSW) (Global Alliance for Incinerator Alternatives, 2013; Mo et al., 2014; Verma et al., 2016; Watson, 2016). The continuous growth rate of PSW is now an issue of concern globally since consistent information on municipal solid waste (MSW) generation is usually scanty or unavailable in low-income countries (LCs). Moreover, in places where MSW data are available, the statistics are usually unreliable; hence regular waste characterization activities on different MSW streams, particularly plastic waste (PW), become very necessary (AbdAlqader & Hamad, 2012; Miezah et al., 2015). From the recent waste composition studies conducted in South Africa, it was observed that the amount of PSW generated was not reported but it was documented that the recycling rate of PSW has been increasing considerably since 2016. Furthermore, a survey conducted between 2013 and 2015 reported that the rate of recycling of PSW grew by 106% but there was very little evidence to support the claim (Infrastructure News, 2017). The significant increase observed in one of their findings was said to be due to public enlightenment on the need for recycling and the need for the rapid creation of end markets (ESI Africa, 2018; Infrastructure News, 2017).

In the same vein, studies conducted by Ayeleru et al. (2016) and Ayeleru et al. (2018) in the city of Johannesburg, South Africa, corroborated the previous reports that there are no records or separate data for the PSW generated from the total volume of PW streams consumed. The studies only reported that approximately 27% of the overall amount of MSW generated was PWs (Ayeleru et al., 2016; Ayeleru et al., 2018). In addition, a study has reported that ~0.5 million tons of PWs were generated in Ghana annually but none of these wastes were collected. These are clear pointers that there are no records for PSW generated in many nations in Africa including South Africa (Kortei & Quansah, 2018). Similarly, some studies in Ghana reported that out of the total volume of PWs generated, only 2% was recycled while the remaining was disposed of in the landfill sites; some was dumped by the roadside and allowed to litter the street, while the remaining was thrown into the drainage (B&FT Online, 2016; Lambert & Sabutey, 2016). Ackah et al. (2012) reported that PWs accounted for about 16.5% of the total MSW generated in Accra, the capital city of Ghana, but there was no statistical data to quantify them into different PW streams (Ackah et al., 2012). Furthermore, a recent study conducted at the University of Lagos, Nigeria, by Adeniran et al. (2017) only focused on the percentage compositions of polythene bags and other plastics (i.e. 24% and 9% respectively); there were no records for other PW streams including the PSW (Adeniran et al., 2017). Adekomaya and Ojo (2016) reported only the total percentage of PWs (57%) from total MSW generated in Lagos State, Nigeria (Adekomaya & Ojo, 2016). Imam et al. (2008), in their study carried out at various locations/outlets of the Federal Capital Territory (FCT) of Nigeria, reported the percentage total amount of PWs generated, there are no data to classify PW into streams (Imam et al., 2008). Abur et al. (2014) in their study also did not report the percentage compositions of different type of PWs (Abur et al., 2014). Gakungu et al. (2012) carried out a study in over 40 tertiary institutions in Kenya, but only revealed the overall compositions of waste generated; there were no data for the amount of PWs generated nor the percentage of PSW (Gakungu et al., 2012). All the above reports have supported other previous studies that there are little or no records for individual PW components in the developing countries (DCs), specifically in South Africa. To the best of our knowledge, there has been little or no study that has focused on the recycling of PSW via chemical recycling techniques (U.S. Environmental Protection Agency, 2015).

11.3 THE 4Rs OF POLYSTYRENE WASTE TO ZERO WASTE

There have been numerous misconceptions about PS waste management. First, there was the 3Rs (reduce, reuse and recycle) concept. The 3Rs were said to be achieved when priorities are rightly placed in a way that can assist in cutting back the amount of PS consumption. The 3Rs are often referred to as the hierarchy of waste management (WM) and have been promoted in many nations of the world, including in Japan (Abdul-Rahman & Wright, 2014; Green Coast, 2019; Organisation for Economic Co-operation and Development, 2019; Wichai-utcha & Chavalparit, 2019; Yolin, 2015). The 3Rs later graduated to the 4Rs (reduce, reuse, recycle and recover) when the former was no longer effective enough to address the issue of acute MSW generation, particularly the PSW. The reduce concept is to decrease waste in the productive stage and to utilize resources efficiently. The reuse idea is to put into use again what may previously have been discarded. This idea was adopted to change the properties of the waste item into a better form. The recover model is employed to recoup resources from materials that ordinarily are to be thrown away, and the innovative method employed is chemical recycling (City of Belmont, 2019; International Institute for Sustainable Development, 2013; Sadi et al., 2012; Skanska, 2012). In recent times, the development of materials with enhanced features and functionalities is gaining wider attention in the field of material science and interface of chemistry (Katagiri, 2017; Pyun & Matyjaszewski, 2001). One of these materials is the hybrid nanocomposite (NC) that can be developed via the incorporation of organic polymer into inorganic materials to form an outstanding novel material (Pyun & Matyjaszewski, 2001). Generally, hybrid organic-inorganic materials depict a boundary between two worlds of chemistry where each of them offers a profound contribution to material science since the properties of each will be of immense benefits to material science (Gomez-Romero, 2001).

Because of the numerous issues confronting many nations of the world with respect to the production and consumption of plastic materials, it has become very crucial to design new and functional hybrid materials or nanocomposites (NC) with enhanced properties from PSWs since this will help to reduce the excessive waste polystyrene in the environment (Krasia-Christoforou, 2015). The NC is comprised of two or more materials with diverse physical and chemical properties to form a novel material, with one of these materials having a grain size range of 1–100 nm (Akpan et al., 2019; Govindaraj et al., 2004; Lateef & Nazir, 2018). NCs have a large surface area that can support interfacial connections alongside polymeric environment when compared to traditional fillers. This brings about the enhancement of the properties of polymers. Recently, organic and inorganic materials have gained attention extensively in the development of NCs owing to their special features (Mallikarjuna et al., 2005; Song et al., 2004). The organic materials include plastic materials with unique properties such as physical, thermal and mechanical properties which make them good candidates for varied applications (Grigore, 2017). In a bid to lessen health and environmental issues connected with poor disposal of waste plastics, several studies have lately incorporated inorganic precursors into polymer materials to produce polymer composites of better properties compared to the original materials (starting materials) (Ali & Venkataraman, 2014). The incorporation of inorganic precursors into the polymer composites to produce NCs is very crucial to numerous applications (Barry et al., 2018). The development of NCs from PSWs can now serve as a sustainable alternative since landfilling and incineration of MSW pose a serious threat to public health and the environment. Hence, PSW which was once regarded as a problem can now begin to realize its potential applications in different fields, thereby providing solutions to the issues of pollution arising from poor PW management (Aslam & Er. Shahan-ur- Rahman, 2009; Bandara et al., 2017; Krasnov et al., 2010; Patil & Khurd, 2015; Yunus & Fauzan, 2017).

11.4 CURRENT METHODS FOR THE DEVELOPMENT OF VALUE-ADDED MATERIALS FROM RECYCLED POLYSTYRENE

NCs as shown in Figure 11.3 are materials of superior value to industries and have been prepared through various routes during the last decades (Oliveira & Machado, 2013). Some of these methods include the sol-gel, melt blending, *in-situ* polymerization, solution blending, direct compounding and melt intercalation (Khan et al., 2016). However, most of these techniques have huge limitations that prevent them from being scaled up for commercialization. Some of the challenges include high temperature for the decomposition of polymers which may affect the processing conditions of the products and could decompose polymer itself, poor dispersion of polymer into the polymer matrix, intricacies encountered in the choice of solvents, formation of aggregates which may not allow homogenous dispersion of polymer into the matrix and sometimes heat is required in some of the methods. All these could also be major drawbacks and could also hinder commercialization of the techniques (Filippi et al., 2007; Paszkiewicz & Szymczyk, 2019; Ravichandran et al., 2018; Ucankus et al., 2018).

11.5 SOLVOTHERMAL AND HYDROTHERMAL METHODS

The hydrothermal technique of material synthesis was established following the Second World War (Buisson & Arnaud, 1994). The first study on hydrothermal processes was conducted by Schalfthaul around the mid-19th century (Suchanek et al., 2004). The hydrothermal method is closely related to the solvothermal process. They are both appropriate for the development of novel materials (Tareen, 2013). The hydrothermal technique is usually described as crystal synthesis/growth that employs single or diverse phase reactions operating under a high temperature and high-pressure water conditions from materials that are not soluble at room temperature and pressure (Hayashi & Hakuta, 2010; Riman et al., 2002). The hydrothermal method is different from the solvothermal technique; while the hydrothermal method utilizes aqueous solvent, the solvothermal synthesis method uses non-aqueous solvents for material synthesis (Riman et al., 2002). The merits of solvothermal/hydrothermal methods over other conventional and non-conventional techniques include the ease of conversion or scaling up to commercial application and cost-effectiveness, and all ceramics such as powders, fibres, polymers, etc., can be synthesized via solvothermal/hydrothermal method (Riman et al., 2002; Suchanek et al., 2004). Hydrothermal/solvothermal reactions are usually done in a sealed reactor (otherwise known as autoclave) or high-pressure bomb. The autoclave comprises a Teflon cup or can be made from Teflon whose purpose is to shield the autoclave from corrosive

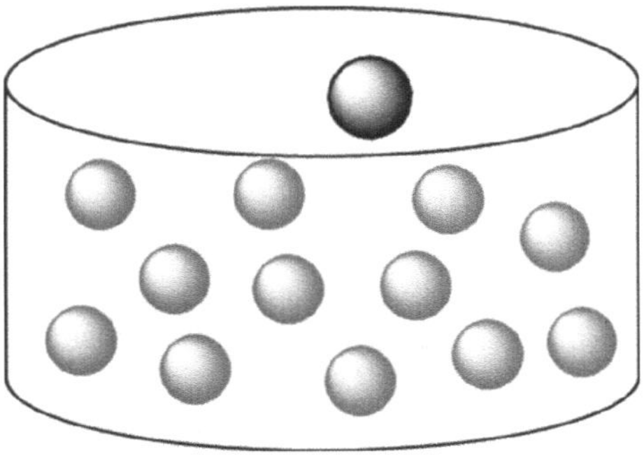

FIGURE 11.3 Organic-inorganic hybrid nanocomposites.

FIGURE 11.4 A hydrothermal autoclave reactor.

TABLE 11.1
Generally Utilized Reactors for Hydrothermal/ Solvothermal Methods (Li et al., 2016)

Type	Temperature (°C)	Pressure (bar)
Quartz tube	250	6
Cone closure	750 to ~1150	4000 to ~13000
Welded closure	400	3000
Modified Bridgman autoclave	500	3700
Piston cylinder	1000	40000
Opposed diamond anvil	>2000	500000

solvents (Li et al., 2016). Figure 11.4 shows the autoclave reactor, and Table 11.1 shows the commonly used reactors for hydrothermal/solvothermal methods.

Currently, there is a need for combinations of diverse techniques which are very common among researchers for optimum results, and the hydrothermal method is not an exception in this drive. Hydrothermal hybrid techniques have been reportedly employed for the synthesis of materials (including nanomaterials, i.e. NCs) and chemical compounds, mainly inorganics. Moreover, in order to enhance the reaction kinetics optimally or for the ability to make new materials, a great amount of work has been done to hybridize the hydrothermal technique with microwaves (MW) (*microwave-hydrothermal processing*), electrochemistry (*hydrothermal electrochemical synthesis*), ultrasound (*hydrothermal-sonochemical synthesis*), mechanochemistry (*mechanochemical hydrothermal synthesis*), optical radiation (*hydrothermal-photochemical synthesis*) and hot-pressing (*hydrothermal hot pressing*) (Suchanek & Riman, 2006).

The hydrothermal method itself, microwave-hydrothermal and microwave-solvothermal methods are, in particular, truly low-temperature methods for the preparation of materials of different sizes and shapes. These methods save energy and are environmentally friendly, because the reactions take place in closed isolated system conditions. The NC materials can be produced in either a batch or continuous process using the above methods (Iwaya et al., 2008; Kharisov et al., 2012). In our recent study by Ayeleru et al. (2020) the solvothermal technique was utilized for the development of nanocomposites from recycled polystyrene materials via the incorporation of metallic precursors into polystyrene matrix, and the results obtained were excellent, showing an economical novel means of recycling PSWs via chemical recycling with little or no impact on public health and the environment (Ayeleru et al., 2020).

11.6 CONCLUSION

The paucity of reliable data can be problematic for researchers in conducting research. However, no matter how complicated the issue might have been, proffering solutions to it goes a long way towards solving the challenge. PS waste generation, application, waste management and the technology involved in their production have been discussed. Emphasis was placed on new methods or process techniques for their production which include hydrothermal/solvothermal processes. More work can be done on detailed analysis on processing methods from different countries alongside detailed waste management methods.

ACKNOWLEDGEMENTS

The authors would like to thank the National Research Foundation (NRF) for the funding provided for this study, the Centre for Nanomechanics and Tribocorrosion (CNT), University of Johannesburg and the Department of Chemical Engineering, University of Johannesburg for providing an enabling environment for research. One of the authors would like to specially thank Dr. Idris Azeez and Mr. Adewale Oladipo of the University of South Africa and Dr. Franklin Olumide Afinjuomo of the University of South Australia for their valuable inputs.

REFERENCES

AbdAlqader, A., & Hamad, J. (2012). Municipal solid waste composition determination supporting the integrated solid waste management in Gaza strip. *International Journal of Environmental Science and Development, 3*(2), 172–177.

Abdul-Rahman, F., & Wright, S. E. (2014). *Reduce, Reuse, Recycle: Alternatives for Waste Management*: NM State University, Cooperative Extension Service.

Abota, C. A. (2012). *Recycling of Plastics Waste in Ghana: A Way to Reduce Environmental Problems/ Pollutions*. (Plastics Technology). Arcada University of Applied Sciences, Finland. 12033.

Abur, B. T., Oguche, E. E., & Duvuna, G. A. (2014). Characterization of municipal solid waste in the Federal Capital, Abuja, Nigeria. *Global Journal of Science Frontier Research: Environmental Earth Sciences, 14*(2), 1–6.

Ackah, R., Carboo, D., & Gyamfi, E. T. (2012). Challenges of plastic waste disposal in Ghana: A case study of solid waste disposal sites in Accra. *Management Arts, 49*, 9879–9885. Retrieved from https://www.elixirpublishers.com/articles/1350467591_49%20(2012)%209879-9885.pdf

Adekomaya, O., & Ojo, K. (2016). Adaptation of plastic waste to energy development in Lagos: An overview assessment. *Nigerian Journal of Technology, 35*(4), 778–784.

Adeniran, A. E., Nubi, A. T., & Adelopo, A. O. (2017). Solid waste generation and characterization in the University of Lagos for a sustainable waste management. *Waste Management, 67*, 3–10.

Akpan, E. I., Shen, X., Wetzel, B., & Friedrich, K. (2019). Design and synthesis of polymer nanocomposites. In *Polymer Composites With Functionalized Nanoparticles* (pp. 47–83). Elsevier.

Al-Maaded, M., Madi, N. K., Kahraman, R., Hodzic, A., & Ozerkan, N. G. (2012). An overview of solid waste management and plastic recycling in Qatar. *Journal of Polymers and the Environment, 20*(1), 186–194.

Ali, T., & Venkataraman, A. (2014). Preparation and characterization of [gamma]-Fe2O3 dispersed polystyrene nanocomposite film. *International Journal of Advances in Engineering & Technology, 7*(1), 122.

Amaniampong, S. B. (2015). *Developing an Option for Sustainable Plastic Waste Management in Ghana: A Case Study of Sunyani Municipality, Ghana*. (Masters). Kwame Nkrumah University of Science and Technology Kumasi, Ghana.

American Chemistry Council, I. (2014). *Polystyrene*. Retrieved from https://www.chemicalsafetyfacts.org/wp-content/uploads/2014/05/082514_ChemSafety_Print-Polystyrene.pdf

Ampofo, S. K. (2015). *The Options for the Effective Management of Plastic Waste in Ghana*. Report on Management of Plastic Waste in Ghana-21-328-STASWAPA. pdf.

Antelava, A., Damilos, S., Hafeez, S., Manos, G., Al-Salem, S. M., Sharma, B. K., … Constantinou, A. (2019). Plastic Solid Waste (PSW) in the context of Life Cycle Assessment (LCA) and sustainable management. *Environmental Management*, *64*(2), 230–244.

Aslam, & Er. Shahan-ur- Rahman. (2009). *Use of Waste Plastic in Construction of Flexible Pavement.* Retrieved from http://www.nbmcw.com/roads-pavements/930-use-of-waste-plastic-in-construction-of-flexible-pavement.html

Awasthi, A. K., Shivashankar, M., & Majumder, S. (2017). *Plastic Solid Waste Utilization Technologies: A Review.* Paper presented at the Materials Science and Engineering Conference Series.

Ayeleru, O. O., Dlova, S., Akinribide, O. J., Olorundare, O. F., Akbarzadeh, R., … Kempaiah, D. M. (2020). Nanoindentation studies and characterization of hybrid nanocomposites based on solvothermal process. *Inorganic Chemistry Communications*, 107704.

Ayeleru, O. O., Ntuli, F., & Mbohwa, C. (2016). *Municipal Solid Waste Composition Determination in the City of Johannesburg.*

Ayeleru, O. O., Okonta, F. N., & Ntuli, F. (2018). Municipal solid waste generation and characterization in the city of Johannesburg: A pathway for the implementation of zero waste. *Waste Management*, *79*, 87–97.

B&FT Online. (2016). *Only 2% of Plastic Waste Generated is Recycled.* Retrieved from https://thebftonline.com/2018/headlines/only-2-of-plastic-waste-generated-is-recycled/

Bandara, W. R. L. N., de Silva, R. M., de Silva, K. M. N., Dahanayake, D., Gunasekara, S., & Thanabalasingam, K. (2017). Is nano ZrO 2 a better photocatalyst than nano TiO 2 for degradation of plastics? *RSC Advances*, *7*(73), 46155–46163.

Barry, S. T., Teplyakov, A. V., & Zaera, F. (2018). The chemistry of inorganic precursors during the chemical deposition of films on solid surfaces. *Accounts of Chemical Research*, *51*(3), 800–809.

Block, C., Brands, B., & Gude, T. (2017). *Packaging Materials: 2. Polystyrene for Food Packaging Applications.* Retrieved from Belgium: http://ilsi.eu/wp-content/uploads/sites/3/2017/12/PS-ILSI-Europe-Report-Update-2017_Interactif_FIN.pdf

Buisson, X., & Arnaud, R. (1994). Hydrothermal growth of quartz crystals in industry. Present status and evolution. *Le Journal de Physique IV,* 4(C2), C2-25–C22-32.

Chemicals and Petrochemicals Manufacturers' Association. (2012). PS – Polystyrene. Retrieved from http://cpmaindia.com/ps_about.php

City of Belmont. (2019). Reduce, reuse, recycle, recover. Retrieved from http://www.belmont.wa.gov.au/Property/Pages/Reduce,Reuse,Recycle,Recover.aspx

Cleanaway. (2018, 20 March). Is polystyrene recyclable? Retrieved from https://www.cleanaway.com.au/about-us/sustainable-future/is-polystyrene-recyclable/

ESI Africa. (2018, 5 June). S.Africa | 1.144m tonnes of recyclable plastic dumped in landfills. Retrieved from https://www.esi-africa.com/s-africa-1-144m-tonnes-of-recyclable-plastic-dumped-in-landfills/

Filippi, S., Mameli, E., Marazzato, C., & Magagnini, P. (2007). Comparison of solution-blending and melt-intercalation for the preparation of poly (ethylene-co-acrylic acid)/organoclay nanocomposites. *European Polymer Journal*, *43*(5), 1645–1659.

Gakungu, N. K., Gitau, A. N., Njoroge, B. N. K., & Kimani, M. W. (2012). Solid waste management in Kenya; A case study of public technical training institutions. *ICASTOR Journal of Engineering*, *5*(3), 127–138.

Geyer, R., Jambeck, J. R., & Law, K. L. (2017). Production, use, and fate of all plastics ever made. *Science Advances*, *3*(7), e1700782.

Global Alliance for Incinerator Alternatives. (2013). *Waste Incinerators: Bad News for Recycling and Waste Reduction*. Retrieved from http://www.no-burn.org/wp-content/uploads/Bad-News-for-Recycling-Final.pdf

Gomez-Romero, P. (2001). Hybrid organic–inorganic materials—in search of synergic activity. *Advanced Materials*, *13*(3), 163–174.

Govindaraj, B., Sastry, N. V., & Venkataraman, A. (2004). Thermal and morphological studies on γ-Fe2O3 polystyrene composites and the affect of additives. *Journal of Applied Polymer Science*, *93*(2), 778–788.

Green Coast. (2019, 21 April). What are the 3R'S of waste management? Retrieved from https://greencoast.org/3rs-of-waste-management/

Grigore, M. (2017). Methods of recycling, properties and applications of recycled thermoplastic polymers. *Recycling*, *2*(4), 24.

Hansen, A. P., da Silva, G. A., & Kulay, L. (2015). Evaluation of the environmental performance of alternatives for polystyrene production in Brazil. *Science of the Total Environment*, *532*, 655–668.

Hayashi, H., & Hakuta, Y. (2010). Hydrothermal synthesis of metal oxide nanoparticles in supercritical water. *Materials*, *3*(7), 3794–3817.

Hussain, Z., Khan, K. M., Perveen, S., Hussain, K., & Voelter, W. (2012). The conversion of waste polystyrene into useful hydrocarbons by microwave-metal interaction pyrolysis. *Fuel Processing Technology*, *94*(1), 145–150.

Imam, A., Mohammed, B., Wilson, D. C., & Cheeseman, C. R. (2008). Solid waste management in Abuja, Nigeria. *Waste Management*, *28*(2), 468–472.

Infrastructure news. (2017). SA's polystyrene recycling growth disproves its 'unrecyclable' status. Retrieved from http://www.infrastructurene.ws/2017/02/15/sas-polystyrene-recycling-growth-disproves-its-unrecyclable-status/

Intelligence Global Business. (2012). Polystyrene (PS) and Expandable Polystyrene (EPS) Global Market to 2020—Continued Development in the Construction (EPS) and Packaging (PS) End-use segments to accelerate future Demand. *End-use Segments to Accelerate Future Demand*. Chesham.

International Institute for Sustainable Development. (2013). The 4Rs - reduction, reuse, recycling and recovery. Retrieved from https://www.iisd.org/business/tools/bt_4r.aspx

Iwaya, T., Tokuno, S., Sasaki, M., Goto, M., & Shibata, K. (2008). Recycling of fiber reinforced plastics using depolymerization by solvothermal reaction with catalyst. *Journal of Materials Science*, *43*(7), 2452–2456.

Kam, C. Z., & Kueh, A. B. H. (2015). Towards sustainable polymeric materials: Zero waste, green and self-healing. *Jurnal Teknologi*, *74*(4).

Katagiri, K. (2017). Organic-Inorganic hybrid nanoarchitecture at mesoscale. In *Supra-Materials Nanoarchitectonics* (pp. 117–133). Elsevier.

Khan, W. S., Hamadneh, N. N., & Khan, W. A. (2016). Polymer nanocomposites–synthesis techniques, classification and properties. In P. D. Sia (Ed.), *Science and Applications of Tailored Nanostructures* (pp. 50–67). One Central Press.

Kharisov, B. I., Kharissova, O. V., & Méndez, U. O. (2012). Microwave hydrothermal and solvothermal processing of materials and compounds. *The Development and Application of Microwave Heating*, *5*, 107–140.

Kortei, N. K., & Quansah, L. (2018, 19 September). Plastic waste management in Ghana - a complete failure and the consequences. Retrieved from https://www.graphic.com.gh/features/opinion/plastic-waste-management-in-ghana-a-complete-failure-and-the-consequences.html

Krasia-Christoforou, T. (2015). Organic–inorganic polymer hybrids: Synthetic strategies and applications. In *Hybrid and Hierarchical Composite Materials* (pp. 11–63). Springer.

Krasnov, A. P., Aderikha, V. N., Afonicheva, O. V., Tikhonov, N. N., Vasil'kov, A. Y., Said-Galiev, E. E., … Nikolaev, A. Y. (2010). Categorization system of nanofillers to polymer composites. *Journal of Friction and Wear*, *31*(1), 68–80.

Kržan, A. (2012). *Biodegradable Polymers and Plastics*. Retrieved from Central Europe: http://www.icmpp.ro/sustainableplastics/files/Biodegradable_plastics_and_polymers.pdf

Lambert, M., & Sabutey, V. K. (2016, 26 July). Ghana's plastic waste management problems - a global issue that needs local awareness. Retrieved from https://www.myjoyonline.com/opinion/2016/july-26th/ghanas-plastic-waste-management-problems-a-global-issue-that-needs-local-awareness.php

Lateef, A., & Nazir, R. (2018). Metal nanocomposites: Synthesis, characterization and their applications. *Science and Applications of Tailored Nanostructures*, 239–256.

Li, J., Wu, Q., & Wu, J. (2016). Synthesis of nanoparticles via solvothermal and hydrothermal methods. *Handbook of Nanoparticles*, 295–328.

Lynwood, C. (2014). *Polystyrene: Synthesis, Characteristics and Applications*. Nova Science Publishers, Inc.

Mallikarjuna, N. N., Manohar, S. K., Kulkarni, P. V., Venkataraman, A., & Aminabhavi, T. M. (2005). Novel high dielectric constant nanocomposites of polyaniline dispersed with γ-Fe2O3 nanoparticles. *Journal of Applied Polymer Science*, *97*(5), 1868–1874.

Miezah, K., Obiri-Danso, K., Kádár, Z., Fei-Baffoe, B., & Mensah, M. Y. (2015). Municipal solid waste characterization and quantification as a measure towards effective waste management in Ghana. *Waste Management*, *46*, 15–27.

Mo, Y., Zhao, L., Wang, Z., Chen, C. L., Tan, G. Y. A., & Wang, J. Y. (2014). Enhanced styrene recovery from waste polystyrene pyrolysis using response surface methodology coupled with Box–Behnken design. *Waste Management*, *34*(4), 763–769.

Oliveira, M., & Machado, A. V. (2013). Preparation of Polymer-Based Nanocomposites by Different Routes. 1–22.

Organisation for Economic Co-operation and Development. (2019). *OECD Environmental Performance Reviews: Japan 2010*. Retrieved from http://www.oecd.org/japan/japan2010.htm

Paszkiewicz, S., & Szymczyk, A. (2019). Graphene-based nanomaterials and their polymer nanocomposites. In *Nanomaterials and Polymer Nanocomposites* (pp. 177–216). Elsevier.

Patil, N. M., & Khurd, V. G. (2015). Utilization of waste plastic in road construction. *International Journal of Inventive Engineering and Sciences (IJIES), 3*(9), 25–28.

Plastic Waste Management Institute. (2009). *An Introduction to Plastic Recycling.* Retrieved from https://www.pwmi.or.jp/ei/plastic_recycling_2009.pdf

Pyun, J., & Matyjaszewski, K. (2001). Synthesis of nanocomposite organic/inorganic hybrid materials using controlled/"living" radical polymerization. *Chemistry of Materials, 13*(10), 3436–3448.

Ravichandran, K., Praseetha, P. K., Arun, T., & Gobalakrishnan, S. (2018). Synthesis of nanocomposites. In *Synthesis of Inorganic Nanomaterials* (pp. 141–168). Elsevier.

Riman, R. E., Suchanek, W. L., & Lencka, M. M. (2002). *Hydrothermal Crystallization of Ceramics.* Paper presented at the Annales de Chimie Science des Materiaux.

Sadi, K., Abdullah, A., Navazandeh Sajoudi, M., Kamal, M. F. M., Torshizi, F., & Taherkhani, R. (2012). *Reduce, Reuse, Recycle and Recovery in Sustainable Construction Waste Management.* Paper presented at the Advanced Materials Research.

Samper, M. D., Garcia-Sanoguera, D., Parres, F., & Lopez, J. (2010). Recycling of expanded polystyrene from packaging. *Progress in Rubber Plastics and Recycling Technology, 26*(2), 83–92.

Skanska. (2012). *4 R Guide: Reduce Reuse Recycle Recover.* Retrieved from https://group.skanska.com/globalassets/sustainability/environmental-responsibility/materials/skanska4rguide.pdf

Song, G., Bo, J., & Guo, R. (2004). The characterization and property of polystyrene compounding of α-Fe 2 O 3 in the nano-scale. *Colloid and Polymer Science, 282*(6), 656–660.

Suchanek, W. L., Lencka, M. M., & Riman, R. E. (2004). Hydrothermal synthesis of ceramic materials. In *Aqueous Systems at Elevated Temperatures and Pressures* (pp. 717–744). Elsevier.

Suchanek, W. L., & Riman, R. E. (2006). *Hydrothermal Synthesis of Advanced Ceramic Powders.* Paper presented at the Advances in Science and Technology.

Tareen, J. A. K. (2013). Hydrothermal synthesis of native elements and simple oxides. In *Handbook of Hydrothermal Technology.* Elsevier, Inc.

Tur, D. A., & Trivedi, M. A. (2014). Design modification and analysis of performance of heat exchanger by replacing copper tubes with polymer tube and composites material- a review. *International Journal for Innovative Research in Science & Technology, 1*(7). Retrieved from http://www.ijirst.org/articles/IJIRSTV1I7092.pdf

Tur, D. A., & Trivedi, M. (2015). Performance analysis of heat exchanger by replacing copper tubes with polymer tube or composites material. *International Journal of Advance Engineering and Research Development, 2*(5). Retrieved from http://www.ijaerd.com/papers/finished_papers/PERFORMANCE%20ANALYSIS%20OF%20HEAT%20EXCHANGER%20BY%20REPLACING%20COPPER%20TUBES%20WITH%20POLYMER%20TUBE%20OR%20COMPOSITES%20MATERIAL-35244.pdf

U.S. Environmental Protection Agency. (2015). *Advancing Sustainable Materials Management: 2013 Fact Sheet: Assessing Trends in Material Generation, Recycling and Disposal in the United States.* Retrieved from Washington, DC: https://www.epa.gov/sites/production/files/2015-09/documents/2013_advncng_smm_fs.pdf

Ucankus, G., Ercan, M., Uzunoglu, D., & Culha, M. (2018). Methods for preparation of nanocomposites in environmental remediation. In *New Polymer Nanocomposites for Environmental Remediation* (pp. 1–28). Elsevier.

UNEP. (2018). *The State of Plastics: World Environment Day Outlook.* Retrieved from https://wedocs.unep.org/bitstream/handle/20.500.11822/25513/state_plastics_WED.pdf?isAllowed=y&sequence=1

Verma, R., Vinoda, K. S., Papireddy, M., & Gowda, A. N. S. (2016). Toxic pollutants from plastic waste-a review. *Procedia Environmental Sciences, 35*, 701–708.

Vilaplana, F., & Karlsson, S. (2008). Quality concepts for the improved use of recycled polymeric materials: A review. *Macromolecular Materials and Engineering, 293*(4), 274–297. doi:10.1002/mame.200700393

Watson, B. (2016). Can incineration and landfills save us from the recycling crisis? Retrieved from https://www.theguardian.com/sustainable-business/2016/jun/14/green-waste-distribution-methods-recycling-plastic-oil-epa

Wichai-utcha, N., & Chavalparit, O. (2019). 3Rs policy and plastic waste management in Thailand. *Journal of Material Cycles and Waste Management, 21*(1), 10–22.

Yolin, C. (2015). *Waste Management and Recycling in Japan Opportunities for European Companies (SMEs Focus).* Retrieved from Japan.

Yunus, M., & Fauzan, R. (2017). *Mechanical Properties of Bioplastics Cassava Starch Film With Zinc Oxide Nanofiller as Reinforcement.* Paper presented at the IOP Conference Series: Materials Science and Engineering.

12 A Review on Leachate Water in BRICS Nations
A Sustainable Perspective

Monika and Abhinav Yadav

CONTENTS

12.1 INTRODUCTION

Uncontrolled population growth encourages the increase in industries and urban agglomeration which have further led to a rise in the amount of municipal solid waste (MSW) (Singh et al. 2011, 2015a). The landfill is a common practice for disposal of MSW, but with a scarcity of appropriate engineering measures it facilitates pollution and affects environmental integrity (Singh et al. 2015b). Inappropriate disposal of solid waste in landfills creates inevitable problems in the synthesis of (landfill) leachate and emission of toxic gases. The production of leachate is a composite result of the unscientific segregation, collection and disposal of MSW. The conjunction of physical, chemical and microbial processes along with several driving factors in the MSW induces transmission of poisonous substances to contaminate as liquid-based leachate which percolates in groundwater and surface water (Harrington and Maris 1986; Christensen and Kjeldsen 1989). These driving factors include waste composition, runoff, ambient temperature, infiltration, transpiration, annual precipitation, evaporation and density, initial moisture content and depth of the landfill (El-Fadel et al. 2002; Renou et al. 2008; Salem et al. 2008). The leachate is produced when rainwater filters through the landfill and absorbs chemical and biological compounds from decaying materials in the landfill. The leachate is one of the several forms of wastewater, and has a mixed effect on both agriculture as well as on human health. Leachate is considered as the toxic soup of many different landfills, and their compaction with biodegradable organic content makes the vicinity oxygen deficient. Moreover, landfill leachate imposes a substantial impact on the utilization and mitigation via complexation of organic ligands and colloidal matters in the soil (Achankeng et al. 2004). Landfill leachate is a water-based solution, and its nature is characterized by the homogenous mixture of four major groups of pollutants: (i) dissolved organic matter, (ii) inorganic macro components, (iii) heavy metals, (iv) xenobiotic organic compounds (Christensen et al. 1994). The generation and dispersal of toxic leachate continue hundreds of years after landfill closures; this poses a potential concern for soil and groundwater quality (Chian and Dewalle 1976; Jorstad et al. 2004). Consequently, further attention has been given to the treatment of landfill leachate as it can pollute larger volumes of groundwater and can make it unusable for domestic and other purposes (Bakare et al. 2000). There are various conventional and nonconventional management techniques used in Brazil, Russia, India, China and South Africa (BRICS nations) for the treatment of leachate. These techniques involve biological treatment, physical-chemical treatment and combinations of both physical and biological treatment. The treatment of leachate improves the quality of wastewater (leachate) and promotes the reuse/recycling of wastewater. The use of wastewater for irrigation has been related with several advantages such as the increase in phosphorus (P), organic carbon (OC), potassium (K), nitrogen (N) and magnesium (Mg) contents of the soil as compared to clean groundwater irrigation (Tiwari et al. 2003). Safe disposal of the MSW is one of the major environmental concerns in BRICS countries. Landfilling and land application of MSW and leachate are mentioned to be the most cost-effective and useful options for MSW disposal.

12.2 COMPOSITION OF LEACHATE

Leachate is a complex composition of concentrated ammonia, phosphorus, nitrogen, suspended solids, heavy metals, phenols, pesticides and aromatic and halogenated compounds (Devare and Bahadir 1994). Other compounds may also exist in very low concentrations in landfill leachate, for instance borate, sulphide, arsenate, selenite, barium, lithium, mercury and cobalt. These hazardous compounds can pose a serious threat to the environment and hinder the management techniques. MSW leachate can be mainly divided into four major groups of pollutants: (i) dissolved organic matter, quantified as chemical oxygen demand (COD) or total organic carbon (TOC), volatile fatty acids (that accumulate during the acid phase of the waste stabilization (Christensen and Kjeldsen 1989)) and more refractory compounds such as fulvic and humic-like compounds. (ii) Inorganic macro components: Calcium (Ca^{2+}), magnesium (Mg^{2+}), sodium (Na^{+}), potassium (K^{+}), ammonium ($NH4^{+}$), iron (Fe^{2+}), manganese (Mn^{2+}), chloride (Cl^{-}), sulphate (SO^{4+}) and hydrogen carbonate (HCO^{3-}) (Kulikowska and Klimiuk 2008). (iii) Heavy metals: Copper (Cu^{2+}), lead (Pb^{2+}), cadmium (Cd^{2+}), chromium ($Cr^{3}+$), nickel (Ni^{2+}) and zinc (Zn^{2+}) (Christensen et al. 2001) (4) Xenobiotic organic compounds (XOCs) originating from household or industrial chemicals and present in comparatively small concentrations (usually less than 1 mg/l of individual compounds). Some other compounds have also included aromatic hydrocarbons, phenols, chlorinated aliphatic, pesticides and plasticizers (Slack et al. 2005; Schiopu and Gavrilescu 2010). The composition of leachate varies significantly, and it depends on the nature of waste buried in the dumpsite, waste age, amount of waste, and landfilling technology. Along with these factors, chemical and biochemical processes are also responsible for the decomposition of waste materials and leachate(wastewater) generation (Fatta et al. 1999).

12.3 AGES OF LANDFILL LEACHATE

The nature of leachate varies in every aspect of its life cycle. It is highly variable and heterogeneous. Generally, it is accepted that landfills experience at least four phases of breakdown, (i) an early aerobic phase, (ii) an anaerobic acid phase, (iii) an early methanogenic phase and (iv) a steady methanogenic phase (recently, an additional aerobic or humic phase of breakdown has been proposed) (Christensen and Kjeldsen 1995; Bozkurt et al. 2000). Some literature concludes that leachate from young landfills has a major share of organics and its chemical oxygen demand (COD) is reported to be 10,000 mg/l, while in leachate from older landfills (more than ten years) their COD values were below 3,000 mg/l (Chang 1989; Chen 1996). Recent literature has demonstrated that acid phase fermented leachate produced from young landfills is characterized by the presence of a significant amount of volatile acids. In mature landfills, the major portion of leachate organics is covered by humic and fulvic-like fractions (Aziz et al. 2007). The possible reason for this contradictory conclusion is the recirculation of waste and leachate as practiced in many landfills.

12.4 GENERATION OF LANDFILL LEACHATE

The total volume of leachate generated is directly dependent on various factors such as evapotranspiration, precipitation, infiltration, surface runoff, groundwater intrusion into the landfill. Approximately one billion tons of waste are produced globally. Hoornweg and Bhada-Tata's (2012) study projected the volume of waste products will exponentially increase manyfold (2 times), i.e. to around 2.2 billion tons by 2025. In many countries including Brazil and India, municipal solid waste is dumped directly in controlled landfills or open dumpsites. ABRELPE (2017) reports that around 215,000 tons of municipal solid waste is produced on a daily basis. If the data is analysed, it is equivalent to 1.033 kg per resident per day in Brazil. Approximately 40.9% of solid waste generated from the Brazilian proximity and out of total 59.1% of MSW of directly and around 80 thousand tons per day has destined inappropriately in open dumpsite or sanitary landfills (Costa et al. 2019). Excessive generation, collection and improper settlement of MSW in uncontrolled landfills are major concerns in various nations. In India, a total of 0.15 million tons per day or 0.2 to 0.6 kg per capita per day of waste is generated (Pande et al. 2015). Approximately 50 million tons of waste are generated annually, and the total moisture content of solid waste is 50–60 % in India (Gupta et al. 1998; Agarwal et al. 2005). Moisture content plays a crucial role in leachate generation. In China 0.2 billion tons MSW were collected and 60.3% of the total MSW was discharged in landfills at the end of 2016. There is a total of 940 solid waste treatment facilities available throughout China. Despite this, the normal concentrations of pharmaceuticals (oxytetracycline, tetracycline and sulfamethoxazole) in MSW from Xi'an city (the capital of Shaanxi province in China) have been reported at 100.9 ± 141.81 mg/kg, 63.8 ± 37.7 mg/kg and 47.9 ± 8.1 mg/kg, respectively, as average concentrations (Song et al. 2016).

12.5 ESTIMATION OF LEACHATE

Leachate quantity is possible to calculate, in accordance with major influencing factors, by water balance equations or hydraulics calculation. Estimation of leachate production from solid waste has been done by software. Models have been developed for calculations. The result will help to fulfil the objective and simplify the simulation of the different scenarios (Serdarevic 2017). As landfill design progresses, the calculations should be refined. The calculation should be of the following form (EPA):

$$\mathrm{L_O} = \left[\mathrm{ER.}(A) + \mathrm{LW} + \mathrm{IRCA} + \mathrm{ER.}\left(\mathrm{A}_{Log}\right)\right]\ \left[\mathrm{a.W}\right]$$

where:

L_o = leachate produced (m^3)

ER = effective rainfall (use actual rainfall (R) for active cells) (m)

A = area of the cell (m^2)

LW = liquid waste (also includes excess water from sludge) (m^3)

IRCA = infiltration through restored and capped areas (m^3). (In areas that have been temporarily capped/restored, an infiltration rate of 20–30% of the effective rainfall (ER) should be used. Infiltration in restored areas would be in the range 2–10% ER in case of a geosynthetic clay liner cap. Infiltration into restored areas (A restored) should be calculated using site-specific information.) The physical, chemical and biological properties of landfill leachate indicate that it is a highly contaminated wastewater. The pH of the solution ranges between 4.0 and 9.0, the value of COD lies from 2,000 to 62,000 mg/l and BOD5 values 60~45,000 mg/l. Leachate has a variant nature throughout the stages of its life cycle; the early stage of the process in landfills reported different COD values up to 90,000 mg/l in the leachate.

12.6 CHARACTERISTICS OF LANDFILL LEACHATE

The colour of leachate may vary from orange-brown or dark-brown; the value of chroma varies around 2,000–4,000. The decomposition of organic matter synthesizes an abundance of organic acid, and it delivers a malodorous smell to the leachate (Williams et al. 2004). The nature of precipitation, age of the leachate quantity or quality of industry waste reduces the intensity of colour from a darker to light shade. In the young stage of landfills, the pH is between 6 and 7, possibly because of the presence of weak acid, and in later stages, it changes to weak alkaline. The pH found in the older stages of landfill can be up to 7–8. The pH will tend to alkaline according to the increasing age of landfills; similarly, these changes also depend on the quality and amount of rainfall and the type of industrial waste. Microbial activity plays a crucial role in the variation of BOD. According to the Serdarevic (2017) Highest BOD value can be reported in acid phase of landfill > initial phase > aged phase > methanogenic phase with 13,000 mg/l > 11,900 mg/l > 260 mg/l > 180 mg/l respectively. The normal land-filling process is from 6 months to 2.5 years, and this time is suitable for maximum BOD. The BOD valuechanges rapidly with the age of landfill waste. The COD value ranges maximum in its recent stage, i.e. >20,000 mg/l, and in mature stage ≤3,000 mg/l. The reduction of COD is relatively delayed and the diminishing of BOD occurs rapidly by the late stage of the leachate life cycle. The reduction of BOD5 or COD proceeds to the reduction in biochemical treatment (Tyre et al. 1997). The ratio of BOD:TOC could reflect the oxidation of organic carbon in leachate. The value of BOD:TOC is high in the later phase of landfills. Leachate has a significant number of suspended solids; it consists of organic matter, inorganic matter, clay, microorganism and high concentrations of salt. The concentration of salts is analogous to variation of conductivity and affects the leachate properties such as TDS, BOD and COD. Concentrations of NO_3, NO_2, NH_4 + N, Na^+, K^+ and Ca^{+2} were found to be much higher in leachate as compared to clean water (Patel et al. 1990).

Landfill leachate consists of clusters of inorganic macro- and micronutrients such as nitrogen, phosphorus, potassium, calcium and trace metals. However, when the landfill is closed, the concentration of phosphorus reduces but the concentration of nitrogen antagonistically rises, possibly because of the slow degradation of complex waste into small fragments in the anaerobic conditions (Rong et al. 2009). The percentage of nitrogen is around 40%~50% in leachate and out of the total nitrogen concentration, ammonium nitrogen constitutes relatively higher portion. The high concentration of ammonium nitrogen brings an imbalance of nutrition to microorganisms. Along with the nutrient constituents, leachate composition constitutes a major portion of heavy metals. The normal heavy metal ions are Cu, Zn, Pb, Cd, Hg, etc. According to Singh et al. (2004) in wastewater, the concentrations of Cd, Cu, Pb and Cr were higher followed by Ni and Zn. The ratio of heavy metals in solution depends on some of the following factors: (i) the amount of waste, (ii) the configuration of waste, (iii) the source of waste such as waste released from either urban, industries, domestic or commercial sectors. Sources define the specific supremacy of any metal in the solution. Only limited studies investigate the presence of microbes in leachate. Leachate plume has several types of microorganism i.e., methanogens, iron reducer, sulphate reducer, denitrify microorganism which responsible for the redox process (Christensen et al. 2001).

12.7 VARIATION IN LEACHATE WATER QUALITY AND ITS IMPACT ON ENVIRONMENT

Time is the main factor that decides the quality of leachate water. Generally, the BOD/COD ratio lies within 0.4–0.75, but the ratio values slightly change with each passing day and become stable at a particular point of time. In older landfills, concentrations of organic matter will reduce, and the ratio of BOD/COD approximately reaches up to 0.1. The water quality of leachate is very fickle, so the technology of the treatment system should be developed strongly. The generation of leachate can promote the layered structures in waste landfills. Leachate navigates its way into the environmental system from the base of the landfill, from where it goes into the groundwater through the unsaturated soil layers which later on blends in with surface water by means of hydraulic-driven associations. The blending of leachate with groundwater causes potential threats to the water reservoir. The pollution potential of leachate depends upon various factors such as the landfill location, leachate concentration, i.e. hydrogeological setting, landfill type (engineered or not) and the quality and volume of the receiving water (groundwater or surface water) (Islam et al. 2013). A layered structure and compactness in the landfill release oxygen and excess moisture from it. Anaerobic conditions stimulates the breakdown of complex products in simpler forms which causes release of methane gas and other GHG (Greenhouse gases). The leakage of gases is also a major problem along with groundwater pollution related to open dumping. Rachel et al.'s (2007) study elucidates the abundant release of greenhouse (mainly CO_2), approximately 818 million metric tons per annum of CO_2 in the form of methane from landfills. Methane is a 25 times more potent greenhouse gas than carbon dioxide that leads to global warming and climate change. The municipal solid waste in landfills is a prominent source of methane emanation. High concentrations of methane ignite very quickly which causes fires and flareups in landfills (Sridevi et al. 2012). There are some treatment methods have been used to diminish the generation of methane gas such as segregation of organic waste, in-vessel composting, tunnel composting and open windrow, and digestion in the absence of oxygen is also used but in confined places so that emitted methane gas is captured easily and transformed into energy. Landfill gases, particularly methane and carbon dioxide, have a great fuel value; hence they can be mined from landfills and exploited as an energy source. So the construction of sanitary landfills having a gas tapping system will prove highly beneficial in the future (Kumar and Sharma 2014). Gupta and Singh (2015) found 16 million metrics (CO_2 equivalents per annum) of methane is released in India from landfill sites (International Energy Agency 2008); potential from landfill gas of different landfills sites of India such as Mumbai (Deonar and Gorai) with 5.6 MW, Ahmedabad (Pirana) 1.3 MW, Delhi (Bhalswa, Ghazipur, and Okhla) 8.4 MW and Pune (Uruli) 0.7 MW annually (Siddiqui and Khan 2011). The methane is finally transformed into primary constituents such as carbon dioxide and water during energy production which decreases its effect on the climate as methane is a 25 times more effective greenhouse gas than carbon dioxide. A report by the Planning Commission (2014) has uncovered the possibilities for the use of waste as an asset and showed that 62 million tons of waste which are created every year can deliver 72 MW of power from landfill gas, 439 MW from the burnable segment and 5.4 million metric ton of fertilizer from natural waste. Application of high-concentration of landfill leachate affects the properties of soil and plants, so only cautious use of leachate can improve the plant and soil health and its integrity with nature. Safe dumping of MSW is one of the major environmental concerns in developing countries. On the other hand, MSW leachate is considered a source of nutrients and water. It can be used as fertilizer in agriculture (Romero et al. 2013). Agricultural use of leachate has multifaceted advantages: (i) it decreases pressure on freshwater, (ii) it increases agricultural production and contributes food safety, (iii) it reduces the uses of synthetic fertilizer, (iv) it decreases the dependency on climate for continuous water supply, (v) it improves agricultural economy, (vi) it doubles farmers' income. The application of leachate in agriculture can be the most cost-effective and convenient option for MSW disposal.

12.8 IMPACT ON SOIL PROPERTIES

On one hand, the application of leachate has multiple advantages but, on the other hand, some major issues arise like salinity, heavy metal, clogging, etc. Treatment with wastewater leaves excessive salts residue on the surface of the soil, and this impairs the soil and plant growth (Leal et al. 2009). Excessive salinity may cause dispersion of clay and reduce the aeration, water infiltration rate, stability of aggregates and porosity. Physical disturbances may also increase the soil global density which elevates the susceptibility to erosion (Mandal et al. 2008; Loncnar et al. 2010). Also, leachate application disturbs the integrity of soil and constrains the microbiological and biochemical processes related to the biogeochemical cycles (Chen et al. 2008; Nakatani et al. 2011; Santos et al. 2013). In a study, landfill leachate was applied in rice and wheat fields (Kuwano et al. 2017). The experimental design was performed in a randomized complete block with four replicates. The treatment was implemented in various depths of 0–10, 10–20, 20–40 and 40–60 cm of Rhodic Kandiudult soil for 4 years. This experiment exolained the enhancement of nitrate (>160 mg kg^{-1}) in the topsoil (down to 40 cm), where landfill leachate increases nitrate (>60 mg kg^{-1}) only at 40 to 60 cm with the highest dose, suggesting leaching. Urea-treated soil had less pH, which might have led to greater retention of nitrate in the topsoil. Sodium (0.02–0.26 cmolc Na kg^{-1}), potassium (0.18–0.82 cmolc K kg^{-1}) and electrical conductivity (0.05–0.14 dS m^{-1}) all increased with leachate dosage. The treatment reduced the dehydrogenase enzyme, and it increased the microbial activity. Panahpour et al. (2011) amended leachate with 25% dilution in a field of acacia and plane. Treatment with leachate affects the properties of soil such as a reduction in pH, enriching soluble salts and increased organic matter percentage. The results also illustrated the improved physical properties, increasing acidic pH, absorption of macronutrients and enhanced concentrations of heavy metals such as phosphorus, iron, zinc, copper and manganese. Leachate is a good source of nutrients; it improves physical properties and plant productivity but in limited use.

12.9 IMPACT ON PLANT BIOCHEMICAL PROPERTIES

The application of leachate is beneficial for plants in lower concentrations on one hand but on the other hand, a higher concentration of leachate reduces the yield and badly affects the biochemical properties. In some studies, crude leachate has been diluted with distilled water to acquire final concentrations of 2.5%, 5%, 10%, 20%, 50% and 100%. The negative control group was exposed to distilled water for the same time period. The application was carried out in replicated sets of three *Zea mays* L. (maize). The results found a higher concentration of leachate reduced chlorophyll (CHL) content but in lower concentrations improved chlorophyll content, for example 2.5% leachate led to a maximum rise of 118.2% but 100% leachate reduce up to 72% maize seedling. Moreover, landfill leachate enhanced the production of lipid peroxidation and protein oxidation in leaf tissues in ten days of treatment, accompanied by an increase in antioxidant status. Variation in leachate concentration is a determinant factor for the function of physiological responses and the growth or inhibition. Higher concentrations exert oxidative stress on plants. Higher concentrations of landfill leachate contain higher levels of pollutants. Their results specify that landfill leachate affected the metabolism of *Zea mays* L. (maize). The risk of hampering metabolism occurred mainly in samples of higher concentration. The application of leachate on plants has reduced constantly producing reactive oxygen species (ROS) during photosynthesis and metabolic activities (Foyer and Noctor 2000). In general, plants comprise composite antioxidant systems (Qureshi et al. 2005) such as catalase (CAT), peroxidase (POD) and superoxide dismutases (SOD) (Verma and Dubey 2003). These antioxidant systems neutralize and scavenge the reactive oxygen species stimulated by environmental stress (Tanyolaç et al. 2007) and also control membrane lipid peroxidation and protein oxidation. Some studies on high plant toxicity under natural and experimental exposure conditions indicate that landfill leachate can stimulate nonbiased impacts on plant growth, genetic stability and biomass production (Bakare et al. 2000; Sang et al. 2006). Studies based on literature suggest that plant enzymes like lipid peroxidase, peroxidase and carotenoids increase with an increase in the concentration of leachate while thiol, protein and chlorophyll decrease with increased concentration of leachate. Therefore, it is important to determine the existence of oxidative damage and the variation in antioxidant enzyme activities in plants to comprehend the tolerance of the plant system to environmental strain (Blokhina et al. 2003). Leachate application on plants has both desirable and undesirable impacts on the growth (Cheng and Chu 2007). Leaf impairment, a reduction in crop yield and premature senescence are the possible effects of leachate irrigation on the plant (Menser et al. 1983). An improvement in the growth of *Salix babylonica*, *Phalaris arundinacea* and *Populous nigra* can be observed on the application of leachate, but some indications of phytotoxicity, like brown and necrotic spots, were witnessed in leaves, while willows had shown deprivation of chlorophyll content and pose chlorosis (Cureton et al. 1991). The amendment of diluted leachate in the irrigation of *Acacia confusa*, *Leucaena leuocephala* and *Eucalyptus torelliana* produces a beneficial response in their growth and stomatal conductance (Liang et al. 1999). Khoshgoftarmanesh and Kalbasi (2002) applied municipal wastewater leachate on rice fields with different treatment concentrations 0, 150, 300 and 600 t. An experiment was performed in Iran with two sets of fields, one treated with municipal wastewater leachate (MWL) ha^{-1} and another control set treated with nitrogen–phosphorus–potassium–zinc (N–P–K–Zn) fertilizer in a randomized complete block design with three replications. Rice seedlings were transferred to plots, which were flooded previously. Municipal wastewater leachate was applied to tiller and panicle stages. The results illustrate that 150 and 300 t MWL ha^{-1} treated plants have shown a rise in grain yield, but 600 t ha^{-1} treatment decreased the yield of grain compared to the control fertilizer treatment. Municipal waste leachate increased the amounts of available macro components (N, P, K) and heavy metals concentration like Fe, Mn, Zn and Cu in soil; these nutrients help in improving soil health and crop productivity up to 4.2 t ha^{-1}. No (Zero) MWL to 16.7 and 6.9 t ha^{-1} in 300 t ha^{-1} MWL treatment, respectively. The amounts of nutrient uptake were highest with 300 t of MWL ha^{-1}. Besides the positive effects, application of MWL significantly increased the levels of EDTA-extractable Ni, Cr, Co and Pb in the soil, and rice at maximum treatment 600 t ha^{-1} MWL. Municipal wastewater leachate is a rich source of nutrients for plants but its noxious characteristics adversely affect plant growth and ultimately contaminate the food chain. Food chain contamination affects the multiple trophic levels of organisms and their health.

12.10 MANAGEMENT AND TREATMENT METHODS IN BRICS NATIONS

Management and treatment techniques are dependent on the age of landfill leachate because the content of leachate will change with time and age. The biological treatment technique is used if a high content of organic matter (>10,000 mg L^{-1}) is present in leachate. Although the treatment is changed if the content replaces with high ammoniacal nitrogen and low biodegradability, this time the appropriate process will be a physical-chemical process or a combination of both physical-chemical and biological techniques. The biological technique is the most cost-effective and highly efficient treatment process in young landfills leachates which are characterized by biodegradable matter (<5 years) (Bulc and Justin 2007). Kurniawan et al.'s (2006) study found the treatment approach varies in mature landfill leachate (more than ten years). In the methanogenic stage of leachate, if the ratio of BOD5 and COD is less than 0.1,it shows high amount of non-degraded complex compounds i.e., fulvic and humic acid. After consequential modification in technology for improving the result of mature landfill leachate treatment, the membrane process was introduced. Reducing the levels of ammonia, COD, BOD and some trace pollutants in leachate is very challenging so the combination treatment only works as the application of combined membrane and biological treatment, carried out in a membrane bioreactor (MBR). Many developing countries are unable to implement progressive leachate treatment technologies because of their expensive implantation, maintenance and the volumes of leachate to be treated (Vital et al. 2014) (Figure 12.1 and Table 12.1).

In the Indian context, the removal of contamination from MSW leachate is done via large-scale industrial units, modular, multistage combined techniques. Most countries prefer biological processes as they are more cost-effective and technically efficient in treating volatile fatty acids in leachates (Chaturvedi and Kaushal 2018). India has treated leachate through the three most conventional leachate treatment processes, namely, the activated sludge process (suspended growth), the fixed film process (attached growth) (Dabi 2015) and an innovative PVA gel treatment technique (combined process) for large-scale leachate treatment. According to Gani et al. (2014), PVA gel acts as an exclusive bio-carrier that has been recognized as an effective biological leachate treatment technology (Chaturvedi and Kaushal 2018). China holds a better position on the list of taking steps for pollution control and environmental protection. Several advanced

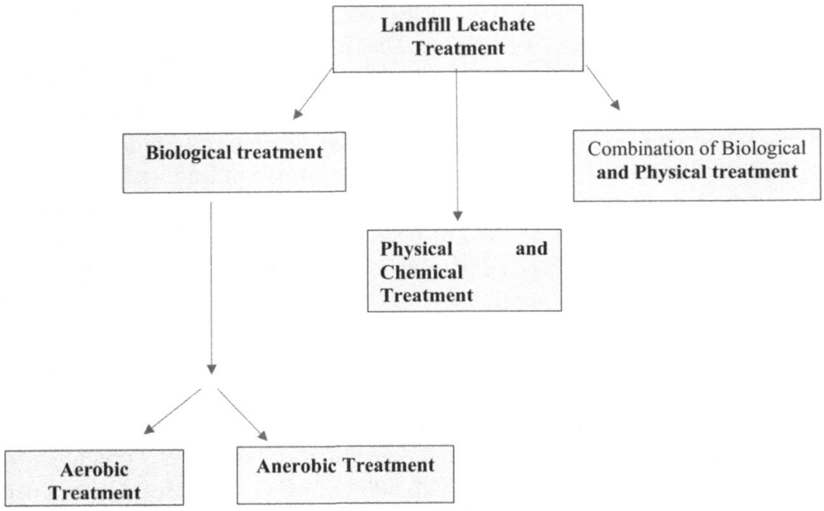

FIGURE 12.1 Different treatment techniques of landfill leachate.

TABLE 12.1
Various Management Technologies for Leachate Treatment

Aerobic Treatment	Anaerobic Treatment	Physical-Chemical Treatment	Combination of Biological and Physical Treatment
Aerated lagoons	Anaerobic filter (AF)	Coagulation -flocculation	Photo- Fenton + biological activated sludge
Aerobic activated sludge	Anaerobic digestion	Chemical precipitation	Agitation + coagulation+ SBR+ filtration
Sequencing batch reactors (SBR)	Up-flow anaerobic sludge	Adsorption	Electro-Fenton and biological treatment
Rotating biological contactors (RBC)	Anaerobic ammonium oxidation (ANAMMOX)	Membrane filtration	
Trickling filters		Ion exchange	
Moving-bed biofilm reactor		Air stripping	
Fluidized bed biofilm reactor (FBBR)		Chemical oxidation/ advanced oxidation	
Membrane biological reactor (MBR)		Electrochemical	
Constructed conventional wetland			
Fungal treatment and phytoremediation			

techniques are used in China such as MBRs, microfiltration/reverse osmosis and photo-Fenton. MBRs have been extensively applied in China (Xiao et al. 2014). The largest 3,200 m^3/d capacity Laogang Leachate Treatment Project, Shanghai, uses MBR treatment techniques (Zhang et al. 2017). Numerous studies found the presence of organophosphorus flame retardants (OPFRs) and metals in municipal landfill leachate in China (Deng et al. 2018). Deng et al.'s (2018) study illustrated that microfiltration/reverse osmosis is the best technique for 98% removal OPFRs and heavy metals in a municipal landfill leachate treatment system set up in Guangzhou, China. Fenton treatment technique used for removal of COD and TOC with efficiency of 65% and 55%

respectively in treatment plants in China (Woi et al. 2010). Brazil has started many programmes such as CONAMA Resolution n° 20 (Brazil 1986), n° 430 (Brazil 2011) for management and improving the condition, parameters and standards of effluent. Generally, biological treatment techniques are followed in Brazil for the removal of organic matter and volatile organic material in solid waste leachate (Costa et al. 2019). Minas Gerais State was the pioneer in Brazil in terms of offering a specific normative deliberation for the discharge of the effluent from leachate treatment systems for sanitary landfills (COPAM/CERH 2008). Lagoon, biological filters, activated sludge and wetlands are the prevalent and conventional examples of biological treatment because they stabilize the leachate properties and, as well as this, are cost-effective (Maia et al. 2015). In the country's largest city Sao Paulo the sanitary landfills commonly perform the treatment of leachate in conjunction with sewage treatment; only 3% of the city's sanitary landfills have a leachate treatment station in their facilities (Sproviere and Contrera 2017). Costa et al. (2019) describe some major cities in Brazil where large landfill sites and adjacent treatment plants have been established like in Seropédica, Sao Gonçalo, Gramacho, Campos, Rio Claro and Osasco, where there was an investment in the outsourcing of services for the disposal and treatment of waste. The advanced technologies like membrane filtration, MBR, ultrafiltration, nanofiltration and reverse osmosis have been introduced for handling wastewater. In Russia, on-site specialized leachate treatment is well-known in many large landfills like in the Perm region (Vaisman et al. 2001). On-site specialized leachate treatment includes aerobic biofilters and biological ponds in Russian landfills. There is a study gap in the management of MSW leachate in South Africa. Management of leachate with advanced and modern technology raised new hopes regarding the recycling of wastewater in the BRICS nation.

12.11 CONCLUSION

A significant amount of municipal solid waste generated in BRICS nations plays a crucial role in leachate release. There are multiple natural factors involved in leachate formation from landfills like precipitation, hydraulic pressure, etc. Leachate has many nutritious macronutrients but its noxious nature contaminates the ground and surface water. Cautious and rational use of leachate promotes the growth of plants and elevates the soil properties. It also reduces the use of synthetic fertilizer and dependency on weather in agriculture via a constant supply of water. But the irrational use of leachate in agriculture invites many difficult problems like heavy metals and salinity and reduces the yield and degrades the soil quality. For the reuse of leachate, BRICS nations have started numerous management programmes and established many treatment plants. Biological treatment techniques are cost-effective and so highly appreciated in most nations like Russia and Brazil. But in the advancement of the last few decades, modern and new technologies have replaced the biological treatment techniques. Advanced technology and a combination of physical, chemical and biological techniques have become most techno-efficient and cost-effective.

REFERENCES

ABRELPE- Brazilian Association of Public Cleaning and Special Waste Companies (2017). Overview of solid waste in Brazil. (In Portuguese). http://abrelpe.org.br/panorama/, Accessed date: 12 October 2018.

Achankeng, E. (2004). Sustainability in municipal solid waste management in Bamenda and Yaounde, Cameroon (Doctoral dissertation).

Agarwal, A., Singhmar, A., Kulshrestha, M., & Mittal, A. K. (2005). Municipal solid waste recycling and associated markets in Delhi, India. *Resources, Conservation and Recycling*, 44(1), 73–90.

Aziz, H. A., Alias, S., Assari, F., & Adlan, M. N. (2007). The use of alum, ferric chloride and ferrous sulphate as coagulants in removing suspended solids, colour and COD from semi-aerobic landfill leachate at controlled pH. *Waste Management & Research*, 25(6), 556–565.

Bakare, A. A., Mosuro, A. A., & Osibanjo, O. (2000). Effect of simulated leachate on chromosomes and mitosis in roots of Allium cepa(L.). *Journal of Environmental Biology*, 21(3), 263–271.

Blokhina, O., Virolainen, E., & Fagerstedt, K. V. (2003). Antioxidants, oxidative damage and oxygen deprivation stress: A review. *Annals of Botany*, 91(2), 179–194.

Bozkurt, S., Moreno, L., & Neretnieks, I. (2000). Long-term processes in waste deposits. *Science of the Total Environment*, 250(1–3), 101–121.

Bulc, T. G., & Justin, M. Z. (2007). Sustainable solution for landfill leachate with a use of phytoremediation. In *Landfill Research Trends* (pp. 103–139). Nova Science Publishers, New York.

Chang, Y. H., & Sheldon, B. W. (1989). Application of ozone with physical wastewater treatments to recondition poultry process waters. *Poultry Science*, 68(8), 1078–1087.

Chaturvedi, H., & Kaushal, P. (2018). Comparative study of different Biological Processes for non-segregated Municipal Solid Waste (MSW) leachate treatment. *Environmental Technology & Innovation*, 9, 134–139.

Chen, P. H. (1996). Assessment of leachates from sanitary landfills: Impact of age, rainfall, and treatment. *Environment International*, 22(2), 225–237.

Cheng, C. Y., & Chu, L. M. (2007). Phytotoxicity data safeguard the performance of the recipient plants in leachate irrigation. *Environmental Pollution*, 145(1), 195–202.

Chian, E. S., & Dewalle, F. B. (1976). Sanitary landfill leachates and their treatment. *Journal of the Environmental Engineering Division*, 102(2), 411–431.

Christensen, T. H., & Kjeldsen, P. (1989). Basic biochemical processes in landfills. In *Sanitary Landfilling: Process, Technology, and Environmental Impact* (pp. 29–49). Academic Press, New York.

Christensen, T. H., & Kjeldsen, P. (1995). Landfill emissions and environmental impact: An introduction. In *5th International Landfill Symposium* (pp. 3–12). Environmental Sanitary Engineering Centre.

Christensen, T. H., Kjeldsen, P., Albrechtsen, H. J. R., Heron, G., Nielsen, P. H., Bjerg, P. L., & Holm, P. E. (1994). Attenuation of landfill leachate pollutants in aquifers. *Critical Reviews in Environmental Science and Technology*, 24(2), 119–202.

Christensen, T. H., Kjeldsen, P., Bjerg, P. L., Jensen, D. L., Christensen, J. B., Baun, A., & Heron, G. (2001). Biogeochemistry of landfill leachate plumes. *Applied Geochemistry*, 16(7–8), 659–718.

Costa, A. M., Alfaia, R. G. D. S. M., & Campos, J. C. (2019). Landfill leachate treatment in Brazil–An overview. *Journal of Environmental Management*, 232, 110–116.

Cureton, P. M., Groenevelt, P. H., & McBride, R. A. (1991). Landfill leachate recirculation: Effects on vegetation vigor and clay surface cover infiltration. *Journal of Environmental Quality*, 20(1), 17–24.

Dabi, N. (2015). Comparison of suspended growth and attached growth wastewater treatment process: A case study of wastewater treatment plant at MNIT, Jaipur, Rajasthan, India. *European Journal of Advances in Engineering and Technology*, 2, 102–105.

Deng, L., Ngo, H. H., Guo, W., Wang, J., & Zhang, H. (2018). Evaluation of a new sponge addition-microbial fuel cell system for removing nutrient from low C/N ratio wastewater. *Chemical Engineering Journal*, 338, 166–175.

Devare, M., & Bahadir, M. (1994). Biological monitoring of landfill leachate using plants and luminescent bacteria. *Chemosphere*, 28(2), 261–271.

El-Fadel, M., Bou-Zeid, E., Chahine, W., & Alayli, B. (2002). Temporal variation of leachate quality from pre-sorted and baled municipal solid waste with high organic and moisture content. *Waste Management*, 22(3), 269–282.

Fatta, D., Papadopoulos, A., & Loizidou, M. (1999). A study on the landfill leachate and its impact on the groundwater quality of the greater area. *Environmental Geochemistry and Health*, 21(2), 175–190.

Foyer, C. H., & Noctor, G. (2000). Tansley review No. 112: Oxygen processing in photosynthesis: Regulation and signalling. *New Phytologist*, 146(3), 359–388.

Gani, K. M., Singh, J., Ali, M., Rose, V., Kazmi, A. A. (2014). A pilot scale study of moving bed biofilm reactor with polyvinyl alcohol gel as biomass carriers for enhanced nutrient removal. In *The 11th International Symposium on Southeast Asian Water Environment* (pp. 1–9). Bangkok.

Gupta, D., & Singh, S. K. (2015). Energy use and greenhouse gas emissions from wastewater treatment plants. *International Journal of Environmental Engineering*, 7(1), 1–10.

Gupta, S., Mohan, K., Prasad, R., Gupta, S., & Kansal, A. (1998). Solid waste management in India: Options and opportunities. *Resources, Conservation and Recycling*, 24(2), 137–154.

Harrington, D. W., & Maris, P. J. (1986). The treatment of leachate: A UK perspective. *Water Pollution Control (Maidstone)*, 85(1), 45–56.

Hoornweg, D., & Bhada-Tata, P. (2012). *What a Waste: A Global Review of Solid Waste Management.*

Islam, M. R., Al Razi, K. A., Hasan, M. R., Hasan, M. H., & Alam, S. (2013). Effect of leachate on surrounding surface water: Case study in rajbandh sanitary landfill site in Khulna City, Bangladesh. *Global Journal of Research In Engineering.*

Jorstad, L. B., Jankowski, J., & Acworth, R. I. (2004). Analysis of the distribution of inorganic constituents in a landfill leachate-contaminated aquifer: Astrolabe Park, Sydney, Australia. *Environmental Geology*, 46(2), 263–272.

Khoshgoftarmanesh, A. H., & Kalbasi, M. (2002). Effect of municipal waste leachate on soil properties and growth and yield of rice. *Communications in Soil Science and Plant Analysis*, 33(13–14), 2011–2020.

Kulikowska, D., & Klimiuk, E. (2008). The effect of landfill age on municipal leachate composition. *Bioresource Technology*, 99(13), 5981–5985.

Kumar, A., & Sharma, M. P. (2014). Estimation of GHG emission and energy recovery potential from MSW landfill sites. *Sustainable Energy Technologies and Assessments*, 5, 50–61.

Kurniawan, T. A., Lo, W. H., & Chan, G. Y. (2006). Physico-chemical treatments for removal of recalcitrant contaminants from landfill leachate. *Journal of Hazardous Materials*, 129(1–3), 80–100.

Kuwano, B. H., Nogueira, M. A., Santos, C. A., Fagotti, D. S., Santos, M. B., Lescano, L. E., ... & Tavares-Filho, J. (2017). Application of landfill leachate improves wheat nutrition and yield but has minor effects on soil properties. *Journal of Environmental Quality*, 46(1), 153–159.

Leal, R. M. P., Herpin, U., da Fonseca, A. F., Firme, L. P., Montes, C. R., & Melfi, A. J. (2009). Sodicity and salinity in a Brazilian Oxisol cultivated with sugarcane irrigated with wastewater. *Agricultural Water Management*, 96(2), 307–316.

Liang, J., Zhang, J., & Wong, M. H. (1999). Landfill leachate used as irrigation water on landfill sites during dry seasons. In *Remediation and Management of Degraded Lands* (pp. 305–317). Lewis Publishers, Boca Raton.

Loncnar, M., Zupančič, M., Bukovec, P., & Justin, M. Z. (2010). Fate of saline ions in a planted landfill site with leachate recirculation. *Waste Management*, 30(1), 110–118.

Maia, S. I., Restrepo, J. J. B., Castilhos Junior, A. B., & Franco, D. (2015). Evaluation of the biological treatment of landfill leachate on a real scale in the Southern Region of Brazil. *Engenharia Sanitaria e Ambiental*, 20, 665–675 (In Portuguese).

Menser, H. A., Winant, W. M., & Bennett, O. L. (1983). Spray irrigation with landfill leachate. *BioCycle*, 24(3).

Nakatani, A. S., Martines, A. M., Nogueira, M. A., Fagotti, D. S., Oliveira, A. G., Bini, D., ... & Cardoso, E. J. (2011). Changes in the genetic structure of bacteria and microbial activity in an agricultural soil amended with tannery sludge. *Soil Biology and Biochemistry*, 43(1), 106–114.

Panahpour, E., Gholami, A., & Davami, A. H. (2011). Influence of garbage leachate on soil reaction, salinity and soil organic matter in east of Isfahan. *Advances in Environmental Biology*, 2336–2343.

Pande, G., Sinha, A., & Agrawal, S. (2015). Impacts of leachate percolation on ground water quality: A case study of Dhanbad city. *Global Nest Journal*, 17(1), 162–174.

Patel, K. P., Singh, B., & Chhibba, I. M. (1990). Crop yields and properties of a sodic soil as influenced by s-containing amendments under draining and nondraining conditions. *Arid Land Research and Management*, 4(3), 173–180.

Planning Commission Report. (2014). https://niti.gov.in

Qureshi, M. I., Israr, M., Abdin, M. Z., & Iqbal, M. (2005). Responses of Artemisia annua L. to lead and salt-induced oxidative stress. *Environmental and Experimental Botany*, 53(2), 185–193.

Rachel, G., Damodaran, N., Panesar, B., Leatherwood, C., & Asnani, P. U. (2007). Methane to markets and landfill gas energy in India. In *Proceeding of the International Conference on Sustainable Solid Waste Management* (pp. 519–525).

Renou, S., Givaudan, J. G., Poulain, S., Dirassouyan, F., & Moulin, P. (2008). Landfill leachate treatment: Review and opportunity. *Journal of Hazardous Materials*, 150(3), 468–493.

Romero, C., Ramos, P., Costa, C., & Márquez, M. C. (2013). Raw and digested municipal waste compost leachate as potential fertilizer: Comparison with a commercial fertilizer. *Journal of Cleaner Production*, 59, 73–78.

Salem, Z., Hamouri, K., Djemaa, R., & Allia, K. (2008). Evaluation of landfill leachate pollution and treatment. *Desalination*, 220(1–3), 108–114.

Sang, N., Li, G., & Xin, X. (2006). Municipal landfill leachate induces cytogenetic damage in root tips of Hordeum vulgare. *Ecotoxicology and Environmental Safety*, 63(3), 469–473.

Santos, C. A., Panchoni, L. C., Bini, D., Kuwano, B. H., Carmo, K. B., Silva, S. M. C. P., ... & Zangaro, W. (2013). Land application of municipal landfill leachate: Fate of ions and ammonia volatilization. *Journal of Environmental Quality*, 42(2), 523–531.

Schiopu, A. M., & Gavrilescu, M. (2010). Options for the treatment and management of municipal landfill leachate: Common and specific issues. *Clean – Soil Air Water*, 38(12), 1101–1110.

Serdarevic, A. (2017, May). Landfill leachate management—control and treatment. In *International Symposium on Innovative and Interdisciplinary Applications of Advanced Technologies* (pp. 618–632). Springer, Cham.

Siddiqui, F. Z., & Khan, M. E. (2011). Landfill gas recovery and its utilization in India: Current status, potential prospects and policy implications. *Journal of Chemical and Pharmaceutical Research*, 3(5), 174–183.

Singh, K. P., Mohan, D., Sinha, S., & Dalwani, R. (2004). Impact assessment of treated/untreated wastewater toxicants discharged by sewage treatment plants on health, agricultural, and environmental quality in the wastewater disposal area. *Chemosphere*, 55(2), 227–255.

Singh, R. P., Tyagi, V. V., Allen, T., Ibrahim, M. H., & Kothari, R. (2011). An overview for exploring the possibilities of energy generation from municipal solid waste (MSW) in Indian scenario. *Renewable and Sustainable Energy Reviews*, 15(9), 4797–4808.

Singh, S., Raju, N. J., & Nazneen, S. (2015). Environmental risk of heavy metal pollution and contamination sources using multivariate analysis in the soils of Varanasi environs, India. *Environmental Monitoring and Assessment*, 187(6), 345.

Slack, R. J., Gronow, J. R., & Voulvoulis, N. (2005). Household hazardous waste in municipal landfills: Contaminants in leachate. *Science of the Total Environment*, 337(1–3), 119–137.

Song, L., Li, L., Yang, S., Lan, J., He, H., McElmurry, S. P., & Zhao, Y. (2016). Sulfamethoxazole, tetracycline and oxytetracycline and related antibiotic resistance genes in a large-scale landfill, China. *Science of the Total Environment*, 551, 9–15.

Sproviere, J. A. S., & Contrera, R. C. (2017). Survey of landfills with treatment of leachate in loco in the state of São Paulo and its technologies. In *V Symposium on Solid Waste (SIRS), September 12th to 15th*. São Paulo, Brazil (In Portuguese).

Sridevi, V., Modi, M., Ch, M. V. V., Lakshmi, A., & Kesavarao, L. (2012). A review on integrated solid waste management. *International Journal of Engineering Science & Advanced Technology*, 2(5), 1491–1499.

Tanyolac, D., Ekmekçi, Y., & Ünalan, Ş. (2007). Changes in photochemical and antioxidant enzyme activities in maize (Zea mays L.) leaves exposed to excess copper. *Chemosphere*, 67(1), 89–98.

Tiwari, K. N., Singh, A., & Mal, P. K. (2003). Effect of drip irrigation on yield of cabbage (Brassica oleracea L. var. capitata) under mulch and non-mulch conditions. *Agricultural Water Management*, 58(1), 19–28.

Tyre, W., Bryan, C., & Dennis, M. (1997). *Site Characterization and Design of On-site Septic Systems – On-sit Sewage Disposal for a Subdivision in a High Groundwater Area.* Editors: M. S. Bedinger, J. S. Fleming, A. Johnson. Now Orleans, P278. ISBN 0-8031-2420-1.

Vaisman, Y. I., Glushankova, I. S., & Rudakova, L. V. (2001). Conceptual approaches for leachate treatment at MSW landfills. In *Abstracts of the 2nd International Congress on Waste Management*, 5–8 June (pp. 526–527). Moscow.

Verma, S., & Dubey, R. S. (2003). Lead toxicity induces lipid peroxidation and alters the activities of antioxidant enzymes in growing rice plants. *Plant Science*, 164(4), 645–655.

Vital, M. H. F., Ingouville, M., & Pinto, M. A. C. (2014). Estimate of investments in landfills to meet targets established by the National Solid Waste Policy between 2015 and 2019. *BNDES*, 40, 43–92. (In Portuguese).

Williams, L. B., & Hervig, R. L. (2004). Boron isotope composition of coals: A potential tracer of organic contaminated fluids. *Applied Geochemistry*, 19(10), 1625–1636.

Xiao, K., Xu, Y., Liang, S., Lei, T., Sun, J., Wen, X., & Huang, X. (2014). Engineering application of membrane bioreactor for wastewater treatment in China: Current state and future prospect. *Frontiers of Environmental Science & Engineering*, 8(6), 805–819.

Zhang, B., Xu, X., & Zhu, L. (2017). Structure and function of the microbial consortia of activated sludge in typical municipal wastewater treatment plants in winter. *Scientific Reports*, 7(1), 1–11.

Part III

Policies and Laws

13 Policy and Governance for Waste Management in Russia

Liliana Proskuryakova

CONTENTS

13.1 INTRODUCTION

Russian waste management policies are partly grounded in the legacy of the Soviet Union, where large-scale waste recycling programmes were in place. In 1979–1985 a programme for the recovery of secondary waste was developed by a research institute. In the 1980s, nearly 30% of all used paper was recycled. Glass bottles were also returned by consumers through glass recycling centres. In the 1990s, after the dissolution of the USSR the recycling programme was closed. Without official support, the programme and infrastructure for the collecting and processing of scrap metal, paper, textile and glass were closed. During the market reforms and the subsequent years nearly all waste was disposed of in authorized and unauthorized landfills. Large-scale projects of extractive industries also resulted in substantial quantities of waste.

According to the Russian Environment Protection Office, by 2018 Russia had accumulated 38.73 billion tons of industrial waste. In 2017 alone the new waste amounted to 6.22 billion tons: A 12.5% increase compared to 2016 and a 66.5% increase compared to 2010 (Russian Ministry of Natural Resources, 2018). Only 2.54 billion tons of waste were recycled for reuse in 2018.

Around 90% of waste comes from industries (primarily extractive) and manufacturing. Other segments of the energy industry that rely on fossil fuels (i.e. power and heat generation) also create significant amounts of specific waste that is hard to process (Krechetov et al., 2019). The volume of municipal solid waste is small compared to the overall amount – 55 to 60 million per year, of which 40% is organic waste, 35% – paper and 6% – plastic (TASS Information Agency, 2019).

Waste generation measured by unit of GDP in 2017 amounted to 98.3 tons per RUB 1 million. In 2011–2017 the indicator increased by 44.6% per RUB 1 million. Before 2012 this indicator had grown by 15% per year; thereafter it was flat until 2015 and showed further growth in 2016 and 2017. From a regional perspective, the largest volume of generated waste was in the Siberian federal district, which accounts for 71% of the country's total. This is explained by the activity of extractive industries, mainly coal mining in Kemerovskaya Oblast, and large volumes of overburden rocks that are classified as type-V hazardous waste, which accounted for 70% of the total waste in this federal district in 2017. All other federal districts also increased the per unit waste generation.

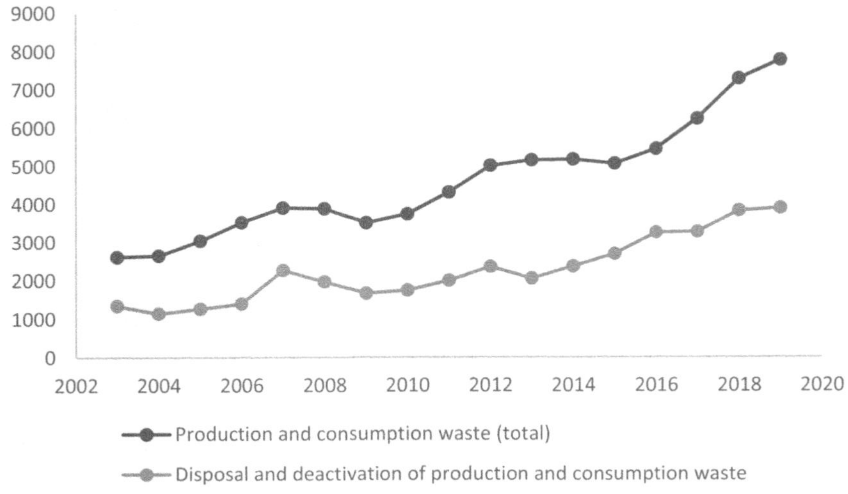

FIGURE 13.1 Generation and disposal of waste from production and consumption of goods (million tons).Source: Russian Federal State Statistics Service (2020).

The accumulated volume of both industrial and municipal waste keeps growing (Figure 13.1). According to different estimates, between 54 and 60 million tons of municipal solid waste (MSW) and 5.5 billion tons of industrial waste are generated in Russia annually. The Russian Ministry of Industry estimates that around 50% of industrial waste and around 10% of MSW is recycled (Maryev, 2019). Experts estimate that the largest share of MSW is disposed of at garbage dumps, 3–4% recycled, 1% burned and 0.3% composted. For comparison, the European Union recovers about 40% of its waste as reusable materials, and 20% as energy. The territory covered by accumulated waste is 4 million hectares, which is comparable to the territory of a medium European country – Switzerland, Belgium or Slovakia. This is similar to the situation in developing countries, where controlled waste disposal covers less than 50% of the total, and in rural areas there is no formal waste disposal system in place (Polprasert & Koottatep, 2017).

The considerable increase of municipal solid waste in the country is partly due to the country's size: A huge territory is not populated. Experts estimate that there are around 17,000 of fly-dumping sites and 700 unauthorized landfills – more than official MSW landfills. Over the last years, these sites have grown fast, causing social tensions and discontent. Official landfills have not been managed properly leading to hazardous gas leaks. Additionally, the economic benefits of resources and materials reuse have not been appreciated.

The most polluted Russian region in terms of MSW is Moscow Oblast, a region neighbouring Moscow city. It handles own waste (around 3.5 million tons) and the bulk of municipal waste of the capital city (around 4.5 million tons). Eight million tons of waste are annually dumped in the landfills of Moscow Oblast, which is around 14% of the total MSW in Russia. However the capacity of regional landfills is only 4.5 million tons (2018).

Making use of waste, such as waste-to-energy solutions, would allow Russia to significantly reduce the amount of harmful emissions from burning more expensive fuel, and save costs in many industries such as metallurgy, cement production, etc. Given that up to 300 litres of synthetic oil can be produced from 1 ton of MSW, the accumulated amount is sufficient to produce up to 3.8 billion tons of synthetic oil, which is equivalent to 46.3% of Russia's internationally confirmed oil reserves. Annually generated domestic and agricultural waste (about 300 million tons in dry weight) is enough to produce up to 90 billion m^3 of gas. This is much more than Russia buys from Turkmenistan (50 billion m^3 a year) (Ermolenko et al., 2019).

The published research on Russia calls for waste policy reforms (Kaazke et al., 2013; Tulokhonova & Ulanova, 2013; Starostina et al., 2014) that would introduce an integrated waste management system instead of focusing on producers' responsibility only (Nevskaya & Marinina, 2015). Other studies also underline the need to offer different frames for the analysis of growth and consumption patterns in the developing economies (Singh et al., 2014). Particularly important is to use the existing indigenous peoples' knowledge about waste management in the policy process (Brednikova & Tkach, 2008; Siraguza & Arzyutov, 2020).

In 2015–2016 a set of federal decrees was adopted to set a renewed waste governance/waste management framework in Russia. Prior to the reform a number of informal waste management practices were developed by the population, civil society organizations and national companies. The informal waste management practices are important for all BRICS countries (da Silva et al., 2019).

In this chapter the Russian national waste policy is presented and analysed. This is done in the context of the ongoing waste management reform and national priority projects. Then, various types of policy instruments applied in Russia are categorized and compared with other BRICS countries, where relevant. Finally, policy implications and recommendations are outlined.

13.2 WASTE MANAGEMENT REFORM

The waste management reform in Russia was launched in 2015–2016 and framed with several interlinked laws and government decrees. Additionally, several national environment and waste management initiatives in the form of projects were launched.

13.2.1 Waste Management Laws and Regulations

In 2015, Government Decree #1520 introduced a comprehensive waste information system that accumulated various data and information for policymaking. The decree established a comprehensive registry of manufacturing companies and importers of goods, waste operators and waste processing capacities, all issued waste processing and waste disposal licenses, data about waste processing technologies and systems for waste handling, and more. As of January 2019, the state registry included 5,526 waste disposal sites. Their cumulative remaining capacity was over 1.7 billion tons. Annually the surface of landfills in Russia grows by 0.4 million hectares (TASS Information Agency, 2019).

Decree #1073 (2015) sets the environmental fee to be paid annually by manufacturing companies and importers. Moreover, Decree #1416 (2015) obliged all manufacturing companies and importers to declare volumes of produced goods and packaging in the previous year. The lists of such goods were set in Decree #1886-p. Decree 2491-p established waste disposal targets by type of goods that should be reconsidered every three years. Decree #1062 set requirements for licensing organizations that collect, transport, process, neutralize, recycle and dispose of hazardous waste.

In 2016 the Government adopted Decree #284 setting fees for waste disposal by manufacturing companies and importers. The fees were set for each category of goods that have to be disposed of after they are no longer in use. This measure allowed the accumulation of resources for waste processing, while also increasing the end prices of goods both produced domestically and imported.

On 1 January 2019 the Federal Law 'On production and consumption waste' (2017) came into force. This signifies a transition to a new MSW management system. The control over waste handling will shift from multiple separate companies to large regional operators, selected through tenders. The new waste handling fees should provide for economic efficiency of waste management. The payment for MSW handling was no longer part of the fixed housing costs calculated for the entire building, it was considered part of flexible utility costs. The new payment should amount to RUB 120–130 per person per month. Finally, on 14 January 2019, President Putin signed a Decree on the establishment of a comprehensive national waste management system.

The 2030 Waste Recycling Industry Strategy was also adopted in 2017. This high-level strategic document sets the following priorities:

- The maximum efficiency of resources and raw materials use
- Preventing waste generation
- Limiting waste generation and lowering the volume of hazardous waste at source
- Waste treatment
- Waste disposal
- Waste detoxification
- Waste dumping (storage and burial) in line with environmental legislation

The 2030 Strategy aims at the creation of a comprehensive waste management system at federal, regional and local levels that is based on the waste management hierarchy above. Conditions are to be put in place for attracting investments into the industry, increasing its resource efficiency and the share of extracted precious fractions. For the entire economy it is planned to gradually substitute non-renewable natural resources with secondary materials obtained from recycled and treated waste and to increase the share of final products manufactured with the use of secondary materials. It is also aimed at lowering these products' prime costs, advancing their technical characteristics, safety and quality so that they are competitive and enjoy high demand. It is planned to start domestic manufacturing of equipment and technologies for waste processing, recycling and detoxification.

One of the tools for the 2030 Strategy implementation is the establishment of eco-industrial parks for various types of waste handling, including mixed industrial-municipal waste. Eco-industrial parks are a measure recommended by UNIDO to attain the Sustainable Development Goals. Essentially, an eco-industrial park is a group of companies bundled together and seeking to achieve better environmental, economic and social performance through cooperation in managing environmental and resource matters. This setting allows companies to switch to sustainable development, while retaining a competitive advantage through exchange of resources – materials, energy, water and by-products (UNIDO, 2020). In Russia the main goals for the establishment of eco-industrial parks in the waste industry is to lower the volume of landfilled waste, increasing the use of secondary materials (recycled resources) in manufacturing and stimulating the use of Russian waste processing technologies. Eco-industrial parks also contribute to lowering the negative social and environmental impact (LECAP, 2019).

The policy tools applied through laws and regulations are summarized in Table 13.1. In addition to laws and regulations described above, there is a set of national projects that are described in Section 13.2.2. National projects are like actions (measures) in the national action plan, and are aimed at the implementation of the legislative and regulatory acts.

As of 2019 most of Russia's regions shifted to the new waste management system. Additional three-year extensions were granted to federal cities (Moscow, Saint Petersburg and Sevastopol). The Ministry of Environment was concerned with the harsh situation in Khabarovsk Region and Jewish Autonomous Region that did not switch to the regional waste operator model, whereby such operator bear responsibility for the entire waste management chain from collection to recycling.

In line with the President's Decree 'On the national targets and strategic development tasks for the Russian Federation until 2024' (2018) the Russian Ministry of Environment initiated the National Project 'Environment'. The project is implemented in five areas, of which 'waste' is one. The overall budget is RUB 4 trillion, and the bulk of it (3.2 trillion) should be allocated from external sources. The national project previews the construction of waste processing infrastructure to be funded by environmental fees collected by Russia's regions. In 2019 these funds will be used to construct 39 facilities in 22 of Russia's regions. The Ministry of Environment projects that the fees will amount to RUB 15 billion by 2024 and will suffice to fund 200 modern facilities for waste sorting, processing and recycling. The main MSW-related indicators for this project are presented in Table 13.2.

Each thematic block of the National Project 'Environment' contains more detailed federal projects in the form of action plans with responsible agencies and decision-makers. In the case of waste there is a project 'Comprehensive management system for municipal solid waste', federal project 'Infrastructure for hazardous waste treatment of I-II class' and federal project 'Clean Country' discussed below.

TABLE 13.1
Main Policy Tools Applied through Legislative and Regulatory Documents in 2015–2019

2015	2016	2017	2018	2019
• Territorial waste management plans for regions of Russia • The establishment of the regional MSW operator	• Introduction of extended producer responsibility in Russia • Disposing of waste that contains useful recyclable components to landfills is prohibited	• 2030 Waste Recycling Industry Strategy • President of Russia's initiatives aimed at sustainable waste management	• Eco-industrial parks start functioning • Regulation of secondary resources and materials circulation • Plan for strategy implementation • Separate collection of MSW	• National Project 'Environment' • Federal Project 'Comprehensive system of MSW management' • Federal Project 'Infrastructure for hazardous waste treatment of I-II class'

Source: Adapted from Maryev, 2019.

TABLE 13.2
Main MSW-Related Indicator for the National Project 'Environment'

		Baseline		Year						
Target Outcomes /Indicators	**Control Level**	**Reverence Value**	**Date**	**2018**	**2019**	**2020**	**2021**	**2022**	**2023**	**2024**
All unauthorized urban landfills, identified before 1 January 2018 are eliminated (units)	Council	0	1 Sept 2018	0	16	42	76	108	146	191
Disposal of all facilities of accumulated environmental damage (units)	Council	10	1 Sept 2018	31	48	57	67	74	74	75
Deployment of facilities for processing, disposal and detoxification of hazardous waste of I and II class (units)	Presidium of the Council	0	25 Jan 2018	0	0	0	0	0	4	7
Share of disposed MSW in the total share of MSW, %	Presidium of the Council	1	1 Sept 2018	3	7	16.2	22.8	24	33	36
Share of processed MSW in the total share of MSW, %	Presidium of the Council	3	1 Sept 2018	7	12	27	38	40	55	60

Source: Russian Ministry of Natural Resources (2018).

In December 2016, the Council of Strategic Development and Priority Projects by the President of Russia approved the project ‘Decreasing the negative impact on the environment by phasing out the facilities with accumulated environmental damage and decreasing the share of MSW burial’ (‘Clean Country’). The project should be implemented in 2017–2025.

The pilot region is Tatarstan Republic that should shift to 100% recycled waste by introducing new approaches to waste management. These approaches include the construction of new waste incineration and waste recycling facilities. In 2017 the construction of four waste incineration plants in Moscow Oblast and one in Kazan began and will cost approximately RUB 140 billion for public budget and investors for a cumulative waste capacity of 3.35 million tons. The target for Moscow Oblast is to decrease waste burial from 9.5 million tons in 2017 to 6.5 million tons in 2025. The following year, the project encompassed 32 activities in 12 of Russia’s regions.

The federal project ‘Comprehensive system of MSW management’ resulted in the establishment of a public company ‘Russian Environmental Operator’ (REO) in January 2019 to coordinate the reform and support the industry actors. Russia’s regions require REO to provide co-funding of waste facility construction. Moreover, REO plans to issue green bonds and co-found small support foundations.

13.2.2 Reform Implementation and First Outcomes

Over 200 regional waste operators have been established that are responsible for waste collection, processing and recycling. In 2019 over 50 MSW companies started operations in 27 regions with cumulative waste processing capacity of 5.5 million tons, and 2.55 million tons waste disposal capacity. REO assesses the waste management infrastructure that has to be put in place at RUB 101 billion. The clear focus is on waste incineration and the construction of modern landfills, while the vision of separate waste collection and waste processing is still vague.

Waste-to-energy solutions through incineration were declared waste processing in the new legislative provisions that were adopted in 2019. This allows the minimizing of separate waste collection and the maximizing of waste incineration. Moreover, with this new classification the environmental fee may also be spent on incineration, while previously it may have been directed only to waste processing/recycling. This signifies a loss of resources and of market opportunities: There are many waste processing enterprises in Russia and they are far from operating at full capacity.

The overall amount of 2019 private investments in municipal treatment facilities is assessed at RUB 20 billion. An example of these is the first automatically operated waste sorting facility that was launched in Belgorod Region. The equipment was manufactured in Russia. The facility staff are only involved in quality control. Part of the secondary raw materials will be recycled on the spot: The polymer waste recycling section can produce 7,000 to 14,000 tons of final products per year. The facility may recycle up to 90% of the overall waste, significantly reducing the volume of dumped waste.

The shift to the separate collection of waste in 70 of Russia’s regions is planned through a two-container system – one for mixed (i.e. organic) waste and one for accumulated waste (metal, plastic, glass, paper). According to the forecast of the Russian Ministry of Environment, by 2024 around 60% of waste will be sorted and 36% processed (recycled). By 2024 in the frame of the National Project ‘Environment’ alone 220 new waste processing facilities will be constructed and over 190 landfills will be remediated.

Eleven policy tools have been adopted to stimulate business involvement in the industry. More specifically, it is planned to offer guarantees to banks, where investors make loans, as well as to compensate part of the lending interest rate. It is also planned to provide 50% co-funding, 25% compensation of capital investments in a facility, provide leasing at a reduced interest rate and other support measures. Some experts assess the reform implementation as satisfactory. Others state that the reform is only aimed at formal attainment of a set of indicators without the establishment of a comprehensive waste management system. Based on the previous experience of the European Union countries the focus on landfills and waste incineration is inefficient (The Village, 2020).

Another policy tool that works towards the reform goals is the stimulation of enterprise green innovation activity. The empirical results of a panel data model show that the impact of environmental innovations on Russian regions' productivity is positive. Aldieri et al. (2019) analysed the relationship between environmental innovation and productivity with the 2010–2015 data from 65 Russian regions. Among other environmental innovations the researchers looked at companies engaged in the recycling of waste, water and materials.

Successful completion of the waste management reform is dependent on several conditions. First, more extensive awareness raising should be done to cover the entire population. This information should include the main features of the reform, rights and obligations (separate waste collection). Discussing the principles of waste processing cost will allow some of the discontent that has accumulated since people were obliged to pay a new waste fee to be addressed. Since the reform launch this fee, as well as the fee collection procedure, varies significantly across regions. For instance, in the Moscow Region the cost of waste disposal (removal) is calculated on the basis of apartment square metres, which led to a considerable price increase. Moreover, in some cases housing services altogether were incorrectly calculated in this region. People should also be informed about the location of waste sorting and processing facilities in each region and their potential social and environmental effects. Should awareness raising measures be supplemented with basic education events the impact may be higher.

A critical success condition is the construction of the entire infrastructure that should be put in operation. If certain facilities are not in place or people do not understand how to use them, the reform will face substantial barriers. Regional operators intend to dismantle trash chutes at multi-apartment buildings to exclude disposable and non-disposable waste mix. These may be replaced with small containers at each floor.

There is also a need to provide stimuli for green manufacturers so that they increasingly use secondary resources (recycled waste) provided by regional operators. An important aspect for the reform is education and vocational training for specialists already working in and newly recruited for the industry.

The waste management reform will give a new development impulse to the machine building industry, as waste processing requires new technologies and equipment. High automation of waste processing allows the number of staff to be limited, increasing the industry efficiency and key performance indicators of companies.

The successful implementation of the waste management reform will limit waste dumping and improve the environmental situation, especially in urban areas, as well making a favourable impact on people's health. It will also provide millions of tons of secondary resources (materials) to the industry, boosting the circular economy.

13.2.3 BRICS Waste Policy Cooperation

Following the fifth BRICS Ministers of Environment Meeting in August 2019 that focused on 'Contribution of Urban Environmental Management to Improving the Quality of Life in Cities', BRICS leaders, in a joint statement, confirmed support for the post-2020 global biodiversity framework and urged all parties involved in the UNFCCC negotiations to come to 'fair and effective' results.

BRICS environment ministers signed the Memorandum of Understanding on Environmental Cooperation at the tenth BRICS Summit in Johannesburg, South Africa (2018). In this document, they agreed to focus on cooperation initiatives through concrete mechanisms and actions. BRICS cooperation in this area should contribute to the increased

> quality of life in cities through the sharing of knowledge and experience on waste management, 'cogeneration,' reverse logistics, sanitation and water quality, better urban air quality and larger urban green areas, and the Partnership on Urban Environmental Sustainability Initiative (PUESI) (SDG Knowledge Hub, 2019).

Brazil proposed to frame the 'environmental quality urban agenda' in line with the following topics: (i) clean rivers and seas; (ii) waste management, resource efficiency and recovery of contaminated areas; (iii) sanitation; (iv) air quality; and (v) urban green areas (IISD, 2019).

The BRICS decision-makers pay special attention to dialogue on hazardous waste handling. In these countries hazardous waste management systems include the construction of facilities, rules and standards for their collection, treatment and transportation (including export and import). Although substantial progress has been achieved, more efforts are required to ensure that 'fully compliant ESM facilities are available in all regions within these countries' (UNEP, 2015).

External trade of steel scrap across national boundaries accounts for less than 20% of the total usage. Russia's share of the export market fell from around 20% in 2005 to 4% in 2010, and rose again to 9% in 2014. Russia is among the leading countries in collecting and consuming recovered paper: Out of 2.6 million tons of collected recovered paper and board, 2.2 million tons were consumed and 0.4 million tons exported in 2012. Russia is also among the top 15 importers of second-hand clothing, part of which may be re-exported as sorted fractions (UNEP, 2015).

Exchanging information and best practices regarding waste management is a priority for business in BRICS, according to the recommendations for the Brasilia summit outlined by the BRICS Business Council in 2019. The Council suggests cooperation in areas related to waste-to-energy solutions, plastic chemical recycling and bioplastic technologies. Minimizing negative impact on the environment is possible through joint efforts of companies and the local authorities, as practical waste management measures are taken at the local level. These include selective waste collection and secondary resources reuse, recycling and recovery that will lead to limiting the amount of waste disposed of in landfills. According to business estimates, solid waste management is still not entirely cost-efficient in some BRICS countries. Therefore, the government should take the lead to kick-start the process, develop business opportunities and provide initial funding to first projects. The BRICS Business Council considers these efforts a contribution to the green economy in the BRICS (BRICS Business Council, 2019).

Russia has provided waste-related development assistance to poor countries. For instance, following Typhoon Haiyan, Philippines, disaster waste management assistance was made available as part of an early recovery plan funded by the Governments of Japan, the Russian Federation and Saudi Arabia and by the Central Emergency Response Fund and UNDP (UNEP, 2015).

Support is also rendered for alleviating hunger and improving nutrition around the world and domestically, for the poor. The not-for-profit organization the Global FoodBanking Network (GFN) helped to establish food banks in over 25 countries, including developed (i.e. Canada, Russia) and developing (i.e. Brazil, Ecuador, India) ones. The Russian branch of GFN, the All-Russian Charity Foundation Foodbank Rus, was registered in 2012. The Foundation distributes food to vulnerable groups of the population across the country. In 2016 the Foundation distributed 5.75 million kg of food and basic commodities. Like in other countries, the Russian FoodBank establishes relations with the largest FMCG manufacturers and other stakeholders (FoodBank Russia, 2020).

13.3 DISCUSSION AND POLICY RECOMMENDATIONS

Although the waste management reform has been launched, its implementation faces difficulties and requires multiple stakeholder efforts. The easiest solutions (i.e. waste incineration) should not be taken for granted. Instead, separate waste collection, waste processing/recycling and other waste-to-energy options should be explored.

13.3.1 Energy from Waste

Second-generation biofuel research has been stepped up in recent years, based on timber, straw, other carbon-containing waste and non-food raw materials; liquid biofuels made by fermentation; solid biofuels produced through torrefaction (low-temperature pyrolysis) and pellet production; and third-generation biofuels – specially grown efficient types of biomass such as algae. As to output, algae is 3.5 times more efficient than palm oil, 8 times more efficient than corn and 40 times more efficient than soybeans (Shchugoreva, 2016). Promising research is also being performed on the disposal of various types of waste. In Russia, organic agricultural waste is the main source of energy biomass (energy content is up to 80 million tons of oil equivalent (toe) a year); forest industry's organic waste if advanced forestry and wood processing technologies are applied (up to 1 billion tons of oil equivalent per year); urban waste (sewage and municipal solid waste); peat (in total 60 billion toe per year); and energy crops (at least 270.9 million toe per year) (REA, 2012).

Energy generation from various types of waste is growing all over the world, including generation by manufacturing companies (in-house generation). Increased amounts of MSW, agricultural and animal farming waste, water treatment sediments and other types of waste (which otherwise just accumulate or get disposed of) all contribute to this trend. The global volume of MSW amounts to about 1.3 billion tons a year, or 1.2 kilos per person per day on average. Specific waste generation figures vary for countries and cities depending on the level of their urbanization and wealth. During the forthcoming decades, the amount of MSW is expected to grow at a higher rate than urbanization, reaching 2.2 billion tons a year by 2025, and 4.2 billion tons a year by 2050 (World Bank, 2012). In the medium to long term, waste may become a profitable energy source comparable with fossil fuels. Unlike hydrocarbons, it is a quasi-renewable resource. Apart from thermal depolymerization, a technology to produce biogas was developed, comprising 55–75% of methane and 25–45% of carbon monoxide, with calorific value ranging between 4,800 and 6,000 kcal/m^3 depending on the methane content. In that respect, biogas is an alternative to natural gas.

In Russia, with its significant hydrocarbon reserves and low-cost natural gas-based power generation, the wider use of RES is aimed at extending the volume while decreasing the costs of energy resources. It also contributes to municipal, industrial and agricultural waste disposal, the diversification of the national energy balance to increase energy security and the reduction of greenhouse gas emissions and environmental pollution. Numerous public and private companies are already working on this market. Specifically, the companies of Rosatom State Atomic Energy Corporation are planning to manufacture equipment for municipal solid waste incineration to generate power and heat (Proskuryakova & Ermolenko, 2019).

13.3.2 A Policy Comparison of Russia with the Other BRICS Countries

Russia is not a unique case among the developing countries in starting the waste reform rather late – after 2005. The country's counterparts among BRICS also adopted national laws and regulations in the same time period (see Table 13.3). With regard to some policy documents, Russia faces similar challenges to other BRICS, and some issues are very distinct for each country.

Extractive industry waste is another common problem for all BRICS countries. China, Russia and Brazil mine oil and gas, and all countries, except Brazil, extract coal (Wilson, 2015). Additionally, the BRICS extract non-energy mineral resources, such as gold.

Similarly common is the problem of MSW. In Brazil, the PNRS (2010) and, in India, the Solid Waste Management Rules (2015) regulate all stages of MSW handling: Collection, separation, storage, transport and final disposal. Despite the waste policy efforts, the adopted standards are often not in line with the local context. For instance, in China the law aimed at preventing environmental pollution by MSW through proper management rules did not achieve the overambitious objectives (Tai et al., 2011). Like most policy areas, inefficient policy implementation hinders the achievement of waste policy goals (Massod et al., 2014).

13.3.3 Policy Recommendations

At the time when there was no comprehensive waste management at the national level some gaps were filled in by non-governmental organizations, civil society initiatives and small businesses.

TABLE 13.3
Waste Laws, Regulations and Related Challenges in the BRICS

	Law, Regulations and Year of Adoption	Challenges
Brazil	National Solid Waste Policy (PNRS) (2010)	Lack of social inclusion and little or no formalization of the waste chain actors; Unemployment, lack of education and awareness on safe practices, licensing and registration hinder transition from informal to formal waste management
Russia	National Law (2015)	Lack of public support; Regional context varies; Local people not consulted, their experience not taken into consideration
India	National Environmental Policy (2006); Solid Waste Management Rules (2015)	Lack of clear statistical data; Lack of law enforcement; Unemployment, lack of education and awareness on safe practices, licensing and registration hinder transition from informal to formal waste management
China	Law on Prevention and Control of Environmental Pollution by Solid Waste (2007); Circular Economy Promotion Law (2008)	Lack of clarity on the responsibilities of different government levels, companies and consumers; The need to define mandatory requests; Regulations are not harmonized with the Circular Economy Law; Lack of policy incentives for recycling
South Africa	National Environmental Management: Waste Act (2008); Waste Management Strategy Implementation (2005)	Uncontrolled access to landfills; Informal settlements near landfills; Lack of monitoring and evaluation of policies; Unemployment, lack of education and awareness on safe practices, licensing and registration hinder transition from informal to formal waste management

Sources: Gutberlet et al., 2013; da Silva et al., 2019; Iyamu et al., 2020.

Now, as the waste management logistics is still not fully in place, these initiatives still function, primarily in cities. Services like Sborbox and Sobirator collect recyclable waste from people's homes. The former functions at no cost for citizens by sending collection workers to pick up even small amounts of waste left by the door at multiapartment buildings. The latter sends vehicles with a fee associated with it, so people often pull together their waste and share the cost. Startups that target various waste-related problems regularly appear in Russia, including Wasteout (sensors and software controlling the occupancy of dumpsters), Ubirator (the aggregator for recyclable material providers, transport companies and recycling centres), Fineplastic (collecting and processing of plastic) and many more (Rusbase, 2019).

Civil society initiatives include the electronic map of Moscow recycling centres "Recyclemap", composed and updated by Greenpeace. This international NGO has other waste-related initiatives, such as 'zero waste' aimed at raising citizens' awareness and extending conscious consumption, undertaking plastic waste studies and promoting statutory bans on single-use plastic (Greenpeace, 2020). To assess the magnitude of the problems volunteers inspect the river and lake strandlines to identify the waste composition. According to this 'plastic watch' assessment, 68% of the waste located at shores is waste; nearly all of this plastic is non-reusable.

To promote the application of waste as an energy resource, it would also make sense to design and approve government support measures for the mass construction of MSW processing plants.

Specifically, we suggest that the Russian Ministry of Energy* step up drafting government regulations on introducing green tariffs to promote the construction of MSW-based generation facilities through selling electricity to the grid in retail markets at regulated prices (tariffs). Moreover, optimal rates should be established for waste disposal services, with part of the revenues co-invested in the construction of waste processing plants. Additionally, regional authorities may decide to grant tax breaks and other benefits to companies implementing such projects.

One policy aspect that is rarely taken into consideration in Russia in any reform is the application of local, including indigenous, people's knowledge. People in rural areas have searched for various ways to use secondary resources. Ethnic groups have long traditions of giving a second life to old goods and using waste to overcome scarcity of resources. The studies of indigenous people's knowledge may help discover 'hidden biases of the notion of circular economy and other "innovative" problem-solving practices' (Siraguza & Arzyutov, 2020). Traditional concepts of circular economy based on the three principles – reduce, reuse and recycle – are not applicable in many development contexts (Jacobi & Bensen, 2011).

A wider multi-stakeholder approach to waste policy design is also underlined by da Silva et al. (2019) for the BRICS nations. However, not all BRICS policies take account of traditional waste collection practices. Arguably the most attentive to local informal practices are Brazilian decision-makers who include support to waste-pickers and their cooperatives through laws and regulations, including the National Solid Waste Policy (2010) and Pro-Catador Program (2010) (Gutberlet et al., 2013).

The actual implementation of the reform will depend on the fine-tuning of the policy interventions at regional and local levels based on different social and economic contexts.

13.4 CONCLUSION

Over the past decades, the waste accumulated and continuously generated in Russia has posed numerous environmental and social problems. The waste management reform was launched in 2015 to tackle these problems. The policy changes started with high-level legislative and legal acts and was further supported with strategies and action plans, and later with national initiatives (projects). Together these measures were aimed at introducing a comprehensive management system for industrial and municipal waste.

Russia's efforts to address waste issues are in general terms comparable with the latest policy interventions in BRICS. The BRICS nations face similar problems with extractive industry and municipal solid waste. The timing for policy changes is also comparable. However, varied national context and history pose different challenges for the BRICS, the most significant of which are related to policy implementation. Some targets are overly ambitious and are not achievable, while others are too low and will not make an impact.

Despite the waste management hierarchy spelled out in the 2030 Waste Recycling Industry Strategy, the clear focus of the reform is on creating more landfills (even though modern ones) and waste incineration that are referred to as 'waste treatment' according to the new legislative provisions. Although waste-to-energy is an important element of waste management, incineration is not the only technology. More attention should be placed on awareness raising, separate waste collection, increasing the efficiency of resource use and various waste recycling options.

One important element for the waste reform success is involving various stakeholders in its implementation. This inclusive process allows making use of informal and local waste collection practices, including those of indigenous peoples. Taking account of cultural and economic contexts improves waste management performance.

---

* In line with the Russian Government Decrees of 13 June 2015 DM-P9-4643r, 23 July 2015 № AD-P9-5014, and 29 September 2015 № AD-P9-6639.

REFERENCES

Aldieri, L. et al. (2019) Environmental innovation, knowledge spillovers and policy implications: A systematic review of the economic effects literature. *Journal of Cleaner Production*, 239, article 118051.

Brazil Federal Decree No. 7,405, 23 December 2010, published in D.O.U. of 23 December 2010.

Brednikova O., Tkach O. (2008) Griaznaia derevnia' i 'zamusorennyi gorod' (obydennye praktiki obrashcheniia s musorom v raznykh soobshchestvakh). *Antropologicheskii Forum*, 8, pp. 338–352.

BRICS Business Council (2019) BRICS Business Council Annual Report. Brasilia, 14 November 2021, URL: https://static.portaldaindustria.com.br/media/filer_public/57/c7/57c7e362-8571-4818-b7ef-b5a91bc00383/brics__annualreport_2020.pdf (accessed 06.04.2021).

da Silva C.L. Weins N., Potinkara M. (2019) Formalizing the informal? A perspective on informal waste management in the BRICS through the lens of institutional economics. *Waste Management*, 99, pp. 79–89.

Economic Growth for an Innovative Future. BRICS Business Council Annual Report 2019. Brazil, Brazilia, November 14, 2019, URL: http://brics2019.itamaraty.gov.br/images/documentos/LowRes_CNI_CEBRICS_Cupula_EN_1.pdf (accessed 03.08.2020).

Ermolenko B., Ermolenko G., Proskuryakova L. (2019) Renewable energy technological potential assessment for evidence-based policy-making. In *Emerging Technologies for Economic Development*, Ed. by D. Meissner, L. Gokhberg, O. Saritas. Springer.

Federal Law No. 12,305, July 2010 and its regulation through Decree No. 7,404 of December 2010.

FoodBank Russia, URL: https://foodbankrus.ru/en/about (accessed 22.07.2020).

Greenpeace (2020) Zero waste, URL: https://greenpeace.ru/projects/zero-waste (accessed 22.07.2020).

Gutberlet J., Baeder A.M., Pontuschka N.N., Felipone S.M.N., dos Santos T.L.F. (2013) Participatory research revealing the work and occupational health hazards of cooperative recyclers in Brazil. *International Journal of Environmental Research Public Health*, 10(10), 4607–4627.

IISD SDG Knowledge Hub (2019) BRICS ministers commit to urban environmental management, global climate and biodiversity issues, URL: https://sdg.iisd.org/news/brics-ministers-commit-to-urban-environmental-management-global-climate-and-biodiversity-issues/ (accessed 21.07.2020).

Iyamu H.O., Anda M., Ho G. (2020) A review of municipal solid waste management in the BRIC and high-income countries: A thematic framework for low-income countries. *Habitat International*, 95, article 102097.

Jacobi P.R., Besen G.R. (2011) Solid waste management in São Paulo: The challenges of sustainability. *Estud. Avançados*, 25, pp. 135–158.

Kaazke J., Meneses M., Wilke B.-M., Rotter V.S. (2013) Environmental evaluation of waste treatment scenarios for the towns Khanty-Mansiysk and Surgut, Russia. *Waste Management Research*, 31, pp. 315–326.

Krechetov P., Kostin A., Chernitsova O., Terskaya E. (2019) Environmental changes due to wet disposal of wastes from coal-fired heat power plant: A case study from the Tula Region, Central Russia. *Applied Geochemistry*, 105, pp. 105–113.

LECAP (2019) Eco-industrial parks of Russia, Dubinchina S. (Ed.), URL: http://lecap.ru/upload/iblock/a50/a50a04d3c10c9731b72c7267fc222499.pdf (accessed 03.08.2020).

Maryev V. (2019) Country report. Russian federation. In *Presentation at the 9th Regional 3R Forum in Asia and the Pacific, 4–6 March, 2019. Research Institute "Environmental Industrial Policy Center", Russian Ministry of Industry and Trade*, URL: https://www.uncrd.or.jp/content/documents/7665Country%20Presentation_Russian%20Federation.pdf (accessed 03.08.2020).

Massod M., Barlow C.Y., Wilson D.C. (2014) An assessment of the current municipal solid waste management system in Lahore, Pakistan. *Waste Management & Research*, 32(9).

National Solid Waste Policy (2010) Brazilian Federal Law No. 12,305, July 2010 and its regulation through Decree No. 7,404 of December 2010.

Nevskaya M.A., Marinina O.A. (2015) Regulatory aspects of mining waste management in the Russian Federation. *Bioscience and Biotechnology Research Asia*, 12, pp. 2619–2628.

Polprasert C., Koottatep T. (Eds.) (2017) *Organic Waste Recycling: Technology Management and Sustainability*. IWA Publishing.

Pro-Catador Program (2010) Brazilian Federal Decree No. 7, 405, 23 December 2010, published in D.O.U. of 23 December 2010.

Proskuryakova L., Ermolenko G. (2019) The future of Russia's renewable energy sector: Trends, scenarios and policies. *Renewable Energy*, 143, pp. 1670–1686.

REA (2012) *Russian Bioenergy Industry in the 21st Century.M.* Russian Energy Agency. (in Russian).

Rusbase (2019) 13 startups in wastetech that solve the waste recycling problems, URL: https://rb.ru/longread/wastetech-startups/ (accessed 30.07.2020) (in Russian).

Russian Federal State Statistics Service (2020) Environment section, URL: https://www.gks.ru/folder/11194 (accessed 03.08.2020).

Russian Ministry of Natural Resources and Environment (2018a) Passport of the national project "Environment", URL: https://www.mnr.gov.ru/activity/directions/natsionalnyy_proekt_ekologiya/ (accessed 23.07.2020).

Russian Ministry of Natural Resources and Environment (2018b) Chapter 8. Production and consumption waste management, URL: http://www.mnr.gov.ru/upload/medialibrary/259-330.pdf (accessed 20.07.2020) (in Russian).

Shchugoreva E. (2016) Nature's gifts: Biofuel// AltEnergiya.ru. URL: http://altenergiya.ru/bio/podarki-ot-pri rody-biotoplivo.html (accessed 28.07.2016) (in Russian).

Singh J., Laurenti R., Sinha R., Frostell B. (2014) Progress and challenges to the global waste management system. *Waste Management Research*, 32, pp. 800–812.

Siragusa L., Arzyutov D. (2020) Nothing goes to waste: Sustainable practices of re-use among Indigenous groups in the Russian North. *Current Opinion in Environmental Sustainability*, 43, pp. 41–48.

Starostina V., Damgaard A., Rechberger H., Christensen T.H. (2014) Waste management in the Irkutsk Region, Siberia, Russia: Environmental assessment of current practice focusing on landfilling. *Waste Management Research*, 32, pp. 389–396.

Tai, J., Zhang, W., Che, Y., Feng, D. (2011) Municipal solid waste source-separated collection in China: A comparative analysis. *Waste Management*, 31, pp. 1673–1682.

TASS Information Agency (2019) Waste treatment in Russia. The birth of the industry, URL: https://tass.ru/info/6000776 (accessed 09.07.2020) (in Russian).

The Village (2020) Authorities refuse to establish a waste processing system, URL: https://www.the-village.r u/village/city/news-city/381437-bez-razdelnogo-sbora (accessed 30.07.2020).

Tulokhonova A., Ulanova O. (2013) Assessment of municipal solid waste management scenarios in Irkutsk (Russia) using a life cycle assessment-integrated waste management model. *Waste Management Research*, 31, pp. 475–484.

UNEP (2015) Global waste management outlook, URL: https://www.uncclearn.org/sites/default/files/inve ntory/unep23092015.pdf (accessed 22.07.2020).

UNIDO (2020) Eco-industrial parks, URL: https://www.unido.org/our-focus-safeguarding-environment-res ource-efficient-and-low-carbon-industrial-production/eco-industrial-parks (accessed 03.08.2020).

Wilson J.D. (2015) Resource powers? Minerals, energy and the rise of the BRICS. *Third World Quarterly*, 36(2), 223–239.

World Bank (2012) *What a Waste. A Global Review of Solid Waste Management*. World Bank.

14 Recycling of Waste and Policy Frameworks in India

C. Ramprasad and S. Rangabhashiyam

CONTENTS

14.1 INTRODUCTION

Over the past few decades the generation of waste (water, solid and gas) has been increasing exponentially due to rapid urbanization, industrialization, economic growth and change of life styles (Wang et al., 2016; Bai et al., 2017; Kumar et al., 2017). Many of the developing countries especially BRICS nations are facing major problems in the management of solid waste (Singh and Sharma, 2016). Effective solid waste management (SWM) is a foremost challenge faced by countries like India where population density is high (Paul et al., 2012; Ramachandra et al., 2018). The culture, climate, physical, geographic, ecological and social features are more diverse than in other developing countries, leading to increased waste generation. According to a World Bank report, 2017, India's population was registered at 1,252 million in the year 2013, and had increased to 1,339.2 million by 2017. Figure 14.1 shows that population growth and the annual waste generation. The growth in the infrastructural projects has created a major impact on environmental health due to mounting of construction and demolition (C&D) waste. The waste is predominantly dumped into rivers, illegally on hill slopes and in low-lying areas, flood-prone areas and with MSW landfill, leading to sudden flooding, resource depletion and groundwater contamination (Abel, 2014; Yahaya et al., 2016; Raju et al., 2018).

In spite of noteworthy growth in the economy, environment and society, the management of MSW, especially of C&D, has remained unchanged. Almost 90–95% of waste is illegally dumped in landfills, and the informal sector is leading in extracting the valuables from the waste (Pathak and Srivastava, 2017; Gupta et al., 2019). There is an urgent need to treat and process the generated solid waste in sustainable ways. The present solid waste treatment systems are incompetent to handle the waste by sustainable methods, and create a negative impact on public health, the environment and the economy (Marshall and Farahbakhsh, 2013; Kumar et al., 2017). The Ministry of Environment

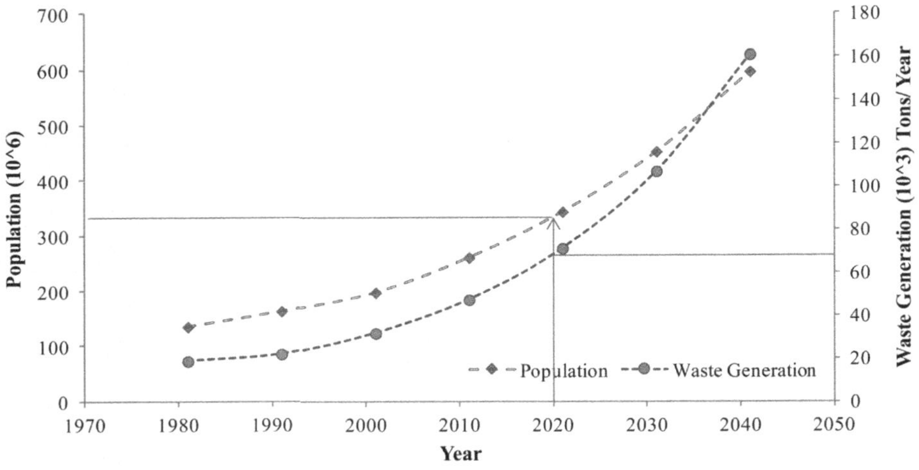

FIGURE 14.1 Population growth trend and waste generation rate from 1981 to 2041.

and Forests (MoEF), Government of India, has initiated the Management and Handling Rules for different types of waste (Sharholy et al., 2008; Sandhu and Sidhu, 2015; Kumar et al., 2017; Gupta et al., 2019), even though the abidance is limited and inconsistent. This chapter discusses the challenges and opportunities associated with two major solid waste contributors, C&D waste and WEEE, and the policies available in India.

14.2 CONSTRUCTION AND DEMOLITION WASTE

14.2.1 Overview

In India, the population growth and continuous industrial growth have forced the local bodies to invest heavily in infrastructure projects such as the construction of roads, flyovers, bridges and residential houses, which lead to huge quantities of C&D waste. The average global generation of C&D waste increased from 20 billion tons in 2007 to 40 billion tons in the year 2014 (Tam et al., 2018). Therefore, the progressive depletion of resources and growing awareness of sustainability has resulted in the recycling and reuse of C&D waste. There are studies showing that a major portion of construction material can be recycled and replaced by reprocessed C&D waste. However, in most developing economies, the lack of regulatory framework and lack of knowledge has stimulated the non-exploitation of value from C&D waste.

Construction waste is bulky and heavy, generated from construction, repair, rehabilitation, demolition and renovation activities. It also includes surplus and damaged commodities arising from construction work that are temporarily used during the construction phase. The waste is generated at different phases of construction and includes concrete, plastic, metal, bricks, glass, tiles, ceramics, wood, plaster and steel, etc. The amount of C&D waste generated during the construction phase was found to be nearly 40–60 kg per sq. m, while during the renovation phase around 40–50 kg per sq. m was estimated to be generated. Similarly, the highest recorded waste generation (300–500 kg per sq. m) is during the demolition of buildings (Patel et al., 2015). The quantities of waste generated based on its individual components or fractions in C&D waste are shown in Figure 14.2. It can be seen from the figure that concrete predominates with 65%, followed by bricks 18%, tiles 7% and the least being from plastic and metals 2%. Concrete is the main constituent of C&D waste and exists in two forms; the first one is mass non-reinforced concrete used for foundations and the second one is reinforced concrete containing steel bars. Therefore, the concrete waste can be processed by various treatment techniques and recycled/reused as aggregates. Concrete that is constructed or made using this kind of aggregate is indicated as recycled concrete aggregate (RCA).

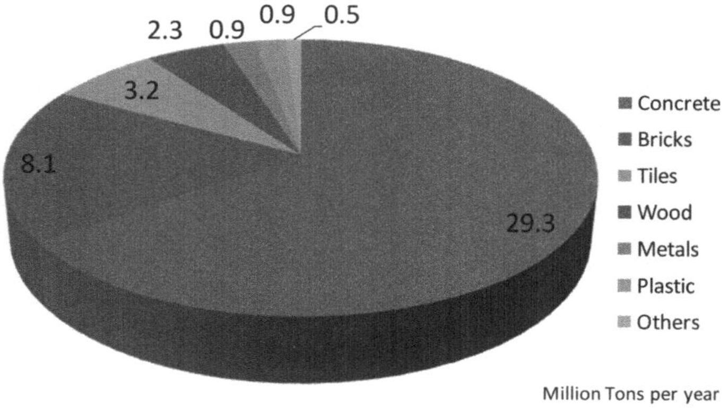

FIGURE 14.2 Construction and demolition waste quantities generated in India.

The recycled aggregates can be used as important constituents of concrete and other recycling operations like pavement making, pit filling and road construction. Thereby, dual benefit can be obtained by preserving the natural resources and reducing landfill disposal. The RCA can be produced by the following steps: Sorting, shredding, separation of ferrous items, sieving and cleaning the demolished concrete into components of different size and quality (Pellegrino and Faleschini, 2016; Singh and Sharma, 2016; Tam et al., 2018), use of manpower/equipment for sorting/shredding are the common methods adopted. The report suggests that RCA has less water absorption, density and specific gravity than normal aggregates (Alengaram et al., 2011; Shahidan et al., 2013; Tosic et al., 2015). Additionally, recycled aggregates are more advantageous as the porosity of the recycled concrete is much higher than that of normal concrete. Table 14.1 shows a comparison of various physical properties of concrete made with normal aggregates and recycled aggregates. Figure 14.3 shows the normal aggregate and recycled aggregate.

14.2.2 Moulded Concrete Cubes

A study was carried out by partially replacing the fine and coarse aggregate by 25% and 50% with recycled aggregate in M50 grade concrete showed a slight variation in the casted concrete cubes and cylinders compressive and split tensile strength respectively (Saravanakumar and Dhinakaran, 2015). The compressive strength on cubes for the 25% replaced recycled aggregate observed after 28 days was 47 megapascal (MPa) (Figure 14.4), while the compressive strength was 50 MPa for

TABLE 14.1
Comparison of Physical Properties of Normal and Recycled Concrete Aggregates

Physical Property	Unit	Normal Concrete Aggregates	Recycled Concrete Aggregates
Absorption	% (w/v)	0.60–0.80	2.03–2.15
Aggregate impact value	% (w/v)	10.0	12.7
Angularity number	NA	7.50	9.50
Bulk density	Kg/ cu. m	1450–1510	1250–1290
Fineness modulus	NA	6.54–6.76	6.79–6.88
Moisture content	% (w/v)	0.17–0.25	1.57–1.88
Porosity	% (v/v)	1.55–1.82	5.03–5.42
Saturated surface specific gravity (dry)	% (w/v)	2.62–2.71	2.53–2.58

NA: Not applicable.

FIGURE 14.3 Photograph of the normal and recycled aggregate.

FIGURE 14.4 Concrete cubes and cylinder with normal and recycled aggregates.

normal concrete (Parthiban and Mohan, 2017). Subsequently, the replacement of 50% of natural aggregate with recycled aggregate showed a maximum compressive strength 55 MPa while the normal concrete showed a compressive strength of 60 MPa (Ramprasad and Kutty, 2016). The shrinkage, split tensile strength and skid resistance test on the recycled concrete aggregate showed satisfactory results. The split tensile strength of the normal concrete was 5.22 MPa whereas the recycled concrete showed a split tensile strength of 4.65 MPa. Therefore, the use of RCA for the construction of non-load bearing structures is advisable in order to protect environmental and natural resources. There is real-time evidence on the use of RCA for making pavements, compound walls, park benches, partition walls, etc.

14.2.3 Recycled Aggregate as Carbon Sequester

As discussed earlier, the recycled aggregates provide substantial benefits to the environment and society. The considerable benefit depends on the types of recycling, reprocessing and reuse supply chain, and the areas are:

(i) The use of recycled aggregates instead of virgin aggregates from quarry
(ii) The processing of quarry aggregates emits a lot of greenhouse gases
(iii) Transportation emissions can be reduced if an onsite re-processing facility is available

A study by Nayana and Kavitha (2017) showed that for the production of 1 ton of natural aggregate around 0.046 tons of CO_2 are emitted, while only 0.0024 tons of CO_2 are emitted for the production of recycled aggregates. There was another study showing that CO_2 emissions from the construction of 1 kilometre of roadway from natural aggregates and recycled aggregates showed a positive effect when utilized (Wijayasundara et al., 2017). Normal aggregates showed CO_2 emissions of approximately 72 tons/kilometre while recycled aggregates emitted approximately 24 tons/kilometre.

14.2.4 Barriers that Prevent Wider Applications of Recycled Aggregate

The use of recycled aggregates in construction has been restricted by many factors like customer perception, limited policies, standard operating procedures, insufficient financial incentives, skilled labour availability and distance/non-availability of waste recycling facilities (Tam et al., 2018). The public perceptions of the use of recycled aggregates are the biggest concern; they think that the quality and life of buildings will be reduced. The standards and specifications for the use of recycled aggregates for construction are lacking in India. The government of India recently initiated a rule on C&D waste in the year 2016, elaborately described in clause 2.5. Additionally, contractors and financial agencies are not supporting the use of recycled aggregates in projects. The concerns according to them are that the use of recycled aggregates in construction is not liked by the public and governments are also not promoting it. Subsequently, the labourers employed in sites are equipped in handling the recycled aggregates and very few recycled units are in existence in India. The ways to overcome the barriers are (i) the Government of India has to encourage the research institutes to prepare an operation manual and guidelines for the usage of recycled aggregates in the construction field; (ii) an awareness campaign needs to conducted to remove the negative perspective on the quality and life of recycled aggregates and (iii) the public needs to accept the use of recycled aggregates and focus on the concept of waste as a resource.

14.2.5 Policies and Framework Followed in India

In India, the first waste management law was enforced in 1986 as the 'Environmental Protection Act', and then a lot of acts and policies were put in place concerning special wastes. The Construction and Demolition Waste Management Rules were approved by the Ministry of Environment, Forest and Climate Change (MoEF&CC), Government of India (GoI) in the year 2016. The rules describe C&D waste collection, sorting, segregation, processing, recycling and reusing techniques in detail. The C&D waste management rule particularly emphasized one methodology named 'Deconstruction' for the demolition of buildings. The demolition of the building using explosives creates a lot of environmental pollution such as air and solid waste. In the deconstruction technique, each individual component such as wooden door and window frames, glasses, electrical items, tiles, bricks, etc., can be pulled down one after the other. Thus, the dismantled items can be recycled or reused as such with slight processing, resulting in less waste generation.

14.3 WASTE ELECTRICAL AND ELECTRONIC EQUIPMENT (WEEE)

The usage of electrical and electronic equipment (EEE) worldwide is augmented due to the technological development and fast economic growth in various sectors consequent to intensive manufacturing practice in various industries related to electrical equipment and electronics. Apart from different positive features of EEE products, the shorter lifespans of EEE products, greater use of EEE products by large consumers, lower rate of recycling of EEE waste and fast growth in waste electrical and electronic equipment (WEEE) are major concerns worldwide (Kiddee et al., 2013; Kumar and Dixit, 2018). WEEE is considered one of the major sources of waste, increasing at a rapid rate compared to other waste generation and reported to amount to more than 40 million metric tons per year (Kumar et al., 2017; Breivik et al., 2014). The global trade, diminution of resource availability, tininess of components and size of EEE products require WEEE rules modification in various countries based on their economic growth and province (Li et al., 2015).

Most countries are involved in the management and regulatory control of WEEE; nevertheless enormous quantities of WEEE from developed countries are illegitimately shipped into developing countries (Chan and Wong, 2013). WEEE transboundary transport is mainly contentious due to the export of WEEE to developing countries (Benin, Ghana, Kenya, Liberia, Nigeria, Senegal, Uganda, South Africa, Cambodia, Pakistan, India, China, Hong Kong, Malaysia, Philippines, Vietnam, Thailand) from developed countries (Canada, United States, EU27, Australia, Belgium, Finland, France, Germany, Japan, Netherlands, Norway, United Kingdom, Singapore, South Korea), studied by the Basel Action Network (BAN), Silicon Valley Toxics Coalition (SVTC), non-governmental organizations (NGOs), United Nations Environment Program (UNEP), Secretariat of the Basel Convention (SBC) and Federal Environmental Agency in Germany (Wong et al., 2007; Ongondo et al., 2011; Jinhui et al., 2013). The transboundary WEEE movement from developed countries is due to the high cost of metal extraction methods, stringent environmental policies for the recycling/recovery of valuable components from WEEE and higher labour costs. Therefore, the unlawful movement of WEEE from developed to developing countries needs to be forbidden in order to prevent metal loss and improper methods of recycling WEEE (Shamim et al., 2015; Zhang et al., 2017). The management of WEEE reported in many studies with the views of recycling/recovery of valuable materials, WEEE hazardous effects in environment and human health.

14.3.1 Composition of WEEE and Hazardous Impacts

WEEE consists of different materials, varies with the waste source generated from electronics and electrical equipment. The different forms of wastes formed as the result of use of different electrical and electronic devices include air conditioning units, refrigerators, televisions, electric blankets, washing machines, microwave ovens, dish washers, toasters, coffee makers, water purification units, vacuum cleaners, iron boxes, computers, mobile phones, laptops, printers, lighting bulbs, medical equipment, electronic monitoring devices, shavers, modems, cameras, toys, etc (Huang et al., 2014; Qinxin et al., 2018; Chen et al., 2015; Hadi et al., 2013). The major materials distributed in WEEE include metals of ferrous and non-ferrous type, plastics, glass, ceramics, resins, etc. Most WEEE consists of iron and steel ferrous metals as the major materials by weight, contributing about half of the WEEE total weight. Next to metals, plastics occur as another major component, occupying one-quarter of the total weight of WEEE. The remaining portions in the WEEE total weight are contributed by metal-plastic mixtures, screens, cables, printed circuit boards, glasses, etc. (Widmer et al., 2005; Yoo et al., 2009). Plastics in WEEE include polycyclic aromatic hydrocarbons, polybrominated diphenyl ethers, polybrominated, dibenzo-p-dioxins, dibenzofurans, etc. The organic material consists of brominated flame retardants, dioxins and dibenzofurans and glass fibres. Large fractions of base metals, rare earth metals, precious metals, hazardous metals and ceramic groups are also distributed in WEEE (Kaya, 2016). The accurate composition of WEEE varies considerably based on their type, period of use and producer (Bigum et al., 2017). Printed circuit boards act as the base of electronics devices and are used in most electronic products. The assemblies of printed circuit boards consist of green or yellow board and many electronic components. The basic organization of printed circuit boards comprised of copper-clad laminates as the substrate material consists of fiberglass-reinforced epoxy resin and materials made of different metals (Zeng et al., 2012). Most of the substrate of printed circuit boards consists of the metallic (30–50%) and non-metallic fractions (50–70%). The non-metallic fractions in printed circuit boards are thermosetting resin (epoxy, cyanate ester and phenol), reinforcing material (fiberglass, ceramics and cloth), substrate type of brominated flame retardants, additives, etc. (Rajagopal et al., 2016; Du et al., 2016). Copper is the important metal in the waste printed circuit boards substrate, and copper contributes 10–20% of the total weight of waste printed circuit boards. Apart from copper, other metals in waste printed

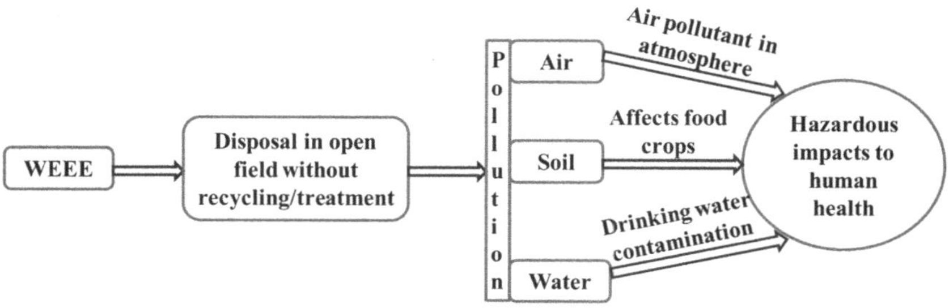

FIGURE 14.5 Impact of WEEE on environment and human health.

circuit boards include aluminium, iron, zinc, lead, arsenic, gold, silver, lead, etc. (Yang et al., 2009; Akcil et al., 2015).

WEEE contains different materials, and most of them are associated with the property of toxicity to human health and potential to cause environmental damage. Nevertheless, some of the materials possess high market value through a proper recycling approach. Figure 14.5 illustrates the improper disposal of WEEE and their pollutants' entry through different routes consequent to the human health hazards (Widmer et al., 2005). Most of the WEEE is transferred to landfill sites/incinerated. The hazardous characteristics of WEEE result in serious environmental pollution if not properly treated/recycled. The incineration of WEEE not only causes the release of toxic ash but also releases hazardous toxic dust into the atmosphere (Mirian et al., 2010). The analysis of the elemental composition of plastics from WEEE has been reported. Less than 1.5% of analysed WEEE showed pollution effects according to Restriction of Hazardous Substances limits. Higher quantities of plastics indicated the inappropriateness of recyclability as 71 parts with a total weight of 11 kg showed bromine concentrations greater than 2,000 mg/kg. WEEE composed of plastics from coloured appliances and plastics of tools group appliances presented high pollution risks (Ayah et al., 2020). Halogenated flame retardants and dielectrics from WEEE are highly persistent, cause bioaccumulation and with toxic potential impose health hazards. Polybrominated diphenyl ethers and their derivatives interfere with the normal function of reproductive and thyroid hormones (Peng et al., 2016). Higher exposure to dechlorane plus stimulates oxidative stress and modifies the activities of detoxification and metabolism (Li et al., 2013). Undesirable health effects have been reported in the workers dismantling WEEE. The exposure of pregnant workers to lead during the separation of electronic waste causes health effects to the foetal development. Children exposed to WEEE work areas along with parents showed higher levels of cadmium in their blood which leads to impaired growth, adaptability and behavioural problems (Zheng et al., 2008).

Polybrominated diphenyl ethers can undergo transformation into OH-polybrominated diphenyl ethers, which have structural resemblance to thyroxine. Due to similarity in structure, OH-polybrominated diphenyl ethers interfere with the production of thyroxine, alter the level of hormones and cause neurotoxicity and dermal problems (Darnerud et al., 2001). Copper, zinc and aluminium are present in large quantities in printed circuit boards, and their excess leaching into the environment poses a serious threat to living systems. The high concentration intake of such metals in humans leads to health issues such as gastrointestinal pain, digestive disorder, anaemia and kidney problems (Raikwar et al., 2008). The levels of polychlorinated biphenyls were analysed in 128 individual human breast milk samples in suspected areas of Ghana. Seven indicator polychlorinated biphenyls were identified in the collected breast milk samples. No major difference was observed in the polychlorinated biphenyls concentrations of milk samples collected from primiparae and multiparae mothers. Polychlorinated biphenyl-28 and polychlorinated biphenyls-180 were confirmed in all the milk samples collected from Agbogbloshie and 99% detection of polychlorinated biphenyl-18, polychlorinated biphenyl-52, polychlorinated biphenyl-101, polychlorinated biphenyl-138 and polychlorinated biphenyl-153. One milk sample collected out of 23 from Kwabenya was found positive for polychlorinated biphenyl-28, and the remaining polychlorinated biphenyls were within detection limits (Anita et al., 2018). Workers recycling WEEE are more exposed to flame retardants, and they are associated with health hazards including damage to DNA, hormonal disorders and undesirable pregnancy results (Grant et al., 2013). Flame retardants (39) were identified in the air of WEEE recycling facilities. The predominant detected flame retardant was polybrominated diphenyl ethers. Higher exposures are reported due to the tasks of dismantling, crushing and bailing. Screens of computers, televisions and cathode ray tubes are connected to higher exposures to flame retardants (Sabrina et al., 2019). Particulate matter and polybrominated diphenyl ethers are major contaminants in fumes emitted during the de-soldering of waste printed circuit boards. The deposition of particulate matter in the respiratory systems through inhalation causes serious health hazards to the workers (Jie et al., 2019). Trace elements such as cadmium, copper and mercury enriched the agricultural soils of WEEE recycling regions. Such enrichment trace elements are due to the activities of WEEE recycling. The contaminations of farmland are due to the trace elements at higher concentrations, which cause carcinogenic effects in the local habitants (Wencheng et al., 2018).

14.3.2 Eco-Friendly Approach to Processing WEEE

In view of major environmental issues and resource reutilization of WEEE, the proper management of WEEE is highly essential. The management strategies include reuse, servicing, remanufacturing, recycling and disposal (Wenzhi et al., 2006). The complex nature of WEEE makes the recycling of heterogeneous components difficult. Nevertheless, to overcome the negative impact of WEEE on the environment and to recover important materials, a suitable process must be employed. The recovery of WEEE constituents like metals cannot be performed in a single approach. The steps include WEEE disassembly, size reduction and segregation; the recovery is through different processes such as the metallurgical type of hydrometallurgical and pyrometallurgical processes (Chauhana et al., 2018). However, these approaches are associated with the drawbacks of inefficiency, higher cost requirements and moreover energy-intensive processes. The bio-based method biosorption is reported as a potential unconventional approach to the recovery of metals from WEEE (Qingping et al., 2016; Zhengsheng et al., 2020).

Biosorption is a physico-chemical process, independent of cellular metabolism, subjecting dead biomass without nutrient requirements and free from the constraints of aseptic conditions. The process is associated with the mechanisms of adsorption, absorption, surface complexation, ion-exchange and precipitation on the biomass (Volesky, 2007; Rangabhashiyam et al., 2014). The practice of biosorption technology has the advantages of low cost, environmental friendliness, higher efficiency, regeneration and reuse, more specificity, freedom from chemical sludge generation and easy control (Vijayaraghavan et al., 2017; Rangabhashiyam and Balasubramanian, 2018; Selvakumar and Rangabhashiyam, 2019). Different biomass types are reported in the biosorption method, including microorganisms (Sakthi et al., 2019; Kshama and Varsha, 2011), macroalgae (Luciana et al., 2018; Rangabhashiyam et al., 2016), lignocellulosic biomass residues (Vijayaraghavan et al., 2016, Herrera-Barros et al., 2020), fish scales (Teshale et al., 2020), bones (Jihoon et al., 2011), feathers (Anantha et al., 2016), industrial sludge (Gurpreet et al., 2017), etc. The biosorption capacities in the removal of metals are shown in Table 14.2.

The macroalgae *Turbinaria conoides* and immobilized *Turbinaria conoides* (polysulfone) were reported to have a biosorption capacity of 200.5 and 157.9 mg/g in thulium removal from aqueous solution. The regeneration of biosorbents for reuse was performed using 0.01 M HCl and usefully tested up to 10 recycles (Rangabhashiyam and Vijayaraghavan, 2019). The biomass of chitin was subjected as biosorbent for the removal of gold from a leachate solution of discarded computer microprocessors. The methods of biosorption and precipitation with biosorption were used for gold recovery from leachates. The maximum uptake capacity using both methods was determined as 35 mg/g. The test indicated 80% gold recovery with 20 g/L chitin (Leticia et al.,

TABLE 14.2
Removal of Metal Ions from Aqueous Solution Using Biosorbents

Biosorbent	Metal Ions	Biosorption Capacity (mg/g)	References
Neem sawdust	Cerium	200.00	Varshini and Das (2014)
Malt spent rootlets	Europium	156.00	Anagnostopoulos and Symeopoulos (2013)
Saccharomyces cerevisiae	Lead	72.50	Shahram et al. (2015)
Aloe vera leaves	Copper	3445.00	Abedi et al. (2015)
Bacillus coagulans	Lead	118.30	Lei et al. (2014)
Ulva lactuca	Praseodymium	69.75	Vijayaraghavan (2015)
Westerdykella sp.	Arsenic	70.00	Jaiswal et al. (2018)
Orange peel	Praseodymium	58.80	Varshini et al. (2015)
Fish scales	Cerium	200.00	Das et al. (2014)
Biomass of coastal seaweed community	Zinc	115.19	Fatih et al. (2017)
Rapeseed waste	Lead	25.13	Irina et al. (2019)
Phanerochaete chrysosporium	Lead	81.93	Ningqin et al. (2020)
Phanerochaete chrysosporium	Cadmium	48.23	Ningqin et al. (2020)
Chitosan	Rhenium	222.00	Wenjing et al. (2020)

2015). Leach liquor from printed circuit board is rich in copper, and the separation of copper was performed with a biosorption method with *Aspergillus oryzae* and yeast. Both *Aspergillus oryzae* and yeast were used as dead biomass in optimal condition for biosorption experiments. *Aspergillus oryzae* showed biosorption efficiency of 88.6 and 91.2% in copper removal from leach liquor and synthetic system. The biosorption efficiency of yeast was 70.9% for copper removal from leach liquor and 83.6% from a synthetic system. The regeneration analysis was performed using the eluent of 0.1 N HCl. Desorption efficiency for both biosorbents was higher in the synthetic system compared to leach liquor (Rachna et al., 2018). The amino-carbamate moiety embedded calcium alginate hydrogel beads were tested for the biosorptive removal of silver ions from aqueous solution. The experimental data better explained the biosorption kinetic model of pseudo-second-order. The equilibrium data best fitted to the biosorption isotherm models of Langmuir (210 mg/g) and Sips. The value of separation factor represented the removal of silver ions as favourable biosorption (Hamza et al., 2020). Biosorbent prepared from banana peel was utilized for the removal of strontium ions from aqueous solution. The characterization of the biosorbent using a scanning electron microscope illustrated that pores of the biosorbent were irregular in shape and exhibited micro rough texture. And these structures favour the binding of strontium ions onto biosorbent. A Freundlich isotherm model and pseudo-second-order kinetic model are found to be the best fitting isotherms for strontium ion sorbing (Komal and Virendra, 2018). Biosorbates such as lanthanum, europium and ytterbium ions removal in single- and multi-component systems were investigated using Ca-loaded *Sargassum polycystum* in a batch system. The mechanism of ion exchange biosorption was involved in the removal of lanthanum, europium and ytterbium ions. In the case of the single-component system, the binding of metal ions to biosorbent increased with the increase of initial solution pH. The multi-component system of biosorption was carried out at initial solution pH 4.0, and the affinity of biosorbate is in the sequence of europium > lanthanum > ytterbium ions (Diniz and Volesky, 2005).

14.3.3 Management of WEEE in India

WEEE generation is increasing worldwide at a rapid rate, and India is the fifth biggest country in the generation of WEEE. The annual generation of WEEE in India is growing at an estimated rate of about 21%. The Basel Action Network and Hazardous Waste Rules (2008) restrict WEEE movement across national borders (Garlapati, 2016; MoEF, 2018). The three main types of WEEE designated in India are from households, information technology and communication equipment and consumer electronics. The Associated Chambers of Commerce of India estimated the annual growth rate of WEEE and generation in 2016. Mumbai tops the list, then Delhi-NCR and Bangalore. Other cities include Chennai, Kolkata, Pune, Hyderabad and Ahmedabad. The responsibilities of stake holders (producer, manufacturer, collection centre, dealer, refurbisher, consumer, dismantler and recycler) in India are defined in the E-waste Rule 2016. The quantity of e-waste generated is about 1,641 metric kilotonnes in India. The responsibilities include extended producer accountability and a deposit refund system. The framework and number of regulatory instruments is six. The most important recycling technology includes mechanical technology with manual dismantling and a recycling rate of 5%. Categories such as information technology, communication equipment and consumer electrical equipment and electronics are considered. Means of solution include motivation of informal recycle sectors to official (Pankaj et al., 2017; Awasthi and Li, 2017).

14.4 CONCLUSION

The growth of population, infrastructural developments and high living standards are major concerns in India for the generation of solid waste. The current situation is that waste is predominantly dumped in an insecure manner and handled by the informal sector. The major issues with waste management are associated with public participation and the lack of responsibility for ecology and the environment. There is an urgent need to cultivate public awareness on sustainable waste management scenarios especially for C&D waste and WEEE. The construction and demolition industries are focusing on sustainable design, processes, operation and maintenance activities. The use of recycled aggregates in construction is one such promising option; many developed countries have enacted policies for the use of recycled aggregates. BRICS nations are at present higher consumers of construction aggregates, and hence have high potential for the recycling and re-use of C&D waste to prolong the life of landfills and natural resources. There are studies showing that partial replacement of normal aggregates with recycled aggregates has not affected the compressive and spilt tensile strength much. Similarly, WEEE can also be treated by various methods to recover the valuable products. Hence, the chapter concludes that C&D and WEEE can be processed wisely by adopting appropriate techniques to recover the valuables, recycle and reuse them.

REFERENCES

Abedi, S. Mousavi, H. Z., & Asghari, A. (2015). Investigation of heavy metal ions adsorption by magnetically modified aloe vera leaves ash based on equilibrium, kinetic and thermodynamic studies. *Desalination & Water Treatment, 57*, 13747–13759.

Abel, D. J. (2014). *Perceptions on illegal dumping in the eThekwini municipality* (Doctoral dissertation, University of the Free State).

Akcil, A., Erust, C., Gahan, C. S., Ozgun, M., Sahin, M., & Tuncuk, A. (2015). Precious metal recovery from waste printed circuit boards using cyanide and non-cyanide lixiviants – A review. *Waste Management, 45*, 258–271.

Alengaram, U. J., Salam, A., Jumaat, M. Z., Jaafar, F. F., & Saad, H. B. (2011). Properties of high-workability concrete with recycled concrete aggregate. *Materials Research, 14*(2), 248–255.

Anagnostopoulos, V., & Symeopoulos, B. (2013). Sorption of europium bymalt spent rootlets, a lowcost biosorbent: Effect of pH, kinetics and equilibriumstudies. *Journal of Radioanalytical and Nuclear Chemistry, 295*, 7–13.

Anantha, R. K., & Kota, S. (2016). Removal of lead by adsorption with the renewable biopolymer composite of feather (*Dromaius novaehollandiae*) and chitosan (*Agaricus bisporus*). *Environmental Technology & Innovation*, *6*, 11–26.

Anita, A., David, K. E., Jens, M., Sergey, V. K., & Erik, G. S. (2018). Assessment of PCBs and exposure risk to infants in breast milk of primiparae and multiparae mothers in an electronic waste hot spot and non-hot spot areas in Ghana. *Science of the Total Environment*, *612*, 1473–1479.

Awasthi, A. K., & Li, J. (2017). Management of electrical and electronic waste: A comparative evaluation of China and India. *Renewable and Sustainable Energy Reviews*, *76*, 434–447.

Ayah, A., Despoina, B., Hugo, T., Yanik, M., & Kerstin, K. (2020). Towards a high quality recycling of plastics from waste electrical and electronic equipment through separation of contaminated fractions. *Journal of Hazardous Materials*, *387*, 121741.

Bai, X., McPhearson, T., Cleugh, H., Nagendra, H., Tong, X., Zhu, T., & Zhu, Y. G. (2017). Linking urbanization and the environment: Conceptual and empirical advances. *Annual Review of Environment and Resources*, *42*, 215–240.

Bigum, M., Damgaard, A., Scheutz, C., & Christensen, T. H. (2017). Environmental impacts and resource losses of incinerating misplaced household special wastes (WEEE, batteries, ink cartridges and cables). *Resources, Conservation & Recycling*, *122*, 251–260.

Breivik, K., Armitage, J. M., Wania, F., & Jones, K. C. (2014). Tracking the global generation and exports of e-waste. Do existing estimates add up? *Environmental Science & Technology*, *48*, 8735–8743.

Chan, J. K. Y., & Wong, M. H. (2013). A review of environmental fate, body burdens, and human health risk assessment of PCDD/Fs at two typical electronic waste recycling sites in China. *Science of the Total Environment*, *463*, 1111–1123.

Chauhana, G., Jadhaob, P. R., Pantb, K. K., & Nigam, K. D. P. (2018). Novel technologies and conventional processes for recovery of metals from waste electrical and electronic equipment: Challenges & opportunities – A review. *Journal of Environmental Chemical Engineering*, *6*, 1288–1304.

Chen, M., Huang, J., Ogunseitan, O. A., Zhu, N., & Wang, Y.-M. (2015). Comparative study on copper leaching from waste printed circuit boards by typical ionic liquid acids. *Waste Management*, *41*, 142–147.

Darnerud, P. O., Eriksen, G. S., Jóhannesson, T., Larsen, P. B., & Viluksela, M. (2001). Polybrominated diphenyl ethers: Occurrence, dietary exposure, and toxicology. *Environmental Health Perspectives*, *109* Supplement 1, 49.

Das, D., Varshini, C. J. S., & Das, N. (2014). Recovery of lanthanum (III) from aqueous solution using biosorbents of plant and animal origin: Batch and column studies, *Minerals Engineering*, *69*, 40–56.

Diniz, V., & Volesky, B. (2005). Biosorption of La, Eu and Yb using Sargassum biomass. *Water Research*, *39*, 239–247.

Du, X., Wang, L., Zhao, W., Wang, Y., Qi, T., & Li, C. M. (2016). Preparation of hierarchical porous carbon from waste printed circuit boards for high performance electric double-layer capacitors. *Journal of Power Sources*, *323*, 166–173.

Fatih, D., & Abdulkerim, K. (2017). Biosorption of heavy metal ions by chemically modified biomass of coastal seaweed community: Studies on phycoremediation system modeling and design. *Ecological Engineering*, *106*, 101–108.

Garlapati, V. K. (2016). E-waste in India and developed countries: Management, recycling, business and biotechnological initiatives. *Renewable and Sustainable Energy Reviews*, *54*, 874–881.

Grant, K., Goldizen, F. C., Sly, P. D., Brune, M. N., & Neira, M., van den Berg, M., & Norman, R. E. (2013). Health consequences of exposure to e-waste: A systematic review. *Lancet Global Health*, *1*, 350–361.

Gupta, J., Tyagi, B., Rathour, R., & Thakur, I. S. (2019). Microbial treatment of waste by culture-dependent and culture-independent approaches: Opportunities and challenges. In *Microbial Diversity in Ecosystem Sustainability and Biotechnological Applications* (pp. 415–446). Singapore: Springer.

Gurpreet, S. D., Guitaya, M. L. R., Surinder, K., Krishnamoorthy, H., Satinder, K. B., Patrick, D., & Mausam, V. (2017). Novel biomaterials from citric acid fermentation as biosorbents for removal of metals from waste chromated copper arsenate wood leachates. *International Biodeterioration & Biodegradation*, *119*, 147–154.

Hadi, P., Gao, P., Barford, J. P., & McKay, G. (2013). Novel application of the nonmetallic fraction of the recycled printed circuit boards as a toxic heavy metal adsorbent. *Journal of Hazardous Materials*, *252–253*, 166–170.

Hamza, S., Ejaz, A., Ahsan, S., Muhammad, I. D., Zahoor, H. F., Imran, N., Rehana, B., & Marrium, I. (2020). Amino-carbamate moiety grafted calcium alginate hydrogel beads for effective biosorption of Ag(I) from aqueous solution: Economically-competitive recovery. *International Journal of Biological Macromolecules*, *144*, 362–372.

Herrera-Barros, A., Tejada-Tovar, C., Villabona-Ortíz, A., González-Delgado, A. D., & Benitez-Monroy, J. (2020). Cd (II) and Ni (II) uptake by novel biosorbent prepared from oil palm residual biomass and Al2O3 nanoparticles. *Sustainable Chemistry and Pharmacy*, *15*, 100216.

Huang, J., Chen, M., Chen, H., Chen, S., & Sun, Q. (2014). Leaching behavior of copper from waste printed circuit boards with Bronsted acidic ionic liquid. *Waste Management*, *34*, 483–488.

Irina, M., Carmen, T., Daniela, F., & Carmen, P. (2019). Simultaneous biosorption of micropollutants from aqueous effluents by rapeseed waste. *Process Safety and Environmental Protection*, *132*, 231–239.

Jaiswal, V., Saxena, S., Kaur, I., Dubey, P., Nand, S., Naseem, M., Singh, S. B., Srivastava, P. K., & Barik, S. K. (2018). Application of four novel fungal strains to remove arsenic from contaminated water in batch and column modes. *Journal of Hazardous Materials*, *356*, 98–107.

Jie, G., Ang, J., Jianbo, W., Oladele, A. O., & Zhenming, X. (2019). Emission characteristics and exposure assessment of particulate matter and polybrominated diphenyl ethers (PBDEs) from waste printed circuit boards de-soldering. *Science of The Total Environment*, *662*, 530–536.

Jihoon, C., Mingcan, C., Min, J., Sang-Hyun, C., Deok, H. M., & Jeehyeong, K. (2011). Kinetic and mechanism studies of the adsorption of lead onto waste cow bone powder (WCBP) surfaces. *Environmental Geochemistry and Health*, *33*(Supplement 1), 81–89.

Jinhui, L., Brenda, N. L. N., Lili, L., Nana, Z., Keli, Y., & Lixia, Z. (2013). Regional or global WEEE recycling. Where to go? *Waste Management*, *33*, 923–934.

Kaya, M. (2016). Recovery of metals and nonmetals from electronic waste by physical and chemical recycling processes. *Waste Management*, *57*, 64–90.

Kiddee, P., Naidu, R., & Wong, M. H. (2013). Electronic waste management approaches : An overview. *Waste Management*, *33*, 1237–1250.

Komal, V. M., & Virendra, K. R. (2018). Utilization of banana peels for removal of strontium (II) from water. *Environmental Technology & Innovation*, *11*, 371–383.

Kshama, A. S., & Varsha, K. V. (2011). Kinetics and equilibrium studies on biosorption of nickel from aqueous solution by dead fungal biomass of Mucor hiemalis. *Chemical Engineering Journal*, *171*, 1234–1245.

Kumar, A., Holuszko, M., & Espinosa, D. C. R. (2017). E-waste: An overview on generation, collection, legislation and recycling practices. *Resources, Conservation & Recycling*, *122*, 32–42.

Kumar, S., Smith, S. R., Fowler, G., Velis, C., Kumar, S. J., Arya, S., … Cheeseman, C. (2017). Challenges and opportunities associated with waste management in India. *Royal Society Open Science*, *4*(3), 160764.

Kumar, V., & Dixit, G. (2018). Evaluating critical barriers to implementation of WEEE management using DEMATEL approach. *Resources, Conservation & Recycling*, *131*, 101–121.

Lei, D.-y. Liu, Z., Peng, Y.-h., Liao, S.-b., & Xu, H. (2014). Biosorption of copper, lead and nickel on immobilized Bacillus coagulans using experimental design methodologies. *Annals of Microbiology*, *64*, 1371–1384.

Leticia, N. C., Eduardo, H. T., Daniel, A. B., & Guilherme, L. D. (2015). Biosorption of gold from computer microprocessor leachate solutions using chitin. *Waste Management*, *45*, 272–279.

Li, J., Zeng, X., Chen, M., Ogunseitan, O. A., & Stevels, A. (2015). "Control-Alt-Delete": Rebooting solutions for the e-waste problem. *Environmental Science & Technology*, *49*, 7095–7108.

Li, Y., Yu, L. H., Zhu, Z. C., Dai, J. Y., Mai, B. X., Wu, J. P., Wang, J. S. (2013). Accumulation and effects of 90-day oral exposure to Dechlorane Plus in quail (Coturnix coturnix). *Environmental Toxicology and Chemistry*, *32*, 1649–1654.

Luciana, P. M., Maria, A. P. C., Selene, M. A., Guelli, U.de S, Rui, A. R. B., & Vitor, J. P. V. (2018). Brown marine macroalgae as natural cation exchangers for toxic metal removal from industrial wastewaters: A review. *Journal of Environmental Management*, *223*, 215–253.

Marshall, R. E., & Farahbakhsh, K. (2013). Systems approaches to integrated solid waste management in developing countries. *Waste management*, *33*(4), 988–1003.

Mirian, C. S., Joaquim, A. N., Nivaldo, B., & Solange, C. (2010). Determination of toxic elements in plastics from waste electrical and electronic equipment by slurry sampling electrothermal atomic absorption spectrometry. *Talanta*, *81*, 1781–1787.

MoEF. (2018). Guidelines for environmentally sound management of E-waste (as approved Ministry of Environment and Forests, MoEF) letter No. 23-23/2007-HSMD; 2008.

Nayana, A. Y., & Kavitha, S. Evaluation of CO_2 emissions for green concrete WITH high volume slag, recycled aggregate, recycled water to build ecol environ.

Ningqin, L., Tianjue, H., Yunbo, Z., Huaqing, Q., Jamila, A., & Hao, Z. (2020). Fungal cell with artificial metal container for heavy metals biosorption: Equilibrium, kinetics study and mechanisms analysis. *Environmental Research*, *182*, 109061.

Ongondo, F. O., Williams, I. D., & Cherrett, T. J. (2011). How are WEEE doing? A global review of the management of electrical and electronic wastes. *Waste Management*, *31*, 714–730.

Pankaj, P., Rajiv, R. S., & Ojasvicz. (2017). Assessment of legislation and practices for the sustainable management of waste electrical and electronic equipment in India. *Renewable and Sustainable Energy Reviews*, *78*, 220–232.

Parthiban, K., & Mohan, K. S. R. (2017). Influence of recycled concrete aggregates on the engineering and durability properties of alkali activated slag concrete. *Construction and Building Materials*, *133*, 65–72.

Patel, S., Pansuria, A., Shah, V., Patel, S., Patel, S., Pansuria, A., … Patel, S. (2015). Construction and demolition waste recycling. *International Journal for Innovative Research in Science and Technology*, *1*(7), 266–286.

Pathak, P., & Srivastava, R. R. (2017). Assessment of legislation and practices for the sustainable management of waste electrical and electronic equipment in India. *Renewable and Sustainable Energy Reviews*, *78*, 220–232.

Paul, J. G., Arce-Jaque, J., Ravena, N., & Villamor, S. P. (2012). Integration of the informal sector into municipal solid waste management in the Philippines–What does it need?. *Waste Management*, *32*(11), 2018–2028.

Pellegrino, C., & Faleschini, F. (2016). *Sustainability improvements in the concrete industry: Use of recycled materials for structural concrete production*. Springer.

Peng, Y., Xia, P., Zhang, J. J., Villeneuve, D. L., Zhang, J. M., Wang, Z. H., Wei, S., Yu, H. X., & Zhang, X. W. (2016). Toxicogenomic assessment of 6-OH-BDE47 induced developmental toxicity in chicken embryos. *Environmental Science & Technology*, *50*, 12493–12503.

Qingping, Y., Ruiyi, F., Feng, X., Huiyu, M., Qinglin, Z., & Zhengrong, L. (2016). Selective recovery of Au(III) and Pd(II) from waste PCBs using ethylenediamine modified persimmon tannin adsorbent., *Procedia Environmental Sciences*, *31*, 185–194.

Qinxin, G., Enci, W., Yongyou, N., & Junyi, S. (2018). Profit or environment? A system dynamic model analysis of waste electrical and electronic equipment management system in China. *Journal of Cleaner Production*, *194*, 34–42.

Rachna, S., Garima, C., Azad, S., Arinjay, K., Sanigdha, A. (2018). A novel eco-friendly hybrid approach for recovery and reuse of copper from electronic waste. *Journal of Environmental Chemical Engineering*, *6*, 1053–1061.

Raikwar, M. K., Kumar, P., Singh, M., & Singh, A. (2008). Toxic effect of heavy metals in livestock health. *Veterinary World*, *1*, 28–30.

Rajagopal, R. R., Rajarao, R., & Sahajwalla, V. (2016). High temperature transformations of waste printed circuit boards from computer monitor and CPU: Characterisation of residues and kinetic studies. *Waste Management*, *57*, 91–101.

Raju, K. V., Ravindra, A., Manasi, S., Smitha, K. C., & Srinivas, R. (2018). Trends and status of environmental resource use in Bengaluru metropolitan area (BMA). In *Urban environmental governance in India* (pp. 109–247). Cham: Springer.

Ramachandra, T. V., Bharath, H. A., Kulkarni, G., & Han, S. S. (2018). Municipal solid waste: Generation, composition and GHG emissions in Bangalore, India. *Renewable and Sustainable Energy Reviews*, *82*, 1122–1136.

Ramprasad, C., & Kutty, M. (2016). Recycling of domestic wastewater by subsurface flow constructed wetland for construction purposes. *International Journal of Environment and Sustainable Development*, *15*(3), 286–292.

Rangabhashiyam, S., & Balasubramanian, P. (2018). Utilization of unconventional lignocellulosic waste biomass for the biosorption of toxic triphenylmethane dye malachite green from aqueous solution. *International Journal of Phytoremediation*, *20*, 624–633.

Rangabhashiyam, S., Suganya, E., Selvaraju, N., & Alen Varghese, L. (2014). Significance of exploiting nonliving biomaterials for the biosorption of wastewater pollutants. *World Journal of Microbiology & Biotechnology*, *30*, 1669–1689.

Rangabhashiyam, S., Suganya, E., Varghese Lity, A., & Selvaraju, N. (2016). Equilibrium and kinetics studies of hexavalent chromium biosorption on a novel green macroalgae Enteromorpha sp. *Research on Chemical Intermediates*, *42*, 1275–1294.

Rangabhashiyam, S., & Vijayaraghavan, K. (2019). Biosorption of Tm(III) by free and polysulfone-immobilized Turbinaria conoides biomass. *Journal of Industrial and Engineering Chemistry*, *80*, 318–324.

Sabrina, G., Jerome, L., Bouchra, B., Miriam, L. D., Liisa, M. J., Jacques, L., Brigitte, R., Marc-Andre, V., Joseph, Z., & France, L. (2019). Halogenated flame retardants and organophosphate esters in the air of electronic waste recycling facilities: Evidence of high concentrations and multiple exposures. *Environment International*, *128*, 244–253.

Sakthi, T. A., Pugazhendhi, A., Rajakumar, S., Rajan, M., & Ayyasamy, P. M. (2019). Batch and column approach on biosorption of fluoride from aqueous medium using live, dead and various pretreated Aspergillus niger (FS18) biomass. *Surfaces and Interfaces*, *15*, 60–69.

Sandhu, V., & Sidhu, A. S. (2015). Environmental governance in India: A systematic review of the initiatives. *Pacific Business Review International*, *8*(4), 49–57.

Saravanakumar, P., & Dhinakaran, G. (2015). Mechanical and durability properties of slag based recycled aggregate concrete. *Iranian Journal of Science and Technology – Transactions of Civil Engineering*, *39*(C2), 271–282.

Selvakumar, A., & Rangabhashiyam, S. (2019). Biosorption of rhodamine B onto novel biosorbents from Kappaphycus alvarezii, Gracilaria salicornia and Gracilaria edulis. *Environmental Pollution*, *255*, 113291.

Shahidan, S., Pullin, R., Holford, K. M., Bunnori, N. M., & Noorsuhada, M. N. (2013). Quantitative evaluation of the relationship between tensile crack and shear movement in concrete beams. In *Advanced materials research* (Vol. 626, pp. 355–359). Trans Tech Publications.

Shahram, A., Madhumita, B. R., & Argyrios, M. (2015). Heavy metals removal from aqueous solutions using Saccharomyces cerevisiae in a novel continuous bioreactor–biosorption system. *Chemical Engineering Journal*, *264*, 863–872.

Shamim, A., Ali, M. K., & Rafiq, I. (2015). E-waste trading impact on public health and ecosystem services in developing countries. *International Journal of Waste Resources*, *5*, 1–12.

Sharholy, M., Ahmad, K., Mahmood, G., & Trivedi, R. C. (2008). Municipal solid waste management in Indian cities–A review. *Waste Management*, *28*(2), 459–467.

Singh, P., & Sharma, V. P. (2016). Integrated plastic waste management: Environmental and improved health approaches. *Procedia Environmental Sciences*, *35*, 692–700.

Tam, V. W., Soomro, M., & Evangelista, A. C. J. (2018). A review of recycled aggregate in concrete applications (2000–2017). *Construction and Building Materials*, *172*, 272–292.

Teshale, F., Karthikeyan, R., & Sahu, O. (2020). Synthesized bioadsorbent from fish scale for chromium (III) removal. *Micron*, *130*, 102817.

Tošić, N., Marinković, S., Dašić, T., & Stanić, M. (2015). Multicriteria optimization of natural and recycled aggregate concrete for structural use. *Journal of Cleaner Production*, *87*, 766–776.

Varshini, C. J. S. Das, D., & Das, N. (2015). Optimization of parameters for praseodymium(III) biosorption onto biowastematerials using response surfacemethodology: Equilibrium, kinetic and regeneration studies. *Ecological Engineering*, *81*, 321–327.

Varshini, J. S. C., & Das, N. (2014). Screening of biowaste materials for the sorption of cerium(III) from aqueous environment. *Research Journal of Pharmaceutical, Biological and Chemical Sciences*, *5*, 402–408.

Vijayaraghavan, K. (2015). Biosorption of lanthanide (praseodymium) using Ulva lactuca: Mechanistic study and application of two, three, four and five parameter isotherm models. *Journal of Environment and Biotechnology*, *1*, 10–17.

Vijayaraghavan, K., Rangabhashiyam, S., Ashokkumar, T., & Arockiaraj, J. (2016). Mono- and multi-component biosorption of lead(II), cadmium(II), copper(II) and nickel(II), ions onto coco-peat biomass. *Separation Science and Technology*, *51*, 2725–2733.

Vijayaraghavan, K., Rangabhashiyam, S., Ashokkumar, T., & Arockiaraj, J. (2017). Assessment of samarium biosorption from aqueous solution by brown macroalga Turbinaria conoides. *Journal of the Taiwan Institute of Chemical Engineers*, *74*, 113–120.

Volesky, B. (2007). Biosorption and me. *Water Research*, *41*, 4017–4029.

Wang, Z., Bao, Y., Wen, Z., & Tan, Q. (2016). Analysis of relationship between Beijing's environment and development based on Environmental Kuznets Curve. *Ecological Indicators*, *67*, 474–483.

Wencheng, W., Yingxin, W., Jiahui, W., Xiaowen, L., Xianbin, C., Xinde, C., & Shixiao, Y. (2018). Regional risk assessment of trace elements in farmland soils associated with improper e-waste recycling activities in Southern China. *Journal of Geochemical Exploration*, *192*, 112–119.

Wenjing, L., Xuelin, D., Lihua, Z., & Heqing, T. (2020). Highly selective separation of Re(VII) from Mo(VI) by using biomaterial-based ionic gel adsorbents: Extractive adsorption enrichment of Re and surface blocking of Mo. *Chemical Engineering Journal*, *387*, 124078.

Wenzhi, H., Guangming, L., Xingfa, M., Hua, W., Juwen, H., Min, X., & Chunjie, H. (2006). WEEE recovery strategies and the WEEE treatment status in China. *Journal of Hazardous Materials*, *B136*, 502–512.

Widmer, R., Oswald-Krapf, H., Sinha-Khetriwal, D., Schnellmann, M., & Böni, H. (2005). Global perspectives on e-waste. *Environmental Impact Assessment Review*, *25*, 436–458.

Wijayasundara, M., Crawford, R. H., & Mendis, P. (2017). Comparative assessment of embodied energy of recycled aggregate concrete. *Journal of Cleaner Production*, *152*, 406–419.

Wong, M. H., Wu, S. C., Deng, W. J., Yu, X. Z., Luo, Q., Leung, A. O. W., Wong, C. S. C., Luksemburg, W. J., & Wong, A. S. (2007). Export of toxic chemicals: A review of the case of uncontrolled electronic waste recycling. *Environmental Pollution*, *149*, 131–140.

Yahaya, N. S., Lim, C. S., Taha, M. R., & Pereira, J. J. (2016). Exposure of municipal solid waste disposal sites to climate related geohazards: Case study of Selangor.

Yang, T., Xu, Z., Wen, J., & Yang, L. (2009). Factors influencing bioleaching copper from waste printed circuit boards by Acidithiobacillus ferrooxidans. *Hydrometallurgy*, *97*, 29–32.

Yoo, J., Jeong, J., Yoo, K., Lee, J. C., & Kim, W. (2009). Enrichment of the metallic components from waste printed circuit boards by a mechanical separation process using a stamp mill. *Waste Management*, *29*, 1132–1137.

Zeng, X., Zheng, L., Xie, H., Lu, B., Xia, K., Chao, K., Li, W., Yang, J., Lin, S., & Li, J. (2012). Current status and future perspective of waste printed circuit boards recycling. *Procedia Environmental Sciences*, *16*, 590–597.

Zhang, S., Ding, Y., Liu, B., & Chang, C.-C. (2017). Supply and demand of some critical metals and present status of their recycling in WEEE. *Waste Management*, *65*, 113–127.

Zheng, L., Wu, K., Li, Y., Qi, Z., Han, D., Zhang, B., Gu, C., Chen, G., Liu, J., Chen, S., Xu, X., & Huo, X. (2008). Blood lead and cadmium levels and relevant factors among children from an e-waste recycling town in China. *Environmental Research*, *108*, 15–20.

Zhengsheng, Y., Huawen, H., Pengya, F., Shuai, Z., Tuoyu, Z., Apurva, K., Saurabh, K., Sabahat, M., & Xiangkai, L. (2020). Recent advances in the recovery of metals from waste through biological processes. *Bioresource Technology*, *297*, 122416.

15 Waste Management Policies and Laws
A Case Study of China

Yunlong Zhou, Zhiyang Xu, Qian Li and Haoqi Wang

CONTENTS

15.1 INTRODUCTION

Since the reform and opening up, China has started an unprecedented modernization process, people's lives have changed dramatically and the material conditions are getting better and better. In the decades since the reform and opening up, China has gone through modernization that took hundreds of years in western countries. Therefore, the modernization of China is named 'compressed modernization' by scholars. 'This modernization not only strengthens the production of risk, but also leaves no time for the systematic expectation and management of risk'. In terms of ecological environment, the problems experienced by western countries in history now co-exist in China, and the domestic waste and industrial waste are also increasing. The increase of garbage is obvious, which brings obvious environmental, health and safety problems. In addition, China's weak economic foundation and backward environmental awareness make the environmental risks and the attached structural risks more serious and difficult treat.

Urban kitchen waste and its treatment is a project that we must pay attention to in the process of urbanization. This is also a sign of whether we enter modern civilization spiritually, and it is also related to the interests of millions of people. Nowadays, a large amount of food waste has caused serious ecological damage and environmental pollution. Food waste is becoming a more and more prominent social problem which needs to be regulated and solved urgently in today's society. The only way to deal with the problem of food waste is to improve the degree of rule of law in the field of food waste supervision. First of all, the random disposal of food waste brings great social harm, including both harm to human health and environmental pollution. To prevent and control the increasingly serious social harm caused by food waste, we need not only soft advocacy and policy guidance, but also governance according to law. Second, the government has no doubt the corresponding responsibility for the treatment of food waste. To promote the legalization of the government's responsibility for the treatment of food waste, so that the government has laws to abide by and must be punished if in violation of laws, is not only conducive to eliminating or reducing the harm of food waste, but also conducive to the government's administration in accordance with laws and achieving the goal of building a government ruled by law. Third, not only is the government responsible for the disposal of kitchen waste, but also the market entities such as restaurants and hotels that produce kitchen waste. At the same time, the treatment of kitchen waste not only consumes resources and costs, but also some of it can be turned into treasures, which can generate benefits through the recycling of kitchen waste. Therefore, through legislation, regulation and policy guidance, the government and relevant enterprises can make joint efforts to develop a circular economy, promote the transformation of food waste to resources and realize green development and ecological development, which can effectively save resources and promote the sustainable development of the national economy. Therefore, it is necessary to strengthen the research on the rule of law guarantee of harmless treatment and resource utilization of kitchen waste, which is also the historical condition and starting point for us to innovate the social governance mode.

With the outbreak of 'gutter oil' and 'swill pig' incidents, the domestic research on the treatment of food waste has increasingly become a hot issue for judicial practice and scholars around the country, which has also attracted the attention of the practical and theoretical circles on the treatment of food waste. In practice, in order to effectively treat kitchen waste, domestic provinces and cities have begun to formulate local laws and regulations and government regulations. Shanghai, Jiangsu, Chongqing, Qinghai and other places have successively issued management methods and regulations for garbage, waste water and waste residue of the kitchen or catering industry. The promulgation of these local laws and regulations has effectively promoted the management of local kitchen waste and achieved good results, which has provided certain experience and laid a good foundation for the formulation of management measures for kitchen waste in China. But kitchen waste is also a recyclable resource, which can effectively alleviate the energy pressure. With more and more attention paid to the recycling of kitchen waste, our country has also formulated some legal systems and policies for the recycling of kitchen waste, such as the collection, transportation and disposal system, regulatory system, charging system and market access system. The problem of food waste treatment has aroused widespread concern in the society. The government has also taken relevant measures, and launched the pilot of resource utilization and harmless treatment of food waste nationwide. The problems in the management of food waste treatment in China have gradually emerged.

15.2 DEVELOPMENT AND CURRENT SITUATION

The premise of the legal treatment of kitchen waste in China is to have a systematic and clear legal basis, so the investigation of legislative status is the premise of legal treatment. There are only 'sporadic provisions' in the national laws, but the local legislation shows the trend of 'a hundred flowers bloom'. In this legislative situation and background, there are multiple practical dilemmas in the legal treatment of kitchen waste that need to be clearly identified, which is the premise of the legal treatment of kitchen waste. At present, there are no special and unified legal provisions for the management of food waste in China. Only vague principled provisions can be found in the legislation of the national level, but there are legal provisions regulating food waste from different perspectives in relevant laws. For example, Article 26 of the Constitution states: 'the state protects and improves the living environment and ecological environment and prevents pollution and other public hazards'. In fact, the provisions of the law have put forward requirements for the treatment of kitchen waste, that is, the treatment of waste cannot cause pollution to the environment and cannot cause damage to the lives of residents. The Environmental Protection Law of the People's Republic of China is a strong backing of the development of environmental protection. The environmental pollution prevention and control regulations put forward in its provisions also play a strong role in the management of kitchen waste. Article 40 of the Environmental Protection Law stipulates that 'the state promotes cleaner production and resource recycling'. This also determines that the treatment process of food waste must be strictly in accordance with the relevant requirements of the Environmental Protection Law, and should pay attention to fundamentally solving the problem of food waste and reducing its production. From the national level, although there is not a law specifically for the problem of kitchen waste in China at present, there are many provisions for the legal regulation of kitchen waste from different perspectives in other relevant laws and regulations, including the Solid Waste Pollution Environment Protection Law, the Circular Economy Promotion Law of the People's Republic of China, etc.

The law of the People's Republic of China on the prevention and control of environmental pollution by solid waste is applicable to the prevention and control of pollution by kitchen waste. According to the definition of domestic waste in the law (article 88.3 of the supplementary provisions), food waste belongs to solid waste, which is a kind of domestic waste. Therefore, the principle of domestic waste treatment in this law is applicable to the treatment of food waste, which is a high-level law in the current law of food waste in China. The regulations on the administration of city appearance and sanitation stipulate the treatment of classified recycling and transportation of urban domestic waste based on the perspective of city appearance and sanitation; article 3 of the regulations on the administration of urban domestic waste stipulates the three principles for the treatment of kitchen waste, article 16.3 stipulates the principle of polluter's payment, and the regulations provide for the treatment of urban domestic waste as a whole. In particular, they stipulate that

> hotels, restaurants, government agencies, colleges and universities and other units shall collect and store their own kitchen waste in accordance with the regulations, and deliver the urban domestic waste collection and transportation enterprises that meet the requirements of these measures to the specified urban domestic waste garbage disposal place. (Song J.F., 2011)

This provides a basic institutional framework for the treatment of food waste in China, and points out that food waste is a special category different from domestic waste, which needs to be treated separately and effectively. The food safety law puts forward special requirements for catering service providers, requiring them to 'formulate and implement raw material control requirements, not to purchase food raw materials that do not meet the food safety standards'. At the same time, the food safety law advocates the implementation of transparent processing, and the disclosure of relevant information through transparent kitchen, raw material tracing and other ways. This provides a clear legal basis to prevent illegal vendors from using kitchen waste to make 'gutter oil' flow into the catering market. In addition to the above-mentioned laws and regulations, China has also formulated and implemented many policies for the treatment of kitchen waste, such as the opinions on strengthening the treatment of gutter oil and the management of kitchen waste issued by the general office of the State Council, the circular economy development strategy and short-term action plan, and the notice on the construction planning of harmless treatment facilities for urban domestic waste during the 12th Five Year Plan, the 2015 circular economy promotion plan released by the National Development and Reform Commission, and the notice on printing and distributing the implementation plan for the pilot city construction of recycling utilization and harmless treatment of kitchen waste with special funds for circular economy development jointly issued by the National Development and Reform Commission and the Ministry of Finance, 16 ministries including the Ministry of Education, the Ministry of Environmental Protection, the Ministry of Housing and Urban Rural Development, the Ministry of Science and Technology and the National Development and Reform Commission jointly issued the notice on further strengthening the treatment of urban domestic waste, the notice on organizing and carrying out the pilot work of resource-based utilization and harmless treatment of urban kitchen waste and other policy documents jointly issued by the National Development and Reform Commission, the Ministry of Agriculture, the Ministry of Housing and Urban Rural Development and the Ministry of Environmental Protection. These policies have introduced requirements on the pilot promotion objectives and main tasks, classified collection and transportation, the emission registration system, resource utilization, management system construction, recycling and other aspects of food waste treatment, and the implementation of these policies provides a policy basis for the legal treatment of food waste in China.

The existing kitchen waste should be regarded as a reusable resource, and its resource transformation should be actively realized, such as using waste oil to produce clean fuel for power generation and civil use. The current laws and policies mainly prevent solid waste from polluting the environment and causing harm to human health. In the process of recycling kitchen waste, it is forbidden to return it to the dining table to avoid harm to the human body. The laws restricting the management of food waste also include the Law of the People's Republic of China on the Quality and Safety of Agricultural Products, Administrative Measures for Municipal Solid Waste and Operating Standards of Catering Enterprise.

In terms of local legislation, local food waste management regulations have been implemented for local food waste treatment. Under the guidance of relevant policies, some big cities with strong economies and a high technical level have emerged with more advanced legal systems for food waste recycling in the process of continuous research on food waste recycling, which effectively solve the local problems caused by food waste. Some experience has been gained in the impact and harm. However, due to the decentralization of local legislation, good legal systems for kitchen waste recycling cannot be adopted and popularized by other places.

For example, Jiangsu's 'legal measures for the management of kitchen waste in Suzhou' clearly stipulate the classification of kitchen waste and non-kitchen waste. Shijiazhuang's 'management measures for the treatment of kitchen waste in Shijiazhuang' classify kitchen waste into kitchen waste and fat waste. In practice, the classification system has an obvious effect on the classification and collection of kitchen waste and non-kitchen waste. Although individual restaurants still mix non-kitchen waste and kitchen waste, the overall effect is good, and most subjects can strictly follow the legal system of the source classification of kitchen waste materials. As the capital of China, Beijing has gained some experience in the treatment of kitchen waste. The main laws and regulations implemented now are the Beijing Municipality Measures for the Administration of Collection, Transportation, and Treatment of Kitchen Waste, Interim Measures for the Registration and Administration of Beijing Kitchen Waste and Waste Oil Discharge Registration. The registration and management measures focus on regulation, and the waste oil discharge registration system is a management method to obtain the discharge registration number through network registration; the catering enterprises that do not register their kitchen waste and waste edible oil without justified reasons will be collected in the enterprise credit information system. Through the information contained in the emission registration number, the whole process of kitchen waste can be traced, and the information management of the source of kitchen waste will be focused on. The implementation of these measures can effectively solve the problem caused by the waste of local kitchens (Li et al., 2009). In Shanghai, the policy of food waste treatment was implemented earlier among the measures of food waste treatment of various places, mainly issuing and implementing Administrative Measures for Disposal of Kitchen Waste in Shanghai and Measures for the Management of Disposal of Waste Oils in Shanghai. It is required that kitchen waste and domestic waste must be recycled by classification. Meanwhile, kitchen waste and waste edible oil should be collected separately. The waste oil should be treated separately and pertinently. The treatment procedure is stricter. The collection, transportation and treatment must be carried out by the collection and transportation enterprises determined by unified bidding. The policy of separate treatment of waste oil implemented in Shanghai is a more prominent and characteristic treatment method, which is worth learning from in the management of waste oil (Huang et al., 2014). The regulations on the management of food waste in Xining city were officially implemented on 1 November 2009, and were the first local regulations on the management of food waste approved by the National People's Congress. The regulations define food waste as 'food residue, food processing waste and slaughtering processing waste produced by units and individuals engaged in catering industry, unit feeding, food processing, slaughtering and processing during the production and operation process and waste edible oils'. The regulation also stipulates that residents have the right to supervise the government's food and kitchen waste disposal activities, and call the government hotline to supervise and report any illegal acts of the government (Hu X.J. et al., 2012). The administrative measures for the treatment of kitchen waste in Urumqi were promulgated by order No. 88 of the People's Government of Urumqi, and came into force on 1 December 2007. The administrative measures stipulate that their legislative purpose is to 'strengthen the management of kitchen waste, ensure food safety and health, and maintain the urban appearance and environment'.

It can be seen from the above policies and regulations that although there are no special laws and regulations on food waste management, it is clear that food waste shall not be fed to pigs or illegally refined gutter oil. In addition, the kitchen waste must be disposed of by the licensed unit, and the violation will be punished: (i) according to the polluter's payment principle, the catering unit must pay for the kitchen waste; (ii) the kitchen waste collection, transportation and disposal unit must obtain the corresponding administrative permission; (iii) the kitchen waste must enter the disposal site recognized by the authority; (iv) for the illegal operator, the relevant authority has the right to punish. In order to adapt to the urgent situation of food waste treatment, in view of the reliability of food waste treatment technology, harmless treatment in resource utilization, safety and quality standards of finished products, under the promotion of policies and laws, relevant ministries and commissions have started to prepare and issue relevant specifications and systems in combination with the actual situation and implementation experience, including Technical Requirements for Utilization of Kitchen Waste Resources, Technical Specifications for Disposal of Kitchen Waste, Technical Regulations for Operation and Maintenance of Kitchen Waste Treatment Plant and Safety and Quality Standards for Recyclable Products of Kitchen Waste.

15.3 CURRENT ISSUES

15.3.1 Defects in the Legal System Structure

At present, the legal system of food waste management in China is mainly composed of national domestic waste and solid waste-related laws and regulations and scattered local food waste management methods (Xu et al., 2011). At the national level, the legislation of food waste is still blank, and local laws are not standardized and adapted to the development. There are still many places, such as small cities, rural areas and other places, whose food waste management only relies on the relevant laws, regulations and policy documents at the national level, and there is a lack of targeted management regulations. As a result, there are numerous problems in the management of food waste in many areas, such as lack of clear legal basis, unclear regulatory responsibilities and irregular collection, transportation and disposal links. On the other hand, with the legislation system mainly relying on local kitchen waste legislation, although there is the advantage of targeted management according to the local collection, transportation and disposal status, due to the lack of the guidance of superior law, some relevant provisions of local legislation have been nonstandard, not adapted to the development status quo and need to be further improved (Li et al., 2016). For example, Chongqing food waste management measures stipulate that the legislative purpose of the measures is to 'strengthen the management of food waste, ensure food safety and people's health, maintain the city's appearance and environmental health, and promote the recycling of resources'. It is not practical to take 'ensuring food health and safety' as one of the legislative purposes. Because there is no system design of food hygiene, the food hygiene department is responsible for the supervision of urban food hygiene and safety, and the urban management department is mainly responsible for food waste. The law enforcement of kitchen waste should focus on the generation, collection, transportation and disposal of kitchen waste. Therefore, it is against the original intention of legislation to take 'food hygiene protection' as the main goal of the kitchen waste treatment (Wang L., 2009). Moreover, with the continuous exploration of the legislation of kitchen waste in various regions, some more advanced and mature management systems of kitchen waste cannot be widely promoted in the whole country due to the decentralization of provincial legislation. Second, national domestic waste- and solid waste-related legislation, or the local food waste management methods, are more focused on the food waste collection and transportation, the laws and regulations on food waste disposal and follow-up utilization are deficient. As one of the prominent characteristics of kitchen waste, the resource value of kitchen waste is far from fully reflected in the legal regulation.

It is found from the national laws and regulations on the recycling of kitchen waste that government policies and industry standards are the common forms of regulations on the recycling of kitchen waste. At present, there is no special law on the recycling of kitchen waste. Although the relevant national laws and regulations have made some provisions on the recycling of kitchen waste, these provisions are more general descriptions or principled provisions, which are not operational. This phenomenon seriously affects the progress of food waste recycling in China, which means the implementation of food waste recycling by the government is in a state of lawlessness.

15.3.2 The Operation Mode under the Guidance of System Affects the Governance Efficiency

At present, in China, the main systems for the treatment of food waste are classified as the recycling system and disposal charging system. The classified recycling of kitchen waste is the premise and foundation of the effective resource utilization of kitchen waste. Through classified collection and classified treatment of kitchen waste, the total amount of final disposal of waste can be effectively reduced and the risk of secondary pollution in the process of waste treatment can be reduced. Special attention should be paid to the classification of waste oil and kitchen waste in the treatment of kitchen waste, because there are great differences in their sources, properties, composition and disposal methods, so it is necessary to classify them strictly (Fang Z.Q. et al., 2007). Although the 'China Agenda 21' and 'law on the prevention and control of solid waste pollution' issued by our country stipulate that garbage should be recycled by classification, these provisions are too principled and vague, and lack specific operation guidance. Therefore, at the legal level, the implementation rules for the classification and recycling of kitchen waste are basically in a blank state, and in real life waste management is a mere formality. Although recyclable and non-recyclable waste bins are set up in the municipal administration, in terms of the actual operation, no matter the residents' living or catering providers, there is no garbage classification.

The implementation of the system depends on a clear reward and punishment system. At present, there is no distinct incentive and punishment system in China, which is one of the important reasons why it is difficult to implement the system (Wang X.Q. et al., 2009). In foreign countries, catering units that actively classify food waste will be rewarded, and those that do not classify food waste will be severely punished. For example, once food and beverage units sell kitchen waste to unqualified food waste collection and transportation units, they will face suspension of business for rectification or even more serious punishment (Pan Y. et al., 2011). However, there are no specific reward and punishment measures in the current treatment of kitchen waste in China. For example, the administrative measures for the treatment of kitchen waste in Shanghai only stipulate that the kitchen waste and other domestic waste should be placed separately, and the waste oil and kitchen waste in the kitchen waste should also be classified and recycled. However, there are no corresponding incentive measures for the units and individuals with good classified collection behaviour, and there is no clear legal responsibility for the classification of kitchen waste. This makes it irrelevant for the catering enterprises to classify the garbage, and in most cases, they will not classify the garbage for convenience, or even directly sell the kitchen garbage to private companies to obtain economic benefits. Driven by the monetary interests, illegal traders sell the processed and refined 'gutter oil' to small catering enterprises and it directly enters people's diet, which is an obvious harm to people's health.

In addition, the kitchen waste charging system is to collect fees from the units or individuals that produce the kitchen waste, so that they can bear the responsibility for the kitchen waste treatment, which is consistent with the principle of 'who pollutes, who pays'. In academic analysis, the value concept of the kitchen waste recycling and charging system is based on the concept of useless kitchen waste, because if society admits that the kitchen waste has use value, the collection, transportation and disposal unit should pay certain compensation to the production unit, which has ownership of the use value of kitchen waste. In daily life, people often say that 'garbage is a misplaced resource', which is to admit the resource and value of kitchen garbage. In fact, this is the identification of whether kitchen garbage is a resource or garbage. The value of general domestic waste is obviously lower than the treatment cost, so it is naturally classified as waste for treatment. However, unlike general domestic waste, food waste contains rich nutrients and fat. If it is reasonably recycled and processed, it can produce high energy and resource value. In practice, although it is unreasonable and unscientific for some catering enterprises to sell food waste directly to private restaurants to feed livestock, as there is a great potential safety hazard, it is an affirmation of the resource and value of food waste, from which catering enterprises can obtain economic income. Therefore, the treatment of kitchen waste in China is based on the concept of waste uselessness, which has become a bottleneck in the treatment of kitchen waste in China. It ignores the economic value of food waste, so that food waste in China is in a vicious circle of production, waste and regeneration (Wang L., 2009).

In short, the collection and charging system of kitchen waste will lead to the unwillingness of the generating unit to hand in kitchen waste. If the quantity of centralized collection is small and the scale is small, the processing unit will not be able to achieve large-scale treatment and profitability. In this way, the vicious circle of food waste treatment cannot be alleviated by relying on the government's financial investment alone. It also deviates from the operation law of market economy, and cannot truly realize the unified recycling of food waste. If the legislative requirements cannot be implemented in practice for a long time, it will damage the legal authority of and public trust in the government.

15.3.3 Excessive Dependence on Government Subsidies for Food Waste Management

For a long time, the management of urban kitchen waste in China has mainly depended on the government. The government is not only an investor but also a manager. However, with this self-management and self-supervision mode it is difficult to effectively manage kitchen waste, and the public's participation in it is bound to question the authority and credibility of the government. Especially for enterprises, due to the lack of external competitiveness and effective interests, in the initial development stage of food waste treatment, China's food waste treatment enterprises rely heavily on the policy support and price subsidies provided by local government (Li et al., 2016);

moreover, they also lack innovation awareness and creativity and enthusiasm for work, which seriously affects the management mode of 'no separation of government and enterprise'. Treatment and environmental protection need to be changed and improved.

15.3.4 Lack of Public Participation in the Treatment of Kitchen Waste Leads to Lack of Social Rationality

People play an important role in the production and treatment of food waste. On the one hand, they produce food waste directly and indirectly through consumption; on the other hand, they will be affected by food waste, and they are the victims of food waste. Finally, they are also participants in the treatment of food waste. The public is widely involved in the treatment process of food waste, which is an effective means to reduce the harm of food waste and improve the utilization of food waste (Li et al., 2016). Therefore, enhancing public participation is conducive to improving the environmental protection awareness of the public, enabling the general public to actively participate in the management of food waste and guiding the public to reflect on their own self-discipline, starting from their own, reducing the production of food waste.

Although there are some regulations on public participation in the measures of local kitchen waste treatment, they are rough, not implemented in detail and lack operability, so that the public cannot really participate in them. For example, 'Xining food waste management measures' stipulate citizens' supervision and reporting rights, that is, citizens can supervise through the reporting hotline when they find illegal behaviours in the treatment of food waste, but there is no corresponding regulation on specific operational content such as what kind of situation can be reported, when to accept and whether there is a reporting reward. This blocks the channels for the masses to report, supervise and obtain relevant information. They know little about the food waste production, recycling and transportation mode, disposal and utilization form, disposal subject and other information in the areas concerned with their own interests. Public supervision and participation are only formal, and cannot mobilize the enthusiasm of public participation.

At present, due to the pursuit of economic interests and the neglect of environmental protection, some Chinese governments are unable to invest more energy in the management of food waste. The hardware facilities and supporting measures for the collection and transportation of kitchen waste are not perfect. This will also make it difficult for citizens to put food waste into different categories and collect food waste, which will undoubtedly discourage citizens from participating actively. In addition, the management of food waste in China relies heavily on the formation of the government, which makes much of the public think that the management of food waste is only the responsibility of the government departments, and their own sense of responsibility is poor. It is difficult for them to realize that they are also responsible for reducing the amount of food waste, and they feel they cannot actively participate in the management of food waste.

15.3.5 Unreasonable Legal Liability

In the local laws and regulations of food waste, there are many defects in the establishment of legal liability. Some only stipulate the legal responsibilities of the production units of kitchen waste, and the responsibilities of the collection, transportation and preservation units and the waste management departments are unclear. There are too few articles about legal liability in some local legislation (Guoyang, 2010). Some local legislation on the establishment of illegal liability is too light, unable to effectively deter illegal enterprises and individuals. Some local legislations do not set the corresponding legal responsibilities for the administrative departments not to perform their legal responsibilities in the part of legal responsibilities, and lack the provisions for the legal responsibilities of the administrative organizations in the management of food waste; this not only fails to reflect the systemic legal responsibilities in this field, but also easily makes some special legal responsibilities of government oversight dereliction of control in the management of food waste. On the issue of food waste, producers, collectors, disposers and regulators should bear the corresponding legal responsibilities. Only by building a complete legal liability system can we ensure that all organizations can strictly perform the corresponding legal obligations, so that the legal regulation of food waste can be implemented smoothly.

15.3.6 Lack of Guarantee Mechanism for the Operation of the Legal System

Whether it's domestic waste or kitchen waste, people cannot sort it out properly (Du J.L., 2017). Either, the kitchen waste is discharged into the sewer or mixed into the domestic waste at will, or, in order to obtain benefits, people directly sell the kitchen waste to some private vendors. Such practices not only pollute the environment and waste resources, but also can bring harmful substances into our food and endanger our life and health. At present, the public's understanding of kitchen waste only stays at the level of waste. On the one hand, the public does not realize the potential threat to our environment, society and life and health caused by the improper treatment of kitchen waste. On the other hand, the public cannot realize the urgency of the lack of resources in China, and does not see the value of resources generated by the correct treatment of kitchen waste. Therefore, the public's awareness of food waste recycling is very weak.

In China, not only is law enforcement insufficient, but law enforcement methods are single. At present, most of the improper behaviours of kitchen waste disposal in China mainly face administrative punishment, and the single means of administrative law enforcement will make the administrative law enforcement unable to achieve the expected effect. In the process of using kitchen waste as a resource, no matter the main body of producing, collecting and transporting, or the main body of processing, if they violate the laws and regulations of kitchen waste as a resource, the government will usually order this main body to make corrections or pay fines within a time limit, so as to guide such main body to regulate its own behaviour. For example, the measures for the management of food waste treatment in Shijiazhuang City stipulate that the government will punish those who violate the law by ordering it to correct within a time limit or by imposing a fine. The two means, of ordering to make corrections within a time limit and fines, are not only weak forms of punishment, but also the design of this punishment system leads to unfair phenomena. Those who are seriously in violation of the law and those who are in violation of the law face the same punishment. Moreover, the improper treatment of kitchen waste causes food safety hazards and environmental problems, which cannot be solved by ordering correction within a time limit and fines. Mild punishment is not conducive to the establishment of the authority of the government, or the guidance of kitchen waste in the direction of resource utilization. Therefore, a with a single means of law enforcement it is difficult to achieve the law enforcement effect of the legislation on the recycling of kitchen waste, and ultimately the implementation of the legal system for the recycling of kitchen waste will be restricted.

15.4 SUGGESTIONS FOR IMPROVEMENT

15.4.1 Establish and Improve Relevant Laws and Regulations on Kitchen Waste Treatment

In recent years, the problems caused by improper treatment of kitchen waste, such as gutter oil, swill pig, etc., have been affecting the public, and have been continuously concerning to the public and the government. Although governments at all levels have issued some normative legal documents on the treatment of kitchen waste, and many provinces and cities have also legislated on the treatment of kitchen waste, there is still no special legislation on the treatment of kitchen waste at the national level (Xu et al., 2011). In order to effectively regulate the activities of food waste treatment, China should enact special legislation on food waste treatment as soon as possible, and set up a scientific and feasible legal system to ensure the standardization and harmlessness of food waste treatment activities (Guoyang, 2010). So that the laws and regulations on food waste at the local level can have unified guidance, so that the food waste management regulations become more systematic and complete. Under the guidance of the basic law, specific laws and regulations shall be promulgated according to specific circumstances. In this way, the law on the management of kitchen waste can form an organic and unified system, which can implement the law according to the actual situation and enhance the operability of the law without violating the upper law.

15.4.2 Define the Responsibilities of the Organization of Food Waste Treatment Supervision

In the process of the management of kitchen waste, the government's positive and correct guidance and effective cooperation between various departments play a crucial role. The management of food waste involves a wide range of aspects, which involves the intersection of the authorities of many departments, such as the industrial and commercial administrative department, the environmental protection administrative department and the health supervision and management department. There are many links and a long process in the work of food waste recycling. These supervision subjects cannot cover the whole process of food waste recycling, so it is easy to shift responsibilities among supervision of organizations, and finally there is a gap in supervision. Often the lack of supervision will cause the kitchen waste to be processed through illegal channels, resulting in the re-pollution of kitchen waste. The effective management of kitchen waste cannot only rely on the exercise of the authority of one department, but also needs the effective cooperation between all departments and clear responsibilities of each department, so that they can manage the kitchen waste within their respective authority. The power and responsibility of each supervision organization should be reasonably stipulated and detailed and clear provisions made for the content requiring joint supervision, so as to avoid the phenomenon of self-governance. The supervision efficiency of food waste recycling should be improved, the supervision strengthened and the utilization of food waste recycling realized. At the same time, we should establish the main leadership responsibility mechanism, lead the smooth implementation of the supervision tasks of all supervision subjects and be responsible for the legalization and effectiveness of the supervision system of food waste recycling. And the network system of the supervision of food waste recycling should be improved, and the classification, recycling and disposal of food waste recycling fully covered. Relevant information in the process of food waste recycling should be included in a comprehensive and timely manner, and the objectivity and authenticity of this information verified. Because the perfect network system can increase the transparency of the work of kitchen waste recycling, curb the emergence of bad vendors and improve the efficiency of supervision.

15.4.3 Establish an Incentive Mechanism for Resource-Based Disposal of Kitchen Waste

In addition, food waste treatment has dual attributes; it is not only a public utility, but also has social and economic attributes. This means that we can use the market-oriented operation mode in this field to encourage more market participants to participate in the collection, transportation and disposal of food waste, which can not only reduce the financial and management pressure on the government, but also make the operation of each link more specialized through careful division of labour, and can further improve the level of food waste recycling through fair competition. The market competition mechanism is introduced in the treatment of kitchen waste to guide enterprises to participate in the collection, transportation, treatment and reuse of kitchen waste. The government is only in the role of process supervisor, which appears in the waste cleaning activities. In order to guide the enthusiasm of the masses and enterprises, it is necessary to establish a clear reward and punishment system and a report reward system, to give material or spiritual rewards to the units with better classification and recovery, to set up severe punishment measures for those who sell kitchen waste to other enterprises and individuals without recovery qualification without permission and to impose punishment on the units of recovery and disposal that do not operate normally.

15.4.4 Improve the Market Access and Fair Competition System of Urban Kitchen Waste

Market access system is the general term for various systems and regulations that allow legal citizens and legal persons to enter the market space, and meet the conditions for engaging in commodity production and operation activities and procedural rules in this space. This is the basic system for the coordination and intervention of national macro-control and market micro mechanisms. It can be regarded as an important link in government management. It is not only the initial starting point for the government to manage the market, but also a basic legal system, which can be implemented under a series of modern market economy conditions. A reasonable market access mechanism should be established. Taking 'high standard' and 'strict requirement' market access system as the ultimate goal, we must guide the powerful new enterprises to join the market. In addition, it is also necessary to strengthen the evaluation and assessment of the collection, transportation and disposal units, urge the production units to sign agreements on the collection and transportation of kitchen waste with the processing enterprises that meet the requirements and strictly prevent the flow of kitchen waste to livestock and poultry breeding and illegal processing of private enterprises.

15.4.5 Establish a Normal System of Public Participation

The treatment of kitchen waste is related to the food safety and health of the public. Without public participation, there is no way to deal with kitchen waste. First, we should do a good job in the information disclosure of kitchen waste management. Information disclosure is the premise of public participation. The public does not have sufficient right to know about the management of kitchen waste, so participation can only be empty talk. The main body of the disclosure of the management information of kitchen waste shall include the kitchen waste management organization, the production unit, the collection and transportation unit and the disposal unit of kitchen waste. Among them, the information disclosure of kitchen waste management is government information disclosure, which is mandatory, while the production unit, collection and transportation unit and disposal unit of kitchen waste are enterprise environmental information disclosure, which is generally voluntary. Second, a smooth channel for public complaints should be established. On the basis of information disclosure, we should establish a smooth channel for citizens to report information, and form a long-term management mechanism for the whole of society to participate in the supervision of food waste treatment. The power of administrative supervision of the government is limited, and it is impossible to cover all aspects. By unblocking the channels of public complaints, we can supervise and manage the production, collection, transportation and disposal units of kitchen waste from multiple perspectives, so as to make the illegal activities more controllable, and inhibit the production of gutter oil and swill pigs. Third, the public's interest expression mechanism on the treatment of kitchen waste should be improved. For some public decision-making involving the vital interests of citizens, we should listen to the public's opinions in the form of hearings, Internet solicitation and so on. For example, decisions on whether food waste treatment should be charged, charging standards and other decisions related to the vital interests of catering enterprises should be published in the mainstream media in advance, and representatives of different types and levels of catering enterprises should be invited to participate in hearings to ensure that public demands and wishes can be effectively expressed.

15.4.6 The Improvement of the Charge and Compensation System for the Treatment of Urban Kitchen Waste

As urban kitchen waste is a recyclable resource, it is 'put in the wrong place'. Hotels, restaurants, enterprise (business) canteens, etc., have the right to dispose of food waste freely. Therefore, in order to prevent the producers of food waste from selling to units or individuals without treatment qualification, the government should give corresponding compensation or reward and other preferential policies to encourage their enthusiasm and consciousness. The government may also consider subsidizing low-income families (http://finance.people.com.cn/n/2013/0703/c1004-22058824.html). For example, Chongqing Municipality has implemented a policy of free garbage treatment for 'minimum living standard families'. In addition, it is also necessary to provide ecological compensation to the food waste treatment units or enterprises. For example, in Nanjing Jiangnan Vein Industrial Park, for each ton of non-local domestic waste processed in the industrial park, the financial subsidies of the beneficiary area will be very different, which will be used to upgrade the technology of production facilities in the industrial park and extend its service life (http://paper.people.com.cn/rmrbhwb/html/2016-01/07/content_164585- 7.htm). Indeed, the kitchen waste treatment industry is a relatively low-profit industry, which needs long-term accumulation or a large amount of capital to enter, which increases the burden on enterprises. Therefore, no matter from the policy level or combined with the actual situation, various subsidies are necessary to promote the enthusiasm of enterprises.

In addition, we should establish and improve the benefit compensation mechanism to effectively promote the marketization of urban kitchen waste treatment. The kitchen waste treatment mechanism requires that the waste treatment unit and the unit or enterprise that produces the kitchen waste shall implement the charging system, and the regulation shall pay the kitchen waste treatment fee in strict accordance with the regulations. This treatment cost will be used in the transportation, collection, treatment and technical update of food and kitchen waste. The charging standard for kitchen waste shall be approved by the competent department and implemented after being approved by the municipal government. At the same time, the government can consider subsidies for the collection, transportation and treatment of kitchen waste. For example, there are clear regulations in the measures for the management of food and kitchen waste in Hangzhou. Garbage collection and transportation are paid services, and the cost standard has unified regulations.

15.4.7 Further Optimize the Classified Management of Kitchen Waste

The classification of kitchen waste is the first step in dealing with it after it is produced. Optimizing the classification management of kitchen waste is the premise of realizing the utilization of kitchen waste as a resource, and also an important part of strengthening the legal regulation of kitchen waste. First of all, the classification and collection standards of kitchen waste should be made clear. We should make it clear in legislation that there are two levels of classification requirements for the main body of food waste production: One is the classified collection of food waste and non-food waste; the other is the classified collection of waste edible oil and other food waste. In particular, it emphasizes the separate collection of waste edible oil. An orderly classification system of kitchen waste is a solid foundation for the later reform of the charging system and standards and the realization of the utilization of kitchen waste as a resource. Second, the supporting measures for the classified collection of kitchen waste should be improved. We should not only make clear the classification and collection of kitchen waste, but also pay attention to the operability and enforceability of the classification system in reality, and at the same time establish relevant supporting measures to ensure its real implementation.

First, in the stage of classified recovery, it is clearly necessary to realize the unified distribution of recovery containers and the forced installation of oil-water separation devices. Promoting the use of food waste collection containers with obvious signs issued by the government can not only effectively realize the supervision of catering enterprises, but also facilitate the control of food waste output and improve the convenience of transportation. In addition, since we need to realize the separate collection of waste edible oil in kitchen waste, we should also put forward the requirements for the installation of the required equipment, i.e. oil-water separators, and stipulate the corresponding legal liabilities (Guoyang, 2010). The installation of oil-water separators can not only effectively recover waste edible oil and improve resource utilization efficiency, but also reduce the difficulty of sewage treatment. We can learn from the foreign practice of monitoring the sewage discharge of catering enterprises, set standards to collect sewage treatment fees and encourage enterprises to actively collect waste edible oil to reduce the payment of sewage treatment fees. We should not only put forward requirements in legislation, but also make detailed provisions on the relevant supporting measures to achieve the requirements, otherwise the relevant classification system is always difficult to truly implement.

Second, in the collection, transportation and disposal stage, it is clearly necessary to continue the classified collection, transportation and separate disposal of kitchen waste. A large part of the problems existing in the classification and recovery of food waste in China are due to the imperfection of relevant supporting measures in the collection, transportation and disposal stage. We require the main body of kitchen waste to classify the waste, so we should make clear the requirements of classification and recycling of kitchen waste in the stage of collection, transportation and disposal in legislation at the same time. How effective the classified collection, transportation, and disposal of domestic waste is and whether it can be processed scientifically is an important criterion for measuring the effectiveness of classified collection, transportation, and disposal of domestic waste. If classified terminal disposal cannot be achieved, the former will be abandoned no matter how well it is divided. Therefore, we must also establish a supporting system for the classified recycling of kitchen waste in the collection, transportation and disposal stage. In terms of legislation, it is strictly required to allocate and use a unified classified recycling container in the collection, transportation and disposal stage. It is necessary to publicize the utilization of classified recycling of waste in the disposal stage, especially the utilization of waste edible oil. Only in the whole collection, transportation and disposal of each link to implement the concept of classified recycling is each link of classified recycling meaningful.

15.4.8 Establish the Risk Prevention System of Food Waste Pollution

Environmental risk prevention refers to the measures taken to prevent possible or serious harm to the environment, aiming at the threat or risk caused by such pollution. There are many provisions on environmental risk in the laws and regulations of environmental protection in our country: Emergency Response Law, National Emergency Response Plan for Environmental Emergencies, etc., but these laws and regulations are only generalizations, and there are few clear uses of the concept and system of environmental risk, and the provisions on food waste are even blank. In order to construct an environmental risk prevention system of food waste in our country, we should first establish a management system of food waste risk prevention, clarify the responsibilities of the national departments for the prevention of waste pollution risks and deal with the coordination between the economic responsibilities and environmental responsibilities of the relevant national departments. The risk information disclosure system shall also be stipulated in the legal system for the prevention of food waste pollution risk, so that the managers of urban waste treatment enterprises shall regularly report the environmental information in the whole process of waste treatment to relevant departments, announce it to the public, guarantee the credibility of the information provided and obtain the trust of the government and the public.

15.4.9 Open up a Green Market for Kitchen Waste By-Products

As a typical representative of the by-products of kitchen waste recycling, biodiesel is facing a series of problems that are hard to solve. It is not suitable for a large regular vehicle oil market. Enterprises can only build their own sales channels and sell them to customers such as fishing boats and transport fleets, which makes it difficult to achieve scale efficiency. It is of great practical significance to promote the use of biodiesel and standardize the development of the biodiesel industry. The specific performance is to make full use of waste oil to produce biodiesel, which can effectively control the 'gutter oil' reflux table. At the same time, it can improve China's energy supply capacity, improve energy structure and realize energy diversification. It is an urgent task to open up a green market for kitchen waste by-products and expand the sales channels of biomass diesel and natural gas terminals. Relevant government departments should take measures as soon as possible to break the monopoly of the oil and gas market by oil enterprises, accept the biodiesel that meets the standards for sale and take the biodiesel blending rate as the assessment index of oil sales enterprises, so as to solve the bottleneck of biodiesel sales. In addition, with more and more enterprises of kitchen waste recycling all over the country, the output of biodiesel is also increasing (Chen et al., 2015). I think that we can specially grant the qualification of biodiesel distribution to kitchen waste disposal enterprises, so that these enterprises can really realize the recycling of kitchen waste.

At present, biofeeds are the second largest type of food waste recycling by-products besides biodiesel, and they have the characteristics of large output and direct feeding of livestock, which is related to food safety issues. Food waste is processed into biological feed after harmless treatment. However, due to the limitations of technology and other conditions, some of the biofeeds produced by food waste treatment enterprises have quality and safety problems, so they can't be fed to livestock directly. However, there is no relevant quality and safety standard for this type of biological feed, and there may be some biological feeds that fail to meet the quality and safety standards entering market circulation, which brings potential food safety risks. The establishment of corresponding quality and safety standards can regulate the production of biological feed, reduce the hidden dangers of food safety, and promote the development of the biological feed market.

Therefore, the Food and Drug Administration, together with the Quality Supervision Department, the Ministry of Agriculture and other relevant departments, should jointly issue the quality and safety standards for food waste recycling feed. It can be divided into different quality grades, and it is stipulated that the inspection shall be carried out by the quality supervision department. Unqualified biological feed must be destroyed and shall not enter the sales market. Production licenses shall be issued for qualified biological feed, and it shall be classified according to different quality levels for the buyer to identify. Establishing corresponding quality and safety standards can maintain the market order of biological feed.

REFERENCES

Beck D.Z.L., Shen G.L. (2010) Risk society and China: a dialogue with Ulrich Beck. *Sociological Study* 05: 212–235.

Chen T., Jin Y., Shen D. (2015) A safety analysis of food waste-derived animal feeds from three typical conversion techniques in China. *Waste Management* 45: S0956053X1530012X.

Du J.L. (2017) *The Investigation of Social Awareness of Food Waste Centralized Treatment and the Study of Management Models in Tianshui Urban Region*. Lanzhou University.

Fang Z.Q., Wu J., Bao L.J. (2007) Discussion on the disposition of food residue. *Journal of South Chine Normal (Natural Science Edition)* 70–74.

Guoyang L.I. (2010) *Discussion on Management of Kitchen Waste. Recyclable Resources & Circular Economy.*

http://finance.people.com.cn/n/2013/0703/c1004-22058824.html

http://paper.people.com.cn/rmrbhwb/html/2016-01/07/content_1645857.htm

Hu X.J., Zhang M., Yu J.F., Zhang G.R. (2012) Food waste management in China: status, problems and solutions. *Acta Ecologica Sinica.*

Huang W., Wang J., Dai X., Li M., Harder M.K. (2014) More than financial investment is needed: food waste recycling pilots in Shanghai, China. *Journal of Cleaner Production* 67: 107–116.

Li Y., Jin Y., Li J., Chen Y., Gong Y., Li Y., Zhang J. (2016) Current situation and development of kitchen waste treatment in China. *Procedia Environmental Sciences* 31: 40–49.

Li Z.S., Yang L., Qu X.Y., Sui Y.M. (2009) Municipal solid waste management in Beijing city. *Waste Management* 29: 2596–2599.

Pan Y., Li H.M. (2011) Current status and solutions of kitchen waste recycling in Tianjin. *Recyclable Resources and Circular Economy* 4: 20–24.

Song J.F. (2011) *A Study on the Disposal of Municipal Solid Waste.* Nanjing Agricultural University.

Wang L. (2009) *Study on the Management of Food Waste Recycling.* Tianjin University.

Xu L.I., Shan Z.B., Chen H.B., Chen G., Xu L.I., Shan Z.B., Chen H.B., Chen G., Zhao L.S., Chen N. (2011) Investigation on the national and regional legislations and regulations of food waste. *Environmental Science & Technology* 32: 405–409.

16 Public Policies and Initiatives for Waste Minimization and Recycling in Brazil

Isabelli Dias Bassin, Luíza Santana Franca, Ingrid R.F.S. Alves and João Paulo Bassin

CONTENTS

LIST OF SYMBOLS AND ACRONYMS

CNI Brazilian National Confederation of Industry
ICMS Tax on Circulation of Goods and Services
EI economic instruments
InpEV National Institute for Processing Empty Packaging
IPEA Institute for Applied Economic Research
MSW municipal solid waste
NGOs non-governmental organizations
NSWP Brazilian National Solid Waste Policy
SNIS National Sanitation Information System

16.1 CONTEXTUALIZATION OF THE WASTE RECYCLING ISSUE IN BRAZIL

Given the worrying scenario of increased solid waste generation, several countries have implemented, in their public policies, measures to reduce the generation of waste with incentives to promote the circular economy through its reuse and recycling – and, in cases in which these activities are not feasible, by energy use (Geissdoerfer et al. 2017). Developed countries still face challenges in waste management, and these challenges are even greater in developing countries, such as Brazil.

Successful waste recycling programmes are known to produce many environmental, health, social, economic and educational benefits. They provide a reduction in the use of raw materials and the amount of waste sent to landfills or incinerators, besides creating opportunities for jobs and income (Grimberg and Blauth 1998).

While European countries, such as Germany and Switzerland, have advanced on the issue of waste for many years, moving from a vision of an environmental problem to that of resources recovery (Silva and Capanema 2019), in Brazil, until 2010, there was no legal instrument to establish general guidelines applicable to solid waste to guide its management appropriately. With legal principles aligned with the most advanced in other countries, the institution of the Brazilian National

Solid Waste Policy (NSWP), by Law 12,305/2010, was an important milestone in the sector within the country (Brasil 2010).

However, the implementation of the NSWP faces difficulties mainly concerning the high cost of the operation process, which involves the collection, treatment and disposal of waste, coupled with the lack of revenue for the sector. Most municipalities in the country do not charge a fee or tariff for services, with only 47% of Brazilian cities having a waste or cleaning fee that is capable of defraying 54% of waste management expenses (Snis 2018). Where the tax is implemented, only residential and commercial buildings are distinguished, but it is not associated with the amount of waste generated. The waste tax is normally charged once a year together with the property tax, and when an entity is exempt from property tax (e.g. public institutions, religious centres, NGOs, etc.), it does not pay the waste tax, although it benefits from its collection and cleaning. All of this makes waste management financially unfeasible (Thaty 2017).

Regarding selective collection, it is estimated that only 38.1% of the municipalities in Brazil have the service, and this index has stagnated in recent years according to data from the National Sanitation Information System (SNIS) (Snis 2018). Regarding recycling programmes, only 32% of Brazilian cities have such programmes active (IBGE 2012). Besides, less than 12% of potentially recyclable materials are collected, a rate similar to other developing countries, such as Russia (11%), China (20%) and South Africa (10%), but below those with similar economies, such as Italy (50%) or France (44%) (Eurostat 2018; Green-Cape 2015; Snis 2015; Snytkova and Salnik 2013; Tai et al. 2011). One reason for this may be the increase in the costs of solid waste transport with the adoption of selective collection, which increases the unit cost of waste management but is not reversed by tax revenue.

The SNIS also calculated the recovery rate of recyclable materials in relation to the total amount collected in the country, reaching a value of only 2.2% (Snis 2019). An estimate by the Institute for Applied Economic Research (IPEA) indicates that the country loses about R$8 billion (around $1.4 billion) a year by not recycling waste (IPEA 2010).

Some Brazilian cities are more advanced when it comes to recycling. For example, the city of Curitiba has had a well-established waste recycling programme for almost 30 years and is considered one of the best in the country. Also, Curitiba achieves better results in coverage and volume of collected waste and has greater investment per capita in the selective collection (higher than the national average). On the other hand, Brasília, the national capital, started its third formal attempt in 2014 to implement a recycling programme (Conke 2018).

In a study to identify barriers to the development of waste recycling in Brazil, Conke (2018) listed obstacles divided into five categories, as shown in Table 16.1. Another issue that hinders the recycling of waste in Brazil is the double taxation on recyclable materials that serve as raw material, which is costly and, therefore, discourages reverse logistics in the country, especially in sectors where it becomes more expensive, such as lamps, tires, lubricants, medicines and pesticides. A survey commissioned by the Brazilian National Confederation of Industry (CNI) points out that the Tax on Circulation of Goods and Services (ICMS) is responsible for 53% of the double taxation of recyclable waste that serves as raw material (Lisboa 2014).

An important initiative in progress is Bill 7,535, of 2017, which proposes the creation of a policy to encourage activities aimed at recycling (Gomes 2017). The proposal has three lines of action, namely: (i) direct incentive to recycling projects; (ii) support fund for actions aimed at recycling (Favorecicle); and (iii) investment funds for recycling projects (ProRecicle). The incipience of the recycling industry in Brazil is reflected in the lack of economies of scale, which causes increased costs and hinders the growth of the sector. In this sense, the creation of economic incentives is one of the pillars to promote an increase in the recycling rate in the country (Xavier 2018).

In addition to recycling, energy generation can also be obtained from solid urban waste; however, it is a process that has not yet become feasible in Brazil, despite being a practice already disseminated in other countries and which is aligned with the need to reduce waste volumes sent to landfills and the urgency of closing dumping sites. Recently, on 30 April 2019, Interministerial Ordinance

TABLE 16.1
Barriers to the Development of Waste Recycling in Brazil

Generation, handling and disposal	• Citizens are unaware of the operational aspects of recycling programmes, such as recyclability of materials, collection schedules and disposal of waste separated at the source • Lack of compassion for environmental appeal (reward, punishment or feedback) • People's aversion to the issue of waste and separation facilities • Illegal dumping • Insufficient number of landfills and questionable location of drop-off sites
Collection, transportation and recovery	• Resistance of waste pickers to the requirements of formal jobs and work attitude distant from cooperative values • Lack of management knowledge in recycling cooperatives • Non-adequate facilities, collecting vehicles and scales • No payments for environmental services
Dealing and recycling	• Inconsistency in the supply of recyclable material • Geographic concentration of the recycling industry • Lack of tax incentives for the recycling industry and for products that use recycled material • Packaging and recoverable parts of products not properly evaluated
Program management	• Unequal sharing of programme costs and responsibilities • Illusory expectations about human behaviour and programme results • Poor orientation in the official communication of recycling programmes (focus on why and not how to recycle) • Unsustainable payment method for third-party collection companies • Absence of adequate waste legislation

Source: Adapted from Conke 2018.

274 from the Brazilian Ministries of Environment, Mines and Energy and Regional Development was published, whose objective is to discipline the energy recovery of urban solid waste (Brasil 2019). It is expected that, in the short term, projects using these technologies will become viable.

An alternative to encouraging management of solid waste is the concession of management services to private providers. Public-private concessions and partnerships (PPP) are considered by many to be the alternative for modernizing public management, giving protagonism to the private initiative in public matters, especially in situations in which the public power does not have the technical and financial conditions to carry them out. In this arrangement, an opportunity is created to stimulate the private market to act more strongly in the sector, leveraging investments with government regulation (Silva and Capanema 2019).

Finally, to insert the country in the context of the circular economy, prioritizing the minimization and reuse of waste, environmental education policies need to be stimulated to raise awareness and educate the population about the importance of correct waste segregation at the source and recycling. A tax policy that promotes economic incentives for the use of waste through reverse logistics, recovery and transformation of waste should also be implemented. Public policies are essential to enhance recycling, whether through taxes, sanctions, incentives, awareness campaigns or other public instruments. For this, a stronger articulation between public power and private initiative is necessary, as well as the mobilization of society.

16.2 PUBLIC POLICIES ADOPTED IN BRAZIL RELATED TO WASTE MINIMIZATION AND RECYCLING

The integration of sustainable development into government policies is a key objective of environmental sustainability (da Cruz et al. 2012; Thabrew et al. 2009). Political support and legislation can encourage the reuse and recycling of materials. Some actions may require public funds, with investments or tax exemptions, and others not, such as sanctions and regulations (Murakami et al. 2015).

The Brazilian Law No. 12,305, which instituted the National Solid Waste Policy (NSWP) in Brazil, regulated by Decree No. 7,404/10, establishes the regulatory framework for the solid waste sector. It defines the guidelines to reduce the generation of solid waste, combating pollution, and refers to all types of waste: Domestic, commercial, construction, electronics, agriculture and cattle waste and hazardous waste, except radioactive waste (Brasil 2010). NSWP is part of the National Environment Policy and is articulated with the National Environmental Education Policy, the National Water Resources and Health Policy, the Federal Basic Sanitation Law and the Public Consortium Law. Law No. 12,305 also includes in its text the responsibilities and penalties of citizens and companies, the requirements for the management of hazardous waste and public incentives, such as economic instruments, associated with solid waste (Brasil 2010).

The economic instruments (EI) provided for by the NSWP have an essential function, since their basic concept is to support practices considered good, but considerably costly. Despite being widely used in several European countries, EI has not been widely explored in Brazil, with few federal, state and municipal laws and regulations for waste. Often the taxes and fees charged on solid waste recycling projects, new technologies for the sector and research on alternative materials are sometimes the same as those charged in the conventional economy, being one of the biggest challenges to overcome current rates of reuse of materials and to reduce demand for natural resources (Sobrinho 2009).

When comparing the EI applied nationally and internationally, it is possible to observe a management proposal that is much more aggressive abroad than in Brazil, as can be seen in Table 16.2. It is possible to note that the concern in the international context is with the generation. In addition to the tax on inputs and products, the increase in the landfill tax is what increases the relative cost of this form of waste disposal.

On the other hand, many gaps are observed in the implementation of EI in Brazil. In many cases, when there is an initiative in this sector, there is no state or municipal regulation that associates such instruments with the goals of reduction, treatment and adequate final disposal of solid waste, generating less visibility as to the results obtained and less effectiveness in changing the behaviour of waste generators (IPEA 2012).

The use of EI is a way to assist the implementation of reverse logistics, which will therefore help to expand recycling in Brazil. According to the NSWP, the federal and local governments will be able to use EI through the institution of inductive measures, credit line and public consortia (Brasil 2010).

TABLE 16.2
Comparison between the Characteristics of National and International Economic Instruments (EI)

Characteristics	National Description	International Description
Proposed use of tax	Proposes to reduce the rate or exemption in case the institutions/companies meet more than the expected goals of the EI	Proposes to increase the prices of products or inputs to reduce the generation
Materials return management	Proposes the implementation of ecopoints and increased selective collection	Institutes the deposit-refund system that rewards the return of recyclable and reusable products
Ways to support solid waste management	Laws that recommend the use of management incentives; school work involving families	Sets the price for each unit collected to raise awareness of the disposal

Source: Batista 2019.

TABLE 16.3
Classification of Laws Found in the Documentary Survey Carried out by Batista (2019)

Group	Economic Instruments (EI) Class	Description	Number of Laws
Directors	Laws recommending available incentives	Corroborate with the NSWP and act as drivers in local plans	14
	Laws that encourage reuse	Prioritize actions towards existing waste	
Economic and social aspects	Laws that establish donations in favour of reverse logistics	Actions in favour of the solid waste management in which some kind of donation by governments is necessary	9
	Law establishing awards	Award for sustainable attitudes	
Tax aspects	Law creating an investment fund	Creation of an investment fund that benefits the reverse logistics chain	17
	Tax incentive exemption	Determine the exemption from any fee or tax as a tax incentive	
	Tax incentive reduction	Determine the reduction of a fee or tax as a tax incentive	
	Laws that provide subsidies to reverse logistics chains	Determine the transfer of investments or the creation of a fund to subsidize something specific	

Batista (2019) investigated the use of EI in Brazil through a documentary survey of municipal laws until the year 2017 that supported reverse logistics in the municipalities of Espírito Santo, Rio de Janeiro and São Paulo. The 40 selected laws were separated into groups, according to Table 16.3.

Despite the signs of international efficiency, the implementation of economic instruments in Brazil has not been representative, with few laws implemented. It is necessary to emphasize that the adoption of these instruments by the Brazilian government is incipient. According to Batista (2019), of the 468 municipalities that presented municipal plans for solid waste, only 6.2% published actions and instruments that benefit good behaviour in waste management. Despite the importance of the management of solid waste and the implementation of reverse logistics, Brazil still has a lot to evolve.

In contrast, the recent Law No. 14,026 enacted in 2020, which institutes the new Regulatory Framework for Sanitation in Brazil, is considered an important instrument for municipalities to meet the requirements of waste management in the country. Law 14,026 provides for the universalization of sanitation services by 2033 and makes it possible to provide more private investments in sanitation services (Brasil 2020). According to the new framework, Brazilian municipalities will be required to implement a specific collection method to fund urban cleaning and urban solid waste management services. Otherwise, municipal management will be unable to receive additional federal funds for this activity. This is a rule that has existed since the NSWP was formulated in 2010, but has never been put into practice. With this, a 50% growth in the sanitation market is estimated in two years, at the same time that it should bring positive environmental impacts such as a reduction in the amount of waste generated and an increase in recycling.

16.3 PERSPECTIVES REGARDING WASTE RECYCLING AND MINIMIZATION PRACTICES IN THE BRAZILIAN CONTEXT

The enactment of the Brazilian Law 12,305/2010, which instituted the National Policy on Solid Waste (NSWP) in Brazil for setting targets for the correct disposal of waste, turns ten in 2020. However, in practice, little has been done by the municipalities on this issue. About 40.5% of the waste collected in the country in 2018 was dumped in inappropriate places by 3,001 municipalities, which means 29.5 million tons of municipal solids waste (MSW) ending up in controlled dumps or landfills. Law 12,305/2010 establishes strategies for preventing and reducing the generation of waste and encouraging recycling. However, the production of solid urban waste in 2018 increased by almost 1% over the previous year. When it comes to recycling, the respective law is also far from reaching its objective, since the country recycles less than 3% of its 79 million tons of waste produced per year, although the potential is up to 30% (ABRELPE 2019).

However, after the institution of the NSWP, some mechanisms became viable in order to stimulate the emergence of initiatives in the scope of the recycling market and, therefore, favour the development of the sector in Brazil. One of these mechanisms is reverse logistics, which guarantees the transportation of waste from the end of one production chain to the beginning of another, thus allowing recyclable waste to return in the form of inputs for the manufacture of new products. It is worth mentioning that the NSWP has made mandatory the reverse logistics of certain residues, such as pesticides, their residues and packaging; batteries; tires; lubricating oils, their residues and packaging; fluorescent, sodium and mercury vapor and mixed light lamps; electronics products and their components; packaging in general. Thus, the process of implementing reverse logistics started to take place through sectoral agreements signed between the government and its manufacturers, importers, distributors or traders of products and packaging.

In this context, the 'Campo Limpo' System is the Brazilian programme for reverse logistics of empty packaging or with post-consumption leftover pesticides. It brings together the industry that manufactures these products, the distribution channel, farmers and public authorities, each with their role and responsibilities defined by legislation, to promote reverse logistics and give this packaging an environmentally appropriate destination. The National Institute for Processing Empty Packaging (InpEV) is the managing entity of this system and the entire process, with more than 400 fixed units, being stations, or receiving centres (INPEV 2019). According to ABRELPE (2019), of the total packaging collected, 94% is reinserted in the production process through recycling, transforming into new packaging and caps for pesticides or other artifacts for civil construction and the automotive and energy industries, among other sectors. The remaining amount (6%) of the material that cannot be recycled is sent to incineration.

As for the reverse logistics of tires that can no longer be reused (waste tires), Reciclanip is responsible for managing and acting with representatives of the national manufacturers of these products. From the beginning of the programme in 1999 to the end of 2017, about 4.5 million tons of waste tires were collected and correctly disposed of, which is equivalent to 916 million passenger car tires. Between 2016 and 2017, there was an increase of 0.22% in the amount of tires recovered: from 457,000 tons in 2016 to 458,000 tons in 2017 (ABRELPE 2019).

Regarding waste lubricating oils and their packaging, 'Jogue Limpo' is the managing body responsible for the reverse logistics of the used plastic packaging and used or contaminated lubricating oil (IJL 2020). In 2018, the programme received 4,774 tons of plastic packaging, equivalent to 95.48 million containers, and 572,114 litres of used or contaminated lubricating oil. Of the total plastic packaging collected, 98% had an environmentally appropriate destination, of which 4,568 were recycled. Regarding the total oil collected, 557,597 litres were destined for re-refining.

In 2014, another national sectoral agreement was signed for the implementation of the reverse logistics system for fluorescent, sodium and mercury vapor lamps and mixed light lamps. The operationalization of the system is under the responsibility of the Brazilian Association for the Management of Reverse Logistics of Lighting Products (Reciclus), a non-profit organization founded in November 2015 by companies linked to the Brazilian Association of the Lighting Industry (Abilux) and the Brazilian Association of Manufacturers and/or Importers of Lighting Products (Abilumi). Reciclus is present in 33 cities in 25 states in the country (except Acre and Tocantins). According to the 2017 Annual Activity and Results Report, the system ended that year with 304 collection points installed in Brazil. In 2017, 2,963,646 lamps were disposed of in an environmentally friendly manner, 65% of which were compact and 35% tube lamps (ABRELPE 2019).

Finally, in the case of reverse packaging logistics in general, since 2013, 391,526 tons of waste were recovered by 144 cooperatives. In 2018 alone, 116,610 tons were recycled, 4.4% more than in 2017 – that is, there was a recovery of 22% of packaging, of the following materials: Paper, plastic, aluminium, scrap and copper, glass and other materials (electronics, residual oils and fats and other unspecified materials) were placed on the market (ABRELPE 2019).

Although the NSWP has been in force for ten years in Brazil, the sector still has some considerable deficits, mainly concerning selective collection, material recovery and final disposal of the collected waste. The implementation of the selective collection is still far from being universal; recycling rates are quite incipient and do not present an optimistic scenario for the return of solid waste to the production chain, which ends up not being disposed of in environmentally friendly locations. Therefore, it is essential to prioritize the sector on the agenda of municipal managers, so that the solid waste management in practice is reformulated, making it sustainable under the economic, environmental and social aspects and guaranteeing compliance with the NSWP.

The new Brazilian Regulatory Framework for Basic Sanitation established in 2020 by Law No. 14,026 appears as an alternative for the universalization of sanitation services, allowing the injection of more private investments in sanitation services until 2033. The law transforms the current contracts into concessions with the private company that takes over the state-owned company and also makes it mandatory to open a bidding process involving public and private companies. In this new scenario, municipalities are expected to be able to meet the requirements of the NSWP, which establishes that goals regarding the selective collection and reverse logistics must be part of the minimum content to be included in the municipalities' integrated solid waste management plans for the public and private sectors. NSWP must be interrelated with urban, industrial, technological and foreign trade policies as well as those that promote social inclusion, considering the important role played by associations or cooperatives of recyclable material collectors that carry out door-to-door collection, one of the most common selective collection approaches in Brazil today.

REFERENCES

ABRELPE. (2019). *Panorama dos Resíduos Sólidos no Brasil 2018/2019.* Associação Brasileira de Empresas de Limpeza Pública e Resíduos Especiais, São Paulo, SP.

Batista, B. (2019). *Análise da Efetividade dos Instrumentos Econômicos da Política Nacional de Resídos Sólidos (PNRS).* Dissertação (Mestrado Profissional em Engenharia de Desenvolvimento Sustentável), Universidade Federal do Espírito Santo.

Brasil. (2010). *Lei nº 12.305, de 2 de agosto de 2010. Institui a Política Nacional de Resíduos Sólidos.* Diário Oficial da União, Brasília, DF.

Brasil. (2019). *Portaria Interministerial nº 274, de 30 de Abril de 2019. Disciplina a recuperação energética dos resíduos sólidos urbanos.* Diário Oficial da União, Brasília, DF.

Brasil. (2020). *Lei nº 14.026, de 15 de julho de 2020. Institui o novo margo legal do saneamento.* Diário Oficial da União, Brasília, DF.

Conke, L. S. (2018). "Barriers to Waste Recycling Development: Evidence from Brazil." *Resources, Conservation and Recycling*, 134, 129–135.

da Cruz, N. F., Simões, P., and Marques, R. C. (2012). "Economic Cost Recovery in the Recycling of Packaging Waste: The Case of Portugal." *Journal of Cleaner Production*, 37, 8–18.

Eurostat. (2018). "Recycling Rate of Municipal Waste." https://ec.europa.eu/eurostat/databrowser/view/sdg_11_60/default/table?lang=en (August 19, 2020).

Geissdoerfer, M., Savaget, P., Bocken, N. M. P., and Hultink, E. J. (2017). "The Circular Economy – A New Sustainability Paradigm?" *Journal of Cleaner Production*, 143, 757–768.

Gomes, C. (2017). *Projeto de Lei nº 7.535/2017, de 2 de maio de 2017. Dispões sobre incentivos para fomentar a indústria da Reciclagem.* Câmara dos Deputados, Brasília, DF.

Green-Cape. (2015). "Market Intelligence Report – Waste Economy." 1–11, Cape Town, South Africa.

Grimberg, E., and Blauth, P. (1998). *Coleta seletiva de lixo: reciclando materiais, reciclando valores.* Instituto Polis, São Paulo, SP.

IBGE. (2012). *Profile of Brazilian Municipalities 2011.* Brazilian Institute of Geography and Statistics, Rio de Janeiro, RJ.

IJL. (2020). "Cartilha Institucional Instituto Jogue Limpo." *Instituto Jogue Limpo*, Instituto Jogue Limpo, https://www.joguelimpo.org.br/institucional/promocional.php (August 24, 2020).

INPEV. (2019). "Relatório de Sustentabilidade 2019." https://inpev.org.br/noticias-publicacoes/relatorio-sustentabilidade/index (August 24, 2020).

IPEA. (2010). *Benefícios Financeiros com a Reciclagem*. Diretoria de Estudos e Políticas Regionais, Urbanas e Ambientais, Brasília, DF.

IPEA. (2012). *Diagnóstico dos Instrumentos Econômicos e Sistemas de Informação para Gestão de Resíduos Sólidos*. Secretaria de Assuntos Estratégicos da Presidência da República, Brasília, DF.

Lisboa, V. (2014). *CNI pede desoneração e diz que reciclados são bitributados em R$ 2,6 bilhões*. Agência Brasil, Rio de Janeiro, RJ.

Murakami, F., Sulzbach, A., Pereira, G. M., Borchardt, M., and Sellitto, M. A. (2015). "How the Brazilian Government Can Use Public Policies to Induce Recycling and Still Save Money?" *Journal of Cleaner Production*, 96, 94–101.

Silva, V. P. M., and Capanema, L. X. de L. (2019). "Políticas públicas na gestão de resíduos sólidos: experiências comparadas e desafios para o Brasil." *Banco Nacional de Desenvolvimento Econômico e Social*, 25, 153–200.

Snis. (2015). *Diagnóstico do Manejo de Resíduos Sólidos Urbanos 2013*. Ministério do Desenvolvimento Regional Secretaria Nacional de Saneamento, Brasília, DF.

Snis. (2018). "Manejo dos Resíduos Sólidos Urbanos –2018." http://www.snis.gov.br/painel-informacoes-saneamento-brasil/web/painel-residuos-solidos (August 19, 2020).

Snis. (2019). *Diagnóstico do Manejo de Resíduos Sólidos Urbanos. Sistema Nacional de Informações sobre Sanreamento (SNIS)*. Ministério do Desenvolvimento Regional Secretaria Nacional de Saneamento, Brasília, DF.

Snytkova, M., and Salnik, V. (2013). "Russia Drowns in 60 Million Tons of Garbage a Year." *Pravda.ru*, https://www.pravdareport.com/russia/124947-russia_garbage/ (August 19, 2020).

Sobrinho, N. (2009). *A reciclagem de resíduos sólidos e a questão tributária no Distrito Federal*. Dissertação (Mestrado em Desenvolvimento Sustentável), Universidade de Brasília.

Tai, J., Zhang, W., Che, Y., and Feng, D. (2011). "Municipal Solid Waste Source-Separated Collection in China: A Comparative Analysis." *Waste Management*, 31(8), 1673–1682.

Thabrew, L., Wiek, A., and Ries, R. (2009). "Environmental Decision Making in Multi-Stakeholder Contexts: Applicability of Life Cycle Thinking in Development Planning and Implementation." *Journal of Cleaner Production*, 17(1), 67–76.

Thaty, M. (2017). *Politica nacional de residuos solidos: dificuldades para a implantação: bloco 1*. Rádio Câmara, Brasília.

Xavier, L. (2018). "Meio Ambiente aprova política de incentivo à reciclagem." *Agência Câmara de Notícias*.

Part IV

Hazardous Waste Management

17 Management of Hazardous Waste in Russia

D. Gribkova Y. Milshina

CONTENTS

17.1 INTRODUCTION

Global population growth, intensive industrialization in emerging markets, rising living standards and the decreasing product life cycles of electronic devices increase the demand for efficient hazardous waste management (Stiehler, 2018). Increasing volumes of hazardous waste and its complexity cause the global shortage of natural resources and escalation of environmental issues as well as stimulating the spread of circular economy and sustainable development principles.

Various kinds of human activities generate hazardous waste: Manufacturing, transportation, mining, medical practice, housekeeping. More than half of the 4 billion tons of waste generated annually worldwide (55%) is considered as industrial waste, including hazardous waste (Mavropoulos et al., 2014). According to various estimates, there are 0.35 to 0.4 million tons of Class I and II industrial waste generated annually in Russia (Voloshina, 2019). So far only 1.5% of Class I and II waste is recycled (Mereminskaya, 2019) by licensed operators. Still, the major part of the 444 types of Class I and II waste is not considered in the final enterprises' reports (Vasileva, 2019), which means that these issues are underestimated in the official statistics.

Over the past 30–40 years, profound changes have taken place in waste management. Social attitude towards waste and waste generation have changed fundamentally: Now waste is being considered as an important source of secondary materials and energy resources. One of the fastest-growing waste markets in the world is the e-waste market. By 2016, the world generated 44.7 million metric tons of e-waste, only 20% of which was recycled properly (Baldé, 2017). Most of it – 18 million tons – was generated in Asia, 12 million tons in Europe, 11 million tons in the Americas, 2 million tons in Africa and 0.7 million tons in Oceania (Stiehler, 2018). E-waste poses serious health and environmental threats when treated improperly, since it contains hazardous components, contaminating air, water and soil. In 2021 the global volume of e-waste may reach 52 million tons (Volkova, 2018).

Hazardous waste management includes such stages as collection, transportation, treatment and disposal of waste materials. In form of solids, liquids, sludges or contained gases, such waste may cause substantial harm to human health and/or to the environment due to inadequate storage, transportation, treatment and disposal (Nathanson, 2016). The global hazardous waste management market is expected to reach 48 billion dollars in 2027, growing at a CAGR of 6% in 2019–2027 (NK Wood Research, 2018). The rise in consumer spending, especially in developing countries, stimulates industrial activities, which, in turn, increase the amount of hazardous waste. Industrial waste was the largest sector of the global hazardous waste management market in 2017 (Technavio, 2018), and this segment is expected to occupy the dominant market share in the future.

Improper waste management, apart from economic environmental issues, leads to a negative social response. Transboundary movements of hazardous waste often inflame more acute public debates than a governmental inability to deal with the dangerous waste inside the country. Hence, the sphere of hazardous waste management needs further research and exploration of more effective and accessible methods and technologies.

The hazardous waste management system in Russia is in the process of transformation and has great potential for further improvement. Analysis of the existing organization (RosRAO) and the strategy for the development of this field shows that an effective system is still to be created. It highly depends on the development of technologies, demand for these technologies based on modern requirements and patterns of waste disposal and governmental support, as well as fostering a waste collection culture in society. However, waste disposal practices vary from region to region depending on the volume of produced hazardous waste and located enterprises. Given long distances through the country, the effectiveness of the system will rely on logistics, which in turn should be formed on timely information dissemination (State Information System).

This study is aimed at describing the current situation of hazardous waste management in Russia, including the existing legislative and practical basis and plans for future development. To achieve this goal several tasks were completed. First, the scientific literature focused on waste management issues was analysed. Second, the Russian legislative base in the field of hazardous waste management was described. Third, the current situation regarding waste generation and treatment, including official statistics, was overviewed. Next, the existing and building infrastructure and waste treatment and disposal methods prevalent in Russia were characterized. Then, the SWOT analysis of an emerging nationwide system for hazardous waste management was carried out. Finally, the case studies were analysed to receive the full picture of existing practices of hazardous waste management in Russia. All of the research was based on analysis of existing legislation, official statistics, reports of specialized organizations and authorized government bodies, scientific papers and professional media.

17.2 LITERATURE REVIEW

The paper analyses the information gathered from the recent (2017–2020) reports released by the industrial organizations and authorized governmental offices, official statistics, scientific articles and professional media that reflect the latest transformations in the sphere. In this section, a variety of studies dedicated to this topic are investigated to identify the important issues highlighted by Russian scholars.

To start with, Russian researchers focus on environmental issues which are closely related to social well-being. For example, Chernyaeva (2013) pays attention to the fact that the impact of the increasing waste generation on ecology is much more than the contaminated lands at the dumpsites, but also includes the global geological and hydrological changes. It can lead to the transformation of non-hazardous waste landfills to hazardous ones. Also, Chernyaeva (2013) notes that the extremely hazardous industrial waste is stored at the economically less developed regions adjacent to the industrially developed ones. The scholar draws attention to the need for a sound logistics system and rational treatment facilities distributed throughout the country. Moreover, Chernyaeva (2013) argues that the effective management of hazardous waste needs the coordinated work of a variety of stakeholders, including the national government institutions, local governments, municipal services and industrial enterprises.

In recent years more attention has been paid to legal responsibility for violations of the rules of hazardous waste treatment. Shumskikh (2012), for example, figures out that despite active legislative development, there is an inconsistency of norms between environmental and economic legislation, as well as between environmental and criminal legislations. In this way, the researcher emphasizes an urgent need to eliminate the existing legislative contradictions.

Kruter (2011), while analysing practices of crime prevention in the sphere of treatment of hazardous waste in Russia and abroad, proves that effective crime prevention measures should include both interdisciplinary and international approaches, especially when the transboundary movements of hazardous waste are considered. The pervasive interdisciplinary approach abroad includes the capabilities of different areas: Legal, environmental, administrative, etc. Kruter (2011) as well as Shumskikh (2012) highlights the importance of elimination of contradictions between various legal acts and provides examples of the developed systems of environmental crime prevention in such countries as Germany and the United States.

Nowadays Russian scholars tend to consider the European waste management system as the most effective one. For instance, Nikulichev (2017) notices that the waste-to-energy concept prevails in the EU's programmes of development and scientific literature. In Germany, Belgium, Sweden, the Netherlands, Austria and Denmark only 1–2% of waste is sent to landfills, while 35–50% of it is burned and 50–60% is recycled and composted (Nikulichev, 2017). These countries put an official ban on waste landfilling without special treatment. Nikulichev (2017) also points out that the concept of waste-to-energy is cost-effective: Municipal solid waste can be a source of fuel comparable to peat and some brown coals. Moreover, this fuel can be generated in the most energy-demanding places, i.e. around large cities, and is predictably renewable. The European Union growth rate of venture capital and investments in this sphere recently reached 200% per year (Nikulichev, 2017).

Both Russian and foreign scientists believe that in the near future incineration with the generation of electric and thermal energy will be the main recycling method. In this regard, Nikulichev (2017) notes that waste-to-energy plants are likely to be a part of an integrated waste management system. The scientist emphasizes the advantages of the Danish waste management model: First, this model covers all types of waste (municipal, industrial and hazardous); second, the local authorities determine how to collect and treat waste, and waste generators should strictly follow the guidance; third, the polluter pays principle is applied to households and industry; fourth, the whole system is based on the concept of separate collection.

Timofeeva et al. (2013) also point out that unused, or improperly disposed of, waste represents billions of tons of irretrievably lost resources and, at the same time, a source of harmful chemical and biological components. The amount of generated and reusable waste shows the level of economic development as well as the level of technological, social and cultural development. Not least, the authors raise a question of high professional risks to those who work with solid waste brought to landfills. Reforms of the waste management systems, especially of the hazardous ones, should be accompanied by improved working conditions.

One of the most recent studies of Russian scholars on the topic is devoted to the waste management issues in the modern green and smart cities (Mingaleva et al., 2019). The authors argue that waste management system development should include both technological, social and cultural aspects (Mingaleva et al., 2019). Waste collection methods directly depend on the methods of waste disposal, while their proper combination leads to a higher degree of coverage of municipal solid waste collection. In this perspective, the waste management system is seen as a part of sustainable development and green city models, which vary among different countries.

Fedotkina et al. (2019) claim that in Russia the tradition of waste collection and processing originates from the Soviet practices. Nevertheless, the new stage of waste management development in Russia implies the ideas of responsible production and consumption, which are consistent with the concept of the circular economy. Authors note that in Russia circular economy is being developed in three areas: Management of existing landfills and dumpsites, optimization of the municipal solid waste system, especially at the stage of collection, adoption of cleaner production patterns with the use of renewable technologies and materials and introducing circular economy education. Also, the researchers stress the significance of a multi-actor and multilevel perspective as inevitable in the construction of a waste management ecosystem. It means that different stakeholders – federal and regional government, municipal administrations, business companies and households – should participate in this process.

The report 'Management of hazardous waste in Russia and some EU countries, 2019' (Bellona, 2019) by the Bellona Foundation, an independent non-profit organization, should be noted as one of the deepest and most comprehensive analyses of the ongoing reform of the Russian hazardous waste management system. Authors analyse the federal project for the development of infrastructure for hazardous waste treatment, point out 'blind spots' and barriers and provide statistics and cases of hazardous waste treatment. Still, according to Bellona (2019), only the framework for this system was created, whereas its performance has to be greatly improved. Findings from the report were used in this paper to form the SWOT matrix.

The Civic Chamber of the Russian Federation conducted a detailed analysis (OP, 2018) of the recent legislative changes in the field of waste management in general, including the sector of hazardous waste management in Russia. This report includes the assessment of the effectiveness of the measures taken to ensure proper waste management, funding, tariffs, proposals regarding the involvement of secondary resources in the economy and development of the solid waste collection system, etc.

Researchers of the Bellona Foundation and the Civic Chamber point out some weaknesses that require special attention. A recent paper by Gilmundinov et al. (2020) highlights that the new legislation substantially clarified the conceptual framework; however, it left some 'blind spots' (Gilmundinov, 2020). The most obvious 'blind spot' is the lack of information on the particular works that are going to be implemented.

Some of the relevant papers are retrospective, so they mostly do not reflect the current stage of Russia's hazardous waste management system development. As a rule, studies in this field examine the cases of particular types of waste, spheres of hazardous waste generation (industrial or municipal), stages of treatment (collection, storage, disposal) or legislative regulation of certain segments (for instance, transportation).

In most of the studies, the researchers consider waste management issues from the ecological point of view, which is consistent with the Russian legislative base. In recent years, waste management issues have been examined from a multi-actor perspective. Research and stakeholders stress the importance of collaboration among government, private and scientific actors.

Attention to the administrative and criminal liability issues, as well as appealing to foreign practices, can be explained by the existing decentralized system of control over waste generation and disposal. The problem of the social separate collection culture is also investigated in comparison with the best long-standing foreign practices, mostly the European ones.

Overall, there is a shortage of academic literature devoted to hazardous waste management in Russia that encompasses the last changes in this sphere and describes the system in general. These papers represent a patchwork of different concepts focused on special aspects of waste management. This can be explained by the relatively recent start of the reform of the hazardous waste management system in Russia and the growing importance of the problem both at the state and public levels. One of the most relevant reports ('Management of hazardous waste in Russia and some EU countries, 2019'), which will be discussed further, was published in 2019, after almost a year from the start of the reform.

17.3 LEGISLATIVE ASPECTS OF HAZARDOUS WASTE MANAGEMENT IN RUSSIA

The problem of extremely hazardous and highly hazardous waste treatment in Russia was pointed out at the meeting of the State Council of the Russian Federation in 2016 on the issue of Russia's environmental development for future generations (President of Russia, 2016). The current reform of the hazardous waste management system was launched in 2018. In May 2018, the President signed Executive Order on National Goals and Strategic Objectives of the Russian Federation through to 2024 (President of Russia, 2018). According to this decree, in the process of development of the National Project 'Ecology', the government was instructed to provide a solution to the issue of modern infrastructure for the secure treatment of hazardous waste and the elimination of dangerous waste facilities. Also, in 2018 the Ministry of Nature proposed State Corporation Rosatom as a federal operator of the future system and initiated necessary legislative procedures.

Waste classification based on hazardousness types was developed much earlier. According to the Russian normative legal acts, hazardous wastes are wastes whose generation or handling potentially poses a danger to human health and life and a major environmental threat (Gosstandart Rossii 2002). The federal law №89 approved in 1998 (Federal Law, 1998) defines five classes of waste, whose classification criteria in terms of the degree of negative environmental impact were formulated in a special legislative act in 2014 (Executive Order N 536, 2014). The paper describes the Class I and II waste management system (Table 17.1).

In Russia, the concept of 'environment' and environmental rights are enshrined in the Constitution, which is not the case, for example, for France, which mostly drives the international environmental protection initiatives (Vasilevskaya, 2018). Russia has a legally fixed definition of the term 'environment' in the federal law "About environmental protection" (Federal Law, 2002). Besides this, the law defines other terms – environmental quality and environmental pollution.

Thus, in 2014, the government approved the criteria of waste classification, based on the degree of negative environmental impact (Table 17.2). The degree of hazardousness for the environment (K) is determined by the sum of the hazardousness degree of every waste component: $K = K_1 + K_2 + \ldots + K_m$ (Executive Order N 536, 2014).

The degree of hazardousness of waste components to the environment (K_i) is calculated as a ratio of the concentration of the waste component (C_i) to the coefficient of its environmental hazardousness degree (W_i): $K_i = C_i / W_i$.

However, there is still inconsistency in the definition and valuation of waste hazardousness because of the inadequate distribution of obligations among the different federal authorities. Indeed, the Federal Service for Supervision of Natural Resources (Rosprirodnadzor) and the Federal Service for the Oversight of Consumer Protection and Welfare (Rospotrebnadzor) use their own classification systems for the same waste, along with two different approval procedures. Rosprirodnadzor states that its classification is developed for environment protection, Rospotrebnadzor – for human wellbeing (Markova et al., 2019). Only in July 2019, did the Ministry of Nature and Rospotrebnadzor adopt a plan for the harmonization of waste classification (Ekopromcentr, 2019). The harmonization of classification still is in process.

Also, Russia has ratified a number of landmark global deals: The Basel Convention on the Control of Transboundary Movements of Hazardous Wastes and Their Disposal (Basel Convention, 1992), the Rotterdam Convention on the Prior Informed Consent Procedure for Certain Hazardous Chemicals and Pesticides in International Trade (Rotterdam Convention, 2004) and the Stockholm Convention on Persistent Organic Pollutants (Stockholm Convention, 2004). Russia follows the hazardous waste classification that unites international classification, including prescriptions of the Basel Convention and OECD.

In 2019, the Intergovernmental Council members of the Eurasian Economic Union have signed an Agreement on the transboundary movement of hazardous waste through the EAEU customs territory to ensure the implementation of international obligations under the Basel Convention. For instance, it regulates the movement of waste, which includes arsenic, mercury or their compounds, the dust of metallurgical products and the production of glass containing beryllium, organochlorine, etc., from one Union country to another (Eurasian Economic Commission, 2019). The Agreement allows transboundary transportation of waste by legal entities and individuals registered as sole proprietors.

This brief overview of the legislative framework shows that measures of its development have only been relatively recently taken and have not removed some inconsistencies (for instance, double

TABLE 17.1
Russian Waste Classification System

Hazard Class	Hazard Description	Waste Stream Project Examples	Degree of Environmental Damage	International Definition
I	Extremely hazardous	Fluorescent light bulbs containing mercury, activated carbon contaminated with mercury sulphide; polychlorinated biphenyls and terphenyls waste, polybrominated biphenyls, as well as waste of products containing them; mercury-containing products; asbestos waste, asbestos dust and fibre, etc.	The ecological system is irreversibly broken with no recovery period.	Hazardous
II	Highly hazardous	Concentrated acids, alkalines, halogenated solvents, lead–acid batteries, dry batteries, used sulphuric acid; solid copper chloride waste; solid lead waste; uncontaminated lead sawdust, etc.	The ecological system is severely disturbed; recovery period is not less than 30 years after the complete elimination of the effect source.	
III	Moderately hazardous	Used lubrication oil, oily sludge, oily rags, used oil filters, non-halogenated solvents, paint wastes, sludge for cleaning pipelines and tanks (barrels, containers, tanks, tarmacs) from oil; diesel fuel with lost consumer properties; aviation, automobile and motor oils with lost consumer properties; cement dust; sand contaminated with gasoline, etc.	The ecological system is broken; a recovery period takes at least ten years after reducing the harmful effects of an existing source.	
IV	Low hazardous	Domestic trash, nonferrous metal scrap, certain chemicals, certain construction waste, treated sewage sludge, treated medical wastes, water-based drilling mud, etc.	The degree of harmful effects of waste for the environment is low; the ecological system is broken, a self-healing period of at least three years.	Non-hazardous
V	Practically non-hazardous	Inert wastes: Plastic, ferrous metal scrap, inert construction wastes, food waste, brushwood, non-treated wood waste.	The degree of harmful effects of waste is very low; the ecological system is particularly not broken.	

Source: Authors with the usage of SOVECO materials and Fedotkina (2019).

waste classifications). Nonetheless, Russia is taking steps to address these issues through the development of a detailed classification of waste hazardousness based on explicit calculations and official standards.

17.4 HAZARDOUS WASTE MANAGEMENT

17.4.1 Overview of Types and Amount of Hazardous Waste Generation

A large part of the sphere of hazardous waste management (75%) in Russia is out of government control (Vasileva, 2019). The Ministry of Nature identified the following main challenges the

TABLE 17.2
The Indicators for the Degree of Waste Danger for the Environment

Hazard Class	Degree of Waste Hazardousness to the Environment
I	$10^6 \geq K > 10^4$
II	$10^4 \geq K > 10^3$
III	$10^3 \geq K > 10^2$
IV	$10^2 \geq K > 10$
V	$K \leq 10$

Source: Executive Order N 536 (2014).

government faces: Lack of reliable data on the amounts of Class I and II waste being produced and a shortage of facilities for the deactivation, disposal and storage of such waste (Markova et al., 2019).

There are 0.35 to 0.4 million tons of Class I and II industrial waste produced annually in Russia (Voloshina, 2019); so far only 1.5% of Class I and II waste is recycled (Mereminskaya, 2019) by operators with licenses. In 2018, of 7 billion tons of Class I–V waste generated by individual entrepreneurs and legal entities in Russia (Rosprirodnadzor, 2019), 278,000 tons were hazardous waste (21,800 tons – Class I and 256,200 – Class II) (Vasileva, 2019), but these figures do not include household waste. Of 85 Russian regions, 20 are the largest generators of Class I–II waste with more than 1,000 tons of hazardous waste per year (Bellona, 2019).

Approximately 10,000 enterprises in Russia generate Class I and Class II waste, and just several thousand enterprises have a license to handle such waste (Mereminskaya, 2019). There are only two landfills where hazardous waste can be stored legally – in Krasnoyarsk region and Tomsk oblast. Enterprises send information about waste generation to Rosprirodnadzor as a notification (it is not a mandatory report). The actual volume and hazard class are underestimated frequently. The existing system of control, accounting and regulation does not reflect the full scale of illegal disposal of Class I–II waste.

The main violations that prosecutors identify are the lack of passports for waste (documents with complete information about the waste – origin, composition, conditions, specific disposal facilities, classification, etc.), improper storage of toxic waste at the enterprises and untimely reporting to Rosprirodnadzor.

For example, in China, more than 80% of hazardous waste is disposed of properly (Stiehler, 2018). Since January 2013, the illegal discharge of hazardous waste is considered as a criminal offence. This measure aims to raise volumes of treated hazardous waste. The European construction industry produces high-quality materials from recycled waste, increasing the sustainability of the sector and economy (Global Recycling, 2019). Among the waste generated in the European Union in 2016, of the 100.7 million tons classified as hazardous waste, 76.8 million tons were treated (European Commission, 2018b). The measures being taken by the Russian government aim to reach 65% of recycled hazardous waste for Class I and 100% for Class II by 2024 (Voloshina, 2019).

The above-mentioned e-waste generation is also an acute issue in Russia: 1.2–1.3 million tons of e-waste are generated annually, and only about 20% is recycled (Volkova, 2018). An improperly developed system of control makes it impossible to identify actual volumes of generated and recycled waste. The insufficient level of waste processing is caused by many factors: A lack of special high-precision and high-tech equipment and highly qualified specialists, low labour cost, heterogeneity of disposed materials according to hazard classes and low market demand, as well as a lack of stable demand for recyclable fractions, etc.

Currently, there are about 80 enterprises involved in e-waste processing in Russia (Mitina et al., 2020). Most of them collect, dismantle and sell only the most commercially attractive fractions (metals, some types of plastic). Only several plants provide deep treatment and extract not only precious metals. At the same time, facilities to treat several types of hazardous e-waste (chemical current sources, lead-bearing glass, chladon) should be built.

Despite the flaws of the existing system of waste management, there has been a sustainable trend in recent years towards declining hazardous waste volumes. The generation of extremely hazardous waste decreased by 6 times in 2016 compared with 2010; highly hazardous waste – by 2.3 times (Volkova, 2018).

17.4.2 Infrastructure for Hazardous Waste Treatment

Market regulation of hazardous waste management has been in force for several decades and has led to a shortage of waste processing capacities. Treatment and control systems of extremely or highly hazardous waste are still under construction. There is a lack of the necessary production capacity for the disposal and recycling of such waste and automatically updated information systems as well.

In 2018, the Russian Government approved the Strategy for development of the industry for sorting, recycling and treatment of waste for the period until 2030 (Decree of the Government, 2018) (further – the Strategy), which prioritizes the following dimensions:

- Using raw materials with maximum efficiency
- Prevention of waste generation
- Decreasing waste generation and reducing the severity of hazardous waste
- Waste recycling
- Waste disposal
- Waste deactivation
- Waste storage and burial in accordance with legislative requirements in the field of environmental protection and the sanitary-epidemiological welfare of the population

The Strategy is planned to be implemented in two stages: in 2018–2021 and 2022–2030. These goals require an effective management system, which would be able to cope with existing problems and has the necessary capacities to control hazardous waste generation and dispose of it. The first stage includes the creation of technological and industrial reserves, the placement of industrial facilities for waste treatment, the establishment of the centre for equipment certification and the improvement of the regulatory legislation.

The Strategy implies the construction of facilities for treatment, deactivation and disposal of hazardous waste using the best available technologies. One of the key issues the Strategy highlights is the development of regional and inter-regional plans for the elaboration of transport infrastructure, increasing the efficiency and optimization of the transportation of hazardous waste by diverse types to industrial facilities.

The Creation of Infrastructure for Safe Handling of Class I and II Hazardous Waste is a part of the National Project 'Ecology', which includes more than 36 billion rubles of spending for the implementation of infrastructure projects (Council for Strategic Development and Priority Projects, 2018). This federal law, adopted by the State Duma in July 2019 (Federal law, 2019), creates a unified management system for extremely hazardous waste and a unified information system. It is also envisaged to create a federal waste management scheme all stakeholders should be included in, to fix the distribution of roles among waste management system stakeholders. From 2022 all enterprises will have two options to dispose of their hazardous waste: Accumulate, deactivate, dispose and store waste on their own in accordance with the federal rules or transfer it to the national operator.

The federal operator should be a cornerstone of this system; according to the law signed by the President in August 2019, RosRAO, State Corporation Rosatom's (State Atomic Energy Corporation Rosatom) subdivision, has been granted the status of Federal State Unitary Enterprise (hereinafter

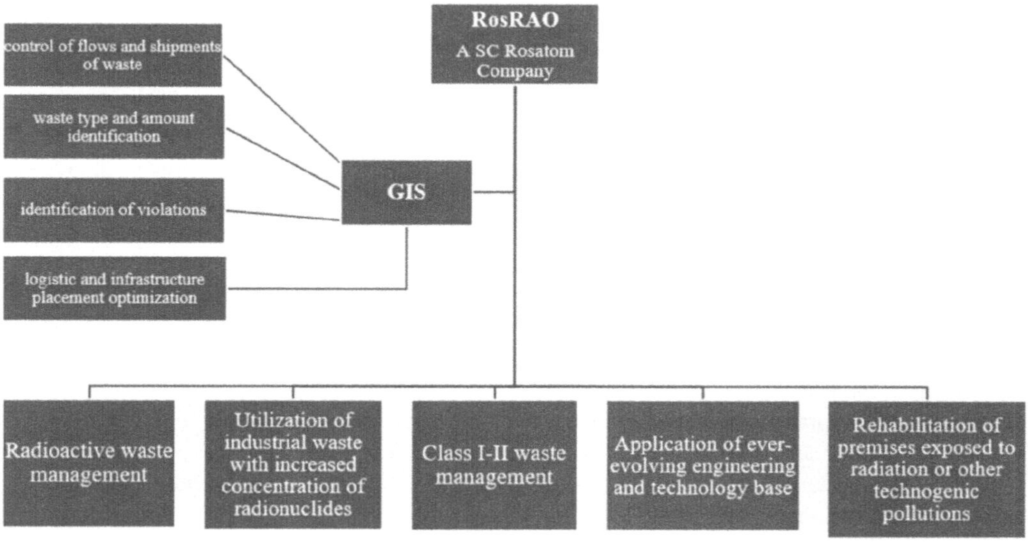

FIGURE 17.1 RosRAO functions. Source: Authors with the usage of RosRAO materials.

Federal operator) for the management of Class I and Class II hazardous waste. RosRAO performs the functions that were assigned to Rosprirodnadzor; the company should make the waste management process more transparent and controlled (Figure 17.1).

Previously, Rosatom has built a similar system for nuclear waste management. Class I and II waste takes the next place after nuclear and radioactive waste in terms of its impact on human health and the environment. Considering Rosatom's high-skilled staff, ever-evolving technology base and specialized authorization in this field, it is reasonable to assume that this practice and infrastructure could be used for hazardous waste handling. Russian atomic specialists have experience in the creation of a radioactive waste management system, which will be taken as a basis for a system of accounting and monitoring of hazardous waste. Rosatom has been working with radioactive waste for decades. Nevertheless, to start work with hazardous waste on a national scale, it needs to build up competencies and attract investments (Burmistrova, 2019).

To strengthen physical facilities, the company intends to build seven industrial engineering complexes. Four of these plants will be based on the former chemical weapons disposal facilities with already existing infrastructure in Kirov, Saratov, Kurgan Oblast and the Udmurt Republic; three more plants should be built by the end of 2024. The capacity of every complex should reach 50,000 tons per year (Burmistrova, 2019). All production processes will be tested for compliance with sanitary and environmental standards. Moreover, RosRAO will cooperate with regional recycling companies with appropriate licenses, capabilities and equipment.

Related to the lack of accurate information on waste generation volume, by 2021 RosRAO intends to create the State Information System (SIS) (Mereminskaya, 2019) for online verification of data, tracking of waste handling and detecting violations. Enterprises will use this system to apply for the disposal of hazardous waste, while the cost of the disposal will be calculated automatically depending on the distance of transportation (Zholobova, 2019).

Data collection is not the only possible application of SIS; it will be integrated with the information systems of taxation services and Rosprirodnadzor to allow automated cross-organizational validation of the information submitted by enterprises. In January 2020, RosRAO reported that companies in Tatarstan are the first to enter the information about hazardous waste into the information system (National Oil and Gas Service Association, 2020). From 2021, SIS should start working in every region of the country.

Among other things, this system will be used to determine the most reasonable places for three recycling plants and logistic centres for the accumulation of hazardous waste and distribution of traffic flows, based on information about the actual volume and composition of waste from enterprises. Waste collection and transportation to final destination will be controlled by Global Navigation Satellite System (Glonass) receivers for trains and trucks.

It should be noted that in 2019 one more organization, the Russian Environment Operator, or REO, was established to ensure the realization of the National Project 'Ecology' and the implementation of legislation in the field of municipal solid waste management. REO is to form an integrated waste management system, to stimulate investment activity and to create a system of secondary processing and involvement of raw materials in the production of energy and new products. On the online platform REO Radar users can report illegal dumping and disposal of mercury thermometers, transformers, mercury-containing devices, waste salts of arsenic, creosol, its residues and medical waste, batteries, tires, oils, oil products or fluorescent lamps.

The level of the processing capacity of the facilities depends on the technologies and methods used for waste treatment and disposal. The waste management system prevailing over decades in Russia requires fundamental technological improvement. According to the Strategy, at the second stage of its implementation, which is to start in 2022, the Russian technological and industrial infrastructure should start functioning on the basis of modern equipment, technologies and mechanisms for waste processing, deactivation and utilization (Decree of the Government, 2018).

17.4.3 Waste Treatment and Disposal Methods

The waste disposal and recycling methods can be divided into two types: Industrial (sorting, composting, incineration and combinations of these operations) and the organization of sanitary landfills for solid waste (Lebedeva, 2015). To a certain extent waste incineration plants can solve the problem of landfills and energy generation issues (waste-to-energy concept). However, when choosing waste incineration, multistage systems of purification of exhaust gases and continuous automatic control of air emissions should be integrated (Gorbulina, 2019). This represents an obstacle for Russian waste incineration plants – the difficulty of neutralizing the gases released from harmful impurities, especially from dioxins and nitrogen oxides. Incineration plants use a single-stage gas purification scheme, which does not allow the complete destruction of the mentioned gases and can cause air pollution (Gorbulina, 2019).

As the world practice shows, the main effective way to solve the problem of waste is recovery, recycling and reuse of waste (Figure 17.2). The 3Rs approach – reduce, reuse and recycle – has been widely integrated, for example, in the Canadian waste management system in the last decade (Pan et al., 2019). In Western Europe, a leader in the field of waste recycling and disposal, the share of recycled waste is already 80–87%, which significantly exceeds the same indicator in Russia (Decree of the Government, 2018). For instance, in Sweden, along with the municipal waste collection system, there are approximately 650 recycling centres for the collection of bulk waste (furniture, old computers, television sets, etc.) and all types of hazardous waste countrywide (Nikulichev, 2017). In 2018 EU countries approved laws formulating new goals in the sphere of waste recycling. Now all EU states are required to recycle at least 55% of municipal waste by 2025, 60% by 2030 and 65% by 2035. Also, the share of waste brought to landfills should be 10% by 2035 (European Commission, 2018a).

Focus on waste recycling leads to an increase of landfills taxes and makes it less attractive for the investors. The level of disposal and reuse of extremely hazardous and highly hazardous waste in Russia is insufficient in comparison with the leading European countries (Germany, Austria, Sweden) and Japan (Decree of the Government, 2018). In Japan, the amount of recirculated resources derived from all waste is more than 50% of the total waste volume, while the amount of waste brought to landfills is 3.4%. These figures are comparable with the European indicators, but the feature of the Japanese approach is that the ratio of reduction amount by dewatering and drying

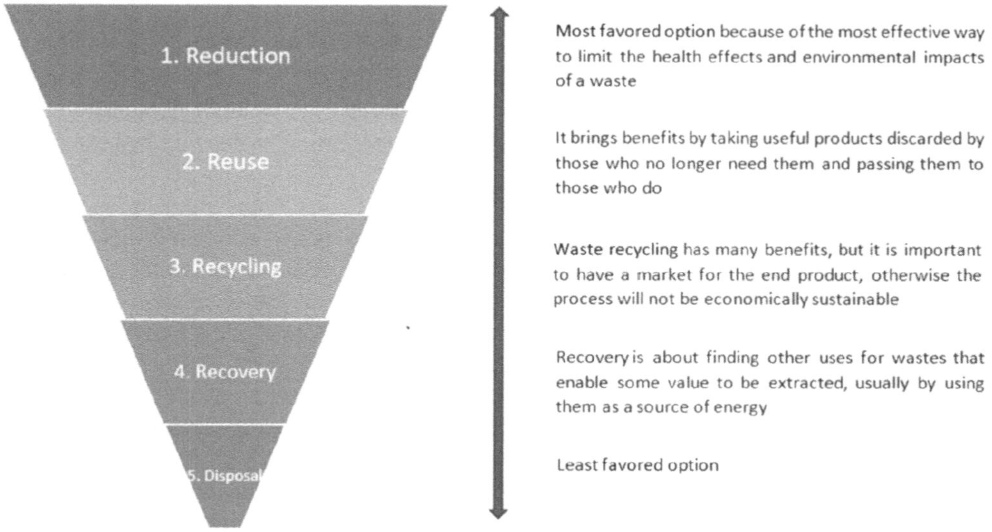

FIGURE 17.2 The waste hierarchy. Source: Open Learn, 2017.

is very large, whereas the ratios of backfilling and final disposal amounts are considerably smaller than in Europe (Takashi, 2018).

'Saudi Arabia's Vision 2030', a strategic framework for future development, includes improving recycling rates (Blaisi, 2019); in 2019 the Kingdom recycled no more than 10% of the 50 million tons of recyclable materials it produces each year (Nereim, 2019). Strategic objectives for recycling by 2035 include the recycling of 81% of the 3.4 million tons of annually produced municipal solid waste and 47% of the 5 million tons of construction and demolition waste per year (Global Recycling, 2019). Recycled materials are to be used in road construction and housing projects.

Landfilling is commonly used in Russia; more than 90% of the waste of all classes is brought to landfills: There are 1,000 municipal solid waste landfills, 15,000 authorized and 17,000 unauthorized landfills and 13,000 unauthorized waste disposal sites with a total surface of 4 million hectares in the country (Fedotkina et al., 2019). A large amount of solid waste is brought to landfills without sufficient treatment. It means the loss of tons of valuable raw materials. The total amount of hazardous waste produced is unclear, so it is difficult to define the share of Class I–II waste brought to landfills.

The Strategy for the development of the industry for the sorting, recycling and treatment of waste for the period until 2030 necessitates a change in the system of waste treatment in favour of recycling and disposal (instead of landfills). For example, currently only several Russian regions have large enterprises (11 in total) enabled to conduct the complex processing, disposal and deactivation of mercury-containing waste, using modern technologies for the refinement of spent mercury and mercury compounds (including ultrapure) production (Decree of the Government, 2018). Of these enterprises four are in the Northwestern Federal District, three in the Southern Federal District, two in the Central Federal District and two in Volga Federal District.

In Russia, the industry for the deactivation and disposal of some liquid Class II waste has not been created yet since the recovery is economically unprofitable. Still, some unusable solvents (alcohols, esters, aromatic hydrocarbons, other solvents containing highly hazardous organic compounds) can be recovered. Enterprises use mainly foreign equipment for the recovery of water-based solvents and flammable solvents. Several companies promote both foreign and domestic solvent recovery equipment in Russia and other countries of the Eurasian Economic Union. Such equipment allows the processing of organic solvents and recovering to consumer standards. More than 70 large Russian enterprises use it (Decree of the Government, 2018).

The collection and disposal of lead-containing Class II waste is the only sector in the sphere of treatment of highly hazardous lead waste, that was recently developed. The reason for this is the collection and disposal of automotive storages which began due to the high content of lead scrap. Other sectors of waste treatment are not economically feasible. Just a small number of domestic enterprises produce equipment for the processing of lead-containing cables and batteries (Decree of the Government, 2018).

Considering the economic inexpedience of the treatment of some types of hazardous waste and lack of economically viable technologies, governmental support and economic mechanisms for stimulation of the production from recoverable resources are of essential importance. They should contribute to the increase of operational characteristics, quality and safety of recycled materials, as well as competitiveness in the internal and external markets (Decree of the Government, 2018).

The Strategy involves the development of scientific and industrial infrastructure for the creation of high-tech environmentally friendly equipment, machinery and operation lines for the treatment of waste with high export potential. A set of government support measures determines the replacement of imported machinery and equipment for waste disposal (Table 17.3).

The Strategy sets ambitious goals comparable to the EU's targets in terms of indicators and deadlines. It determines the necessary conditions for ensuring the environmental security of the country, resource saving and reducing waste generation, thus changing the structure of the waste management system in favour of disposal and recycling. The Strategy stresses the importance of improvement of collection methods and technologies and the transition to effective resource-saving methods. One of the key objectives is the development of domestic methods and technologies for Class I–II waste treatment.

The Strategy describes two possible scenarios for the further development of the waste disposal industry in Russia. The conservative one assumes preservation of current trends, standards and conditions of the waste treatment system, slow changes of economic, regulatory, management frameworks without infrastructure development and implementation of domestic technologies. In this case, the lack of support from the government will affect the implementation of ever-evolving technologies in waste processing and disposal. All indicators will remain at the current level or will

TABLE 17.3
Target Indicators of the Strategy for the Development of the Industry for Sorting, Recycling and Treatment of Waste for the Period until 2030

Title	Unit Measure	2016	2025	2030
The share of disposed and neutralized waste in the total amount of generated waste	%	59.6	75	86
The share of municipal solid waste to be treated in the total amount of accumulated waste	%	8.9	50	80
Eco-technologies parks	unit	-	30	70
Industrial and technical facilities for waste treatment, disposal and recycling	unit	6	101	226
Solid waste sorting complexes	unit	60	210	310
Multifunctional complexes for industrial disposal of waste	unit	10	50	110
Multifunctional sorting complexes	unit	10	50	110
Contribution of the waste treatment industry to the GDP	%	0.08	0.1	0.11
Waste reduction rate	%	-	–1.8	–3.7
The share of imported equipment for waste treatment	%	60	20	10

Source: Decree of the Government, 2018.

decrease due to the import dependence. Landfilling will prevail over recycling. This scenario is also characterized by a low level of investment in this field.

The conservative option may have several negative consequences: Increase in waste generation and consumption of natural raw materials, growth in the number of landfills, environmental degradation, widening gap between real and key indicators of waste disposal and loss of industry's stability and efficiency.

The innovative scenario assumes a comprehensive, balanced development of the industry, return of recycled materials to economic turnover and the attraction of financing. It supposes the use of domestic high-tech resource-saving equipment and technologies and the development of effective infrastructure. According to this scenario, the successful development of disposal infrastructure depends on the integration of the waste management system with different sectors of the economy. The waste management system should cover all stages of the waste treatment, production and sale of domestic equipment and quality, environmentally friendly, competitive products from recycled materials.

As a result, a reduction in the amount of produced waste and the number of landfills should be achieved. Waste reduction will support efforts to promote a safer environment. Implementation of this scenario will facilitate the transition to sustainable socio-economic and industrial development, ensuring environmental security and improving the quality of life of the population.

17.5 SWOT ANALYSIS OF RUSSIAN HAZARDOUS WASTE MANAGEMENT SYSTEM

The ongoing reform of the sphere of hazardous waste management in Russia is in the initial stage, but some experts and stakeholders note 'blind spots' in the emerging system, which need to be improved, and the advantages of the system. The advantages and disadvantages as well as opportunities and threats of the developing hazardous waste management system are visualized through a SWOT matrix. Strengths and weaknesses are internal factors of the system, and opportunities and threats are external factors to the system, presented by the environment.

This SWOT analysis is based on information reported by the Bellona Foundation (Bellona, 2019), the Civic Chamber of the Russian Federation (OP RF, 2018), RosRAO and specialists of this sphere (Figure 17.3).

The existing waste management system can be described only as a framework, while its performance is still unclear. For example, the National Project 'Ecology', which includes a wide range of waste management issues, should be expanded by adding the information on technologies for the conversion of the chemical weapons disposal facilities, as well as for the construction of the new ones.

Moreover, the methods of collaboration of the Federal operator with manufacturers of waste disposal equipment are not defined (Bellona, 2019). At the same time, as the Strategy points out, there is a significant number of unused facilities of bankrupt industrial enterprises, which are potential waste recyclers, as well as of the best domestic technologies. The availability of accurate data on hazardous waste quantity and types, especially in regions, makes it possible to formulate their needs for disposal facilities, and therefore to plan funding. Also, it is reasonable to reflect information on available technologies for waste disposal on an industrial scale.

The State Information System should provide reliable and full information on the amount of waste being produced countrywide to clarify the process of logistical optimization. However, some experts already note that most enterprises produce a small amount of hazardous waste, so its transportation over long distances can be unprofitable for the operator (Bellona, 2019).

The Strategy for development of the industry for sorting, recycling, and treatment of waste for the period until 2030 describes the problem of e-waste treatment, but this description needs to be more detailed and encompass indicators of its processing. Given the specificity of this type of waste, it is

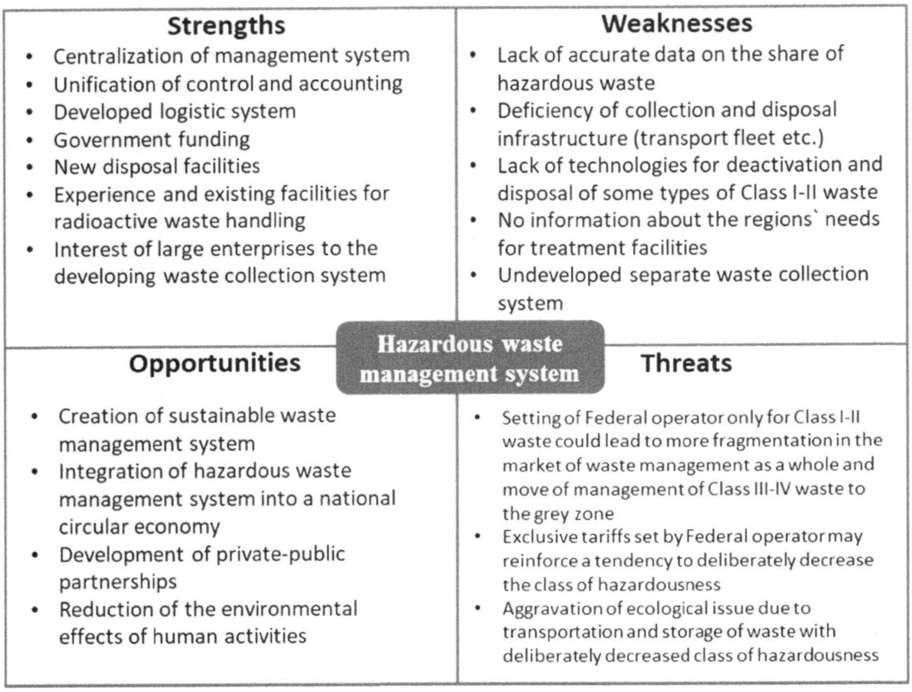

FIGURE 17.3 SWOT analysis of existing Russian hazardous waste management system. Source: Authors.

reasonable to consider it separately and create additional facilities for e-waste treatment (Volkova, 2018). The Federal operator is supposed to manage only Class I and Class II waste. Class III–V waste management system, as well as disposal tariffs. It means higher monopoly tariffs on Class I–II waste and poses the risk of a deliberate decrease of the class of waste hazardousness.

To sum up, most of the acute issues depend on the lack of information on hazardous waste generation. Hence, the effectiveness of the ongoing reform depends critically on the effectiveness of SIS aimed at controlling the enterprises' reports and illegal landfills. Hazardous waste management highly depends on waste sorting culture and existing infrastructure in households. In Russia, this sphere has potential for development and enlargement because societal initiatives in this field are local, not national or even regional as a rule. For instance, in the USSR, the population got payment for the collection of certain types of waste, which represented an economic stimulus. Today responsible garbage collection is based on personal convictions and feelings of personal obligation toward the environment. For this reason, the reform should also consider the creation of country-wide separate garbage collection infrastructure, which will allow setting apart the hazardous waste from municipal solid waste.

17.6 CASE STUDIES OF HAZARDOUS WASTE MANAGEMENT IN RUSSIA

Table 17.4 contains some of the best practices of hazardous waste management in Russia, provided by companies which produce the full cycle of hazardous waste management, equipment and technologies or ecological consulting, and by companies whose main activities do not involve waste management.

It is reasonable to assume that the collaboration of RosRAO with these companies could drive the efficiency of the building management system. These companies already have experience and essential technologies, as well as information about waste generation in their specific field of activity and regions. The use of their experience and best practices can accelerate positive changes.

TABLE 17.4
Hazardous Waste Management Companies

	Company	Legal Entity	Functions/Products	Customers
1	Megapolisresource	Limited liability company	Disposal of batteries, accumulators and electronic scrap	Companies Households
2	SOVECO	Limited liability company	Hazardous waste disposal, waste collection from national industrial enterprises; ecological consulting	Companies
3	Ecorecycle	Limited liability company	Collection and disposal of fluorescent lamps, mercury-containing lamps, bactericidal lamps, mercury-containing products and devices, including thermometers	Companies Households
4	EcoPromTechnologia	Limited liability company	Collection and disposal of mercury-containing waste	Companies
5	RusUtilit	Limited liability company	E-waste collection and processing	Companies
6	Vtorsintez	Limited liability company	E-waste collection, transportation and processing; extraction of precious metals	Companies
7	Ecoutilizatsiya	Limited Liability Company	E-waste collection, transportation and processing; extraction of precious metals	Companies Individual entrepreneurs
8	Eco Group	Limited liability company	Consulting on eco-friendly technologies issues and Class I–V waste management	Companies
9	RETECO	Limited liability company	Industrial waste disposal	Companies
10	EcoBox	Limited liability company	Production of containers for energy-saving lamps, batteries and thermometers	Companies Households
11	Green city	Closed joint-stock company	Disposal of I–III hazardous waste, first of all mercury and energy-saving lamps Provide social initiatives for waste collection	Companies
12	SIBUR (petrochemical company)	Public joint-stock company	Processing and disposal of own production waste Provide social initiatives for waste collection	Companies Households
13	Poligon	Joint-stock company	Disposal of toxic industrial waste	Companies
14	Kubantsetmet	Joint-stock company	Processing and disposal of mercury-containing waste	Companies
15	Gazenergostroy Corp.	Joint-stock company	Production of equipment for neutralization and processing of more than 3,500 types of waste with production of secondary raw materials; ecological consulting	Companies

Source: Authors.

Selected cases represent the current situation in Russia's hazardous waste management market. Cases include companies with a wide scope of functions, counting collection, transportation, storage and disposal, and highly specialized companies, which collect, transport or utilize waste, or produce technologies and equipment. In terms of types of hazardous waste, they also can be divided into those that specialize in specific types of waste (for instance, mercury-containing waste or e-waste), and those that process all types of Class I–II waste. Some of them, for example, SIBUR, are not recycling companies, but recycling is a part of the production process. Most of them collaborate with national enterprises, governmental organizations or businesses. There are fewer companies whose activities are aimed at working with households.

Below are descriptions of some noteworthy cases.

1. Public joint-stock company SIBUR is the largest vertically integrated petrochemical company in Russia. It implements the 'Green Wings' model, which assumes the 3Rs approach (reduce, reuse, recycle) throughout the value chain. Within the framework of this model, the company has been making progress in the following areas: Environmental safety based on the commodity model, eco-friendliness of production processes and products' eco-friendliness and recyclability. SIBUR procures and supplies petrochemistry's best available technologies designed to bring about further reduction in CO2 emissions and decreased consumption of natural resources (SIBUR, 2017).

   SIBUR's waste management strategy focuses on optimizing the generation, processing and disposal of its own production waste and on helping local communities implement measures to foster waste collection and recycling. Also, SIBUR actively supports Russia's transition to the circular economy principles through the support of the national project 'Sort it right', initiatives to install containers and automatic machines for separate solid waste collection in Moscow, etc. (MDPI, 2019).

   Waste management is a part of SIBUR's industrial processes. The feature of the company's activity is that SIBUR's petrochemical enterprises are mainly provided with raw materials of their production, obtained on the basis of hydrocarbon production by-products purchased from oil and gas companies.
2. Sovremennaya ecologia (SOVECO) has been engaged in waste management since 1994 and has experience in the field of ecology and industrial waste disposal. Waste disposal is the company's target and is carried out in coordination with Rosprirodnadzor and the Ministry for Civil Defense, Emergencies and Elimination of Consequences of Natural Disasters. The company collects, transports, deactivates and disposes of industrial waste, carries out the regeneration, purification and separation of complex mixtures of organic solvents, disposes of chemical and galvanic waste, paint-and-lacquer materials, etc.

   Among the company's customers are national industrial enterprises, for instance, Khrunichev State Research and Production State Center, the Almaz-Antey Corporation and medical centres, for instance, Bakulev National Medical Research of Cardiovascular Surgery.
3. Joint-stock company Poligon is a company responsible for waste management and landfill near Tomsk that works with enterprises all over the country. In 1984 the Ministerial Council of the USSR decided to organize a toxic waste landfill near Tomsk; its construction encompassed the experience of creating similar landfills in Sweden, Finland and Italy. In 1992 the landfill began accepting Class II–III waste, and in 1994 Class I waste. Currently, this landfill has facilities for the burying of Class I–IV waste, except liquid noncombustible waste and oil. The activity of the landfill decreased environmental threats, including to the Ob River basin.

   All of the hlandfill's facilities for disposing of toxic industrial waste have strong environmental isolation. The waterproofing used is inert to the effects of most aggressive liquids, precipitation, microorganisms, highly oxidizing agents, acids, mineral salts and alkalis. Moreover, environmental protection from Class I waste is achieved by the external and internal insulation of containers with 10 mm metal walls. Thus, the landfill's bunker with containers has five degrees of protection, which completely eliminate environmental pollution by hazardous waste. Monitoring of the landfill's impact on the environment is carried out twice a year and includes evaluation of the impact on underground and surface water, soil, vegetation and atmospheric air.

4. Kubantsetmet is more specialized and accumulates, deactivates and disposes of waste which includes mercury. A few years ago, the disposal facility processed up to 10,000 tons of waste per year; the volume decreased to 200–300 tons in 2018 because of the closure of several large chemical plants. The mercury-containing waste processing market is continue to decline; in 2018 Russia started to limit the use of mercury lamps, the use of compact fluorescent lamps was prohibited and restrictions have affected the use of high-pressure mercury lamps.
5. Gazenergostroy Corp. produces and sells equipment for the neutralization and processing of more than 3,500 types of waste with the production of secondary raw materials: Chemical industry waste, oil and gas production and processing waste, household garbage and transport companies' waste. The corporation's technologies can be applied to neutralize and process all types of hazards: Solid, liquid and slurred production and consumption waste, medical waste (excluding mercury-containing items) and biological waste.
6. Ecorecycle since 1992 has collected and disposed of fluorescent lamps, mercury-vapor lamps, bactericidal and other mercury-containing lamps of all configurations and sizes and mercury-containing products and devices, including thermometers, sells special containers for the collection, storage and transportation of fluorescent lamps and thermometers and transports collected waste by special vehicles. Among those who use its services are the State Duma, the Government of the Russian Federation, the Accounts Chamber of the Russian Federation, libraries, pharmacies and others in Moscow and the Moscow region (Yuzbekova, 2019).

   Every month Ecorecycle collects and disposes of 150,000–300,000 lamps (Yuzbekova, 2019); in Ecorecycle's small production facility mercury-containing lamps are separated into glass and mercury. The latter is transferred to Kubantsetmet, the only mercury-recycling facility with state-of-the-art technologies and equipment for recovering mercury and lead in Russia.
7. Megapolisresource is a group of companies, the leading Russian processor of batteries, accumulators and electronic scrap. It was the first company in Russia to begin to utilize batteries. The company's specialists integrate related technologies to dispose of waste: From organic chemistry to pyro- and hydrometallurgy and electrochemistry. Of the disposed materials of manganese-zinc batteries, 94.4% become resources again. Megapolisresource disposes of batteries and accumulators, non-ferrous scrap, etc.
8. Green City supplies containers for the storage and transportation of mercury lamps, and a convenient scheme for the reception of energy-saving lamps. The company uses Ecotrom-2, one of the most effective types of equipment, which allows the removal of at least 95–97% of the mercury from the lamps. From waste lamps the company gets glass and socles that can be reused. Green City has its own disposal site, a unique complex, the only one in the region with professional installations for the disposal of highly toxic waste.
9. EcoPromTechnologia – the main sphere of the company's activity is the production of non-metallic mineral products. Also, EcoPromTechnologia transports and recycles mercury-containing items, recycled materials.
10. The main area of RusUtilit's activity is the processing of computers and consumer electronics. More than 350 companies in Russia are partners of RusUtilit.
11. Vtorsintez processes computers, laptops, office equipment, printers, network equipment, consumer electronics, audio and video equipment, means of communication, medical equipment, bank equipment and cash registers.
12. Ecoutilizatsiya is a specialized company created for the development of e-waste management (computers, office technics, industrial equipment and other electrical scrap). Also, the company processes waste containing precious metals.
13. EcoBox is a specialized company producing containers for collecting energy-saving lamps, batteries and thermometers.

14. Eco Group provides a full range of environmental services, including ecological consulting, the development of recommendations and standards for waste generation and disposal for companies producing hazardous waste, assignment of waste to a specific hazard class, etc.
15. RETECO is a group of companies, certified for the dismantling of equipment, transportation, storage and disposal of industrial hazardous waste. RETECO has implemented more than 250 projects throughout Russia.

17.7 CONCLUSION

The development of an efficient hazardous waste management system is a part of sustainable development and the shift to a green and circular economy which implies that products are reused, remanufactured and recycled at the end of their life cycle. Nowadays Russia is moving towards the Industry 4.0 technologies, which enable the development of a circular economy and steps towards remanufacturing, reusing and recycling.

The hazardous waste management system in Russia is in the process of transformation and has great potential for further improvement. One of the most important advantages of the system is a tendency to centralize and streamline the system by the creation of the Federal State Unitary Enterprise RosRAO. Still, there are some weaknesses, such as tariffs and regulation, that need close collaboration among governmental offices on the federal, regional and municipal levels, private stakeholders and population. Moreover, the Russian regions are not equal in the amount of hazardous waste generated as well as resources needed for the establishment of a new waste disposal system.

One of the key elements of hazardous waste management in Russia should be the SIS for online verification of data, tracking of waste handling and detecting violations, which will be created by 2021 by RosRAO. It will also help to build efficient logistics, which is especially important given the geographical features of Russia and different situations with hazardous waste generation in regions. The main efforts will be directed to the modernization and construction of the processing technologies and facilities, the development of high-tech transport control mechanisms and modern recycling technologies.

One of the most potential directions of the hazardous waste treatment relates to production technologies of secondary raw materials from waste. The following fields are the top scientific and technological priorities: Development of energy-saving technologies for the deep processing of mineral and technogenic raw materials, technologies for the environmentally safe recycling of waste, the environmentally friendly deactivation of toxicants, the burning of combustible materials, the use of biomass as a source of energy and the processing of solid fuels. Another promising area of scientific and technological development is the production of domestic high-tech equipment for the disposal of sorted and unsorted waste, environmentally friendly and resource-saving waste handling and the disposal of extremely hazardous and highly hazardous waste (Decree of the Government, 2018).

Taking into account current volumes of produced e-waste, as well as national plans to develop the digital economy, which potentially will lead to an increase of e-waste, this sphere requires special attention.

In terms of societal activity in the waste management system, separate waste collection is the most obvious and basic way to improve this system. The separation of hazardous waste from the total volume of waste plays a key role. To make waste recycling environmentally friendly, it is necessary to ensure waste separation and sorting. Public participation needs to be supported by information and education activities.

In case of the implementation of the above-mentioned measures (support from the government, development and use of domestic technologies, return of recycled materials to economic turnover, attraction of financing), the innovative scenario described in the Strategy will be probable. Its implementation will require closer cooperation among departments from different sectors of the economy and already existing recycling companies. An effective up-to-date system should cover all stages of waste treatment, and the production of quality, environmentally friendly, competitive products from recycled materials.

ACKNOWLEDGEMENTS

The book chapter was prepared within the framework of the Basic Research Program at the National Research University Higher School of Economics.

BIBLIOGRAPHY

Baldé, C.P., Forti, V., Gray, V., Kuehr, R., Stegmann, P. (2017) The global E-waste monitor. United Nations University (UNU), International Telecommunication Union (ITU) & International Solid Waste Association (ISWA), Bonn/Geneva/Vienna, 1–109. URL: https://www.itu.int/en/ITU-D/Climate-Change/Documents/GEM%202017/Global-E-waste%20Monitor%202017%20.pdf (date of access: 02.04.2020)

Basel Convention. (1992) URL: http://www.basel.int/ (date of access: 05.04.2020)

Bellona. (2019) Management of hazardous waste in Russia and some EU countries. URL: https://network.bellona.org/content/uploads/sites/4/2019/02/waste_07.pdf (date of access: 10.04.2020)

Blaisi, N.I. (2019) Construction and demolition waste management in Saudi Arabia: Current practice and roadmap for sustainable management. *Journal of Cleaner Production*, 221, 167–175. URL: https://www.sciencedirect.com/science/article/pii/S0959652619306730?casa_token=NKPXJXtq67AAAAAA:MP_ZvQHQXyL8-FRZTb8BAMGwvUJt2RACbcrvXLsWOWGKihoFbaSVPB5MRKdLCNZSkVU5oFQEKVw (date of access: 19.04.2020)

Burmistrova, S. (2019) V pravitel'stve odobrili peredachu «Rosatomu» utilizaciyu opasnyh othodov. *RBK*. (The Government approved the transfer of the hazardous waste disposal to Rosatom) URL: https://www.rbc.ru/business/20/05/2019/5cdebaba9a79478f4c644fdd (date of access: 26.03.2020)

Chernyaeva, T.K. (2013) Actual`nye problem vliyania othodov proizvodstva i potreblenia na ob`ekty okruzauschey sredy (obzor). *Gigiena i sanitaria*. (Actual issues of waste production and consumption influence on environment and health of population) URL: https://cyberleninka.ru/article/n/aktualnye-problemy-vliyaniya-othodov-proizvodstva-i-potrebleniya-na-obekty-okruzhayuschey-sredy-i-sostoyanie-zdorovya-naseleniya-obzor (date of access: 27.04.2020)

Council for Strategic Development and Priority Projects. (2018) Pasport nacional'nogo proekta "Ekologiya". (National project "Ecology") URL: http://www.mnr.gov.ru/activity/directions/natsionalnyy_proekt_ekologiya/ (date of access: 30.03.2020)

Decree of the Government. (2018) Decree of the Government of the Russian Federation of 25 January 2018 N 84-r "On the approval of the strategy for development of industry for sorting, recycling, and treatment of waste for the period until 2030". URL: https://www.garant.ru/products/ipo/prime/doc/71767672/ (date of access: 05.04.2020)

EcoBox. Official website. URL: https://ecobox.ru/ (date of access: 19.04.2020)

Eco Group. Official website. URL: https://www.ecogrouponline.com/

EcoPromTechnologia. Official website URL: http://ecologytech.ru/

Ecorecycle. Official website. URL: http://www.rtutinet.ru/ (date of access: 19.04.2020)

Ecoutilizatsiya. Official website URL: http://www.ecoutilization.ru/company/

Ekopromcentr. (2019) Plan garmonizacii ekologicheskih trebovanij k opredeleniyu klassa opasnosti othodov. (Harmonization plan on environmental claims on definition of the hazard class) URL: https://ecopromcentr.ru/plan-garmonizacii/ (date of access: 05.04.2020)

Eurasian Economic Commission. (2019) Intergovernmental Council has adopted decisions in implementing the EAEU digital agenda, interaction of the Union countries in industrial and trade cooperation. URL: http://www.eurasiancommission.org/en/nae/news/Pages/09-08-2019-3.aspx (date of access: 25.03.2020)

European Commission. (2018a) Circular economy: New rules will make EU the global front-runner in waste management and recycling. URL: https://ec.europa.eu/commission/presscorner/detail/en/IP_18_3846 (date of access: 26.03.2020)

European Commission. (2018b) Hazardous waste treatment URL: https://ec.europa.eu/eurostat/statistics-explained/index.php/Waste_statistics#Hazardous_waste_treatment (date of access: 25.03.2020)

Executive Order N 536. (2014) On approval of criteria of classification od wastes as I-V hazard classes by the degree of negative impact on the environment. URL: http://docs.cntd.ru/document/420240163 (date of access: 28.03.2020)

Federal Law. (1998) Federal law of 24.06.1998 N 89-FL On production and consumption wastes. URL: http://www.consultant.ru/cons/cgi/online.cgi?from=342041-166&rnd=1515EA0A47A86F6923B192D9108929ED&req=doc&base=LAW&n=349705&REFDOC=342041&REFBASE=LAW#7hx3blvci1w (date of access: 05.04.2020)

Federal Law. (2002) Federal Law of 10.01.2002 N 7-FL On environmental protection. URL: http://www.consultant.ru/document/cons_doc_LAW_34823/ (date of access: 05.04.2020)

Federal Law. (2019) Federal Law of 26.07.2019 N 225-FL On amendments to the Federal Law "On production and consumption wastes" and the Federal Law "On the State Atomic Energy Corporation "Rosatom" N 225-FL. URL: http://www.consultant.ru/document/cons_doc_LAW_329988/ (date of access: 26.03.2020)

Fedotkina, O., Gorbashko, E., Vatolkina, N. (2019) Circular economy in Russia: Drivers and barriers for waste management development./*Sustainability*. URL: https://webcache.googleusercontent.com/search?q=cache:40nKHqVAZ8kJ:https://www.mdpi.com/2071-1050/11/20/5837/pdf+&cd=2&hl=ru&ct=clnk&gl=ru (date of access: 29.03.2020)

Gazenergostroy. Official website. URL: https://gazenergostroy.ru/en/unichtozhenie-otxodov.html (date of access: 19.04.2020)

Gil`mundinov, V.M., Tagaeva, T.O., Boksler, A.I. (2020) Analysis and forecasting of waste management processes in Russia. *Studies on Russian Economic Development*, 31, 92–98. URL: https://link.springer.com/article/10.1134/S1075700720010074 (date of access: 05.04.2020)

Global Navigation Satellite System. Official website. URL: https://www.glonass-iac.ru/en/ (date of access: 20.04.2020)

Global Recycling. (2019) Recycled construction waste in Europe. URL: https://global-recycling.info/archives/2964 (date of access: 05.04.2020)

Global Recycling. (2019) Saudi Arabia: Integrated waste management and recycling plan for Riyadh. URL: https://global-recycling.info/archives/3215 (date of access: 17.04.2020)

Gorbulina, A.R. (2019) Approaches to waste management in Russia p. 100–104. URL: https://elibrary.ru/item.asp?id=40392126& (date of access: 20.04.2020)

Gosstandart Rossii. (2002) Resources saving. Waste treatment. Terms and definitions. URL: http://docs.cntd.ru/document/gost-30772-2001 (date of access: 05.04.2020)

Government of the Russian Federation. (2014) Dmitry Medvedev approved the forecast prepared by the Ministry of Education and Science of Russia for the scientific and technological development of the Russian Federation for the period until 2030. URL: http://government.ru/news/9800/ (date of access: 26.03.2020)

Green City. Official website. URL: www.z-gorod.ru (date of access: 05.04.2020)

Japan-guide.com. (2019) Waste disposal. URL: https://www.japan-guide.com/e/e2222.html (date of access: 17.04.2020)

Kruter, M.S. (2011) Ugolovno-pravovoe I criminologicheskoe preduprezdenie narushenii pravil obraschenia opasnukh othodov. *Publichnoe i chastnoe parvo* 1(9), 80–91. (Legal and criminological prevention of violations of rules for hazardous waste management) URL: https://www.elibrary.ru/item.asp?id=16903472 (date of access: 05.04.2020)

Kubantsetmet. Official website. URL: http://kcvm.info/index.html (date of access: 19.04.2020)

Lebedeva, A.A. (2015) Metodika kompleksnoj ocenki sistemy obrashcheniya s othodami v naselennyh punktah Rossijskoj Federacii. RGGMU. (Methodology of integrated assessment of waste management system in Russia) URL: http://www.rshu.ru/university/dissertations/files/169/%D0%94%D0%B8%D1%81%D1%81%D0%B5%D1%80%D1%82%D0%B0%D1%86%D0%B8%D1%8F%20%D0%9B%D0%B5%D0%B1%D0%B5%D0%B4%D0%B5%D0%B2%D0%B0.pdf (date of access: 26.03.2020)

Markova, V., Morozenko, Y., Vakhrusheva, K. (2019) Hazardous waste management in Russia and the EU. *Bellona*, p. 19. URL: https://bellona.org/publication/hazardous-waste-management-in-russia-and-the-eu (date of access: 11.04.2020)

Mavropoulos, A., Wilson, D.C., Appelqvist, B., Velis, C., Cooper, J. (2014) Globalisation and waste management. *ISWA TASK Force*, 1–35. URL: https://webcache.googleusercontent.com/search?q=cache:_UZhSjXISKwJ:https://www.iswa.org/knowledgebase/tfgfinal+&cd=1&hl=ru&ct=clnk&gl=ru (date of access: 25.03.2020)

McCurry, J. (2019) Fukushima: Japan will have to dump radioactive water into Pacific, minister says. *The Guardian*. URL: https://www.theguardian.com/environment/2019/sep/10/fukushima-japan-will-have-to-dump-radioactive-water-into-pacific-minister-says (date of access: 21.04.2020)

MDPI. (2019) Circular economy in Russia: Drivers and barriers for waste management development. URL: https://www.mdpi.com/2071-1050/11/20/5837/htm (date of access: 05.04.2020)

Megapolisresource. Official website URL: https://eco2eco.ru/ (date of access: 05.04.2020)

Mereminskaya, E. (2019) Top-meneger "dochki" "Rosatoma": Vsego proizvoditsya 444 vida opasnykh othodov. *Vedomosti* (Rosatom`s top-manager: In total, 444 types of hazardous waste are produced) URL: https://www.vedomosti.ru/economics/characters/2019/11/26/817162-vsego-proizvoditsya-vida-opasnih-s-ezhednevno (date of access: 02.04.2020)

Mingaleva, Z., Vukovic, N., Volkova, I., Salimova, T. (2019) Waste management in green and smart cities: A case study of Russia. *Sustainability*, 12(1). URL: https://www.mdpi.com/2071-1050/12/1/94/htm (date of access: 17.04.2020)

Ministry of the Environment. (2018) 一般廃棄物の排出及び処理状況等(平成28年度)について. URL: https://www.env.go.jp/press/105322.html (date of access: 17.04.2020)

Mitina, N.N., Gnetov, E.M. (2020) Utilizaciya promyshlennyh othodov v Rossii i v mire: problemy i resheniya. *Neftegaz.ru* (Industrial waste disposal in Russia and in the world: Problems and solutions). URL: https://magazine.neftegaz.ru/articles/ekologiya/536780-utilizatsiya-promyshlennykh-otkhodov-v-rossii-i-v-mire-problemy-i-resheniya/ (date of access: 05.04.2020)

Nacional'naya Associaciya neftegazovogo servisa. (2020) Predpriyatiya Tatarstana pervymi v RF nachnut vesti uchet opasnyh othodov v 2020 godu. (National Oil and Gas Service Association (2020) Tatarstan enterprises will be the first to keep records of hazardous waste in 2020) URL: https://nangs.org/news/ecology/predpriyatiya-tatarstana-pervymi-v-rf-nachnut-vesti-uchet-opasnyh-othodov-v-2020-godu (date of access: 26.03.2020)

Nathanson, J.A. (2016) Hazardous waste management. *Encyclopedia Britannica*. URL: https://www.britannica.com/technology/hazardous-waste-management (date of access: 28.03.2020)

Nereim, V. (2019) Saudi Arabia plans new recycling facility to build up industry. *Bloomberg*. URL: https://www.bloomberg.com/news/articles/2019-09-01/saudi-arabia-plans-new-recycling-facility-to-build-up-industry (date of access: 19.04.2020)

Nikulichev, U.V. (2017) Upravlenie othodami. Opyt Evropeiskogo souza. Analiticheskii obzor. *RAN. INION*, 55. (Waste management: EU`s experience. Analytical review) URL: http://inion.ru/site/assets/files/1109/nikulichev_upravlenie_otkhodami.pdf (date of access: 05.04.2020)

NK Wood Research. (2018) Global hazardous waste management market forecast 2019–2027. URL: https://www.inkwoodresearch.com/reports/hazardous-waste-management-market/ (date of access: 28.03.2020)

OP RF. (2018) Obshchestvennaya palata RF Doklad «Analiz effektivnosti mer po obespecheniyu pererabotki tverdyh kommunal'nyh othodov i predlozheniya po obespecheniyu ucheta mneniya grazhdan Rossijskoj Federacii pri stroitel'stve ob"ektov, ispol'zuemyh dlya pererabotki ukazannyh othodov». 173. URL: https://www.oprf.ru/files/1_2019dok/Doklad_TKO_OPRF25012019.pdf (date of access: 05.04.2020)

Open Learn. (2017) Solid waste reduction, reuse and recycling. URL: https://www.open.edu/openlearncreate/mod/oucontent/view.php?id=80574&printable=1 (date of access: 26.03.2020)

Pan, C., Bolingbroke, D., Tsun Wai, N.K. (2019) The use of waste diversion indices on the analysis of Canadian waste management models. *Journal of Material Cycles and Waste Management*, 21, 478–487. URL: https://link.springer.com/article/10.1007/s10163-018-0809-3 (date of access: 26.03.2020)

Poligon. Tomsk. Official website. URL: http://poligon.tomsk.ru/ (date of access: 19.04.2020)

President of Russia. (2016) State Council meeting on Russia's environmental development for future generations December 27, 2016. URL: http://en.kremlin.ru/events/president/news/53602 (date of access: 17.04.2020)

President of Russia. (2018) The President signed executive order on national goals and strategic objectives of the Russian Federation through to 2024. URL: http://en.kremlin.ru/events/president/news/57425 (date of access: 17.04.2020)

REO Radar. Official website. URL: https://radar.reo.ru/ (date of access: 26.03.2020)

RETECO. Official website URL: http://reteco.ru/

Romanov, V.I. (2016) Othody Rossii vnachale XXI veka: obzor, analiz, prognozy: spravochnoe posobie. Moskva: Rossiskaia ecologicheskaia academia, 2016, 127–132. (Waste in Russia in the beginning of XXI: Overview, analysis, forecasts)

Rosprirodnadzor. (2019) Svedeniya o deyatel'nosti v oblasti obrashcheniya s othodami. 1–14. URL: https://rpn.gov.ru/upload/iblock/789/789106f5651e63e4ad6b4ae61c51e53e.pdf (date of access: 05.04.2020)

RosRAO. Official website. URL: https://reo.ru/ (date of access: 26.03.2020)

Rotterdam Convention. (2004) URL: http://www.pic.int/ (date of access: 05.04.2020)

RusUtilit. Official website. URL: https://rusutilit.ru/

Sant'ana, L.P. (2015) Hazardous waste management in Japan: A report. *UNISANTA BioScience*, 4. URL: https://www.researchgate.net/publication/303445578_Hazardous_Waste_Management_in_Japan_A_report (date of access: 17.04.2020)

Shumskikh, Yu. (2012) Current state of hazardous waste and substances law / Intelligence. *Innovation. Investments*. Buzuluk humanitarian and technological institute, branch of Orenburg State University, 4, 164–166 URL: https://www.elibrary.ru/item.asp?id=18780638 (date of access: 29.03.2020)

SIBUR. (2017) Sustainability openness responsibility report. URL: https://www.sibur.ru/upload/iblock/627/627b3f529f629a5c994abb40be4a687c.pdf (date of access: 19.04.2020)

Sistema obrashcheniya s othodami v Yaponii. (2019) (Waste management system in Japan) URL: https://zen.yandex.ru/media/axe_machinery/sistema-obrasceniia-s-othodami-v-iaponii-5c7f866638124300b42bd7d2 (date of access: 17.04.2020)

SOVECO. Official website. URL: http://www.soveco.ru/ (date of access: 19.04.2020)

State Atomic Energy Corporation Rosatom. URL: https://www.rosatom.ru/en/ (date of access: 26.03.2020)

Stiehler, A. (2018) Longer term investments Waste management and recycling. *UBS*, 1–10. URL: https://www.ubs.com/content/dam/WealthManagementAmericas/documents/waste-management.pdf (date of access: 25.03.2020)

Stockholm Convention. (2004). URL: http://pops.int/ (date of access: 05.04.2020)

Takashi, A. (2018) Current state and trend of waste and recycling in Japan. *International Journal of Earth & Environmental Sciences*, 1–11. URL: https://doi.org/10.15344/2456-351X/2018/155 (date of access: 19.04.2020)

Technavio. (2018) Global hazardous waste management market. URL: https://www.technavio.com/report/global-hazardous-waste-management-market-analysis-share-2018 (date of access: 27.03.2020)

Timofeeva, S.S., Timofeev, S.S., Sheshukova, L.V. (2013) Professional`nye riski personala poligonov po zahoroneniu tverdyh bytovyh othodov. *Vestnik Irkutskogo gosudarstvennogo tehnicheskogo universiteta*, 2(73), 84–88. URL: https://www.elibrary.ru/item.asp?id=18848819& (date of access: 17.04.2020)

UN. (2010) Japan Waste management. Sustainable development goals. URL: https://sustainabledevelopment.un.org/content/documents/dsd/dsd_aofw:ni/ni_pdfs/NationalReports/japan/Waste_Management.pdf (date of access: 17.04.2020)

Vasileva, A. (2019) Naiti I obezvredit`. *Kommersant* (To find and defuse. *Kommersant*) URL: https://www.kommersant.ru/doc/4048063 (date of access: 02.04.2020)

Vasilevskaya, D. (2018) V ozhidanii ratifikacii. *Bellona*. (Pending ratification / Bellona) URL: https://bellona.ru/2018/07/06/v-ozhidanii-ratifikatsii/ (date of access: 10.04.2020)

Volkova, A.V. (2018) Rynok utilizatsii othodov. *NIU VSHE, Tsentr rasvitia*, 1–87. (Waste disposal market) URL: https://dcenter.hse.ru/data/2018/07/11/1151608260/%D0%A0%D1%8B%D0%BD%D0%BE%D0%BA%20%D1%83%D1%82%D0%B8%D0%BB%D0%B8%D0%B7%D0%B0%D1%86%D0%B8%D0%B8%20%D0%BE%D1%82%D1%85%D0%BE%D0%B4%D0%BE%D0%B2%202018.pdf (date of access: 23.04.2020)

Voloshina, V. (2019) Ne pererabotany I ochen opasny. *Strana, Rosatom*. (Non-recycled and highly dangerous). URL: http://strana-rosatom.ru/2019/07/02/%D0%BD%D0%B5-%D0%BF%D0%B5%D1%80%D0%B5%D1%80%D0%B0%D0%B1%D0%BE%D1%82%D0%B0%D0%BD%D1%8B-%D0%B8-%D0%BE%D1%87%D0%B5%D0%BD%D1%8C-%D0%BE%D0%BF%D0%B0%D1%81%D0%BD%D1%8B/ (date of access 12.04.2020)

Vtorsintez. Official website. URL: http://vtorsintez.ru/

Xinhua. (2019) Feature: Garbage sorting, a basic survival skill of Japanese. URL: http://www.xinhuanet.com/english/2019-07/08/c_138209510.htm (date of access: 22.04.2020)

Yuichi, I., Kenji, T. (2018) Where did our NIMBY go? Graduate School of Economics, Kobe University. URL: http://www.econ.kobe-u.ac.jp/activity/publication/dp/pdf/2018/1818.pdf (date of access: 13.04.2020)

Yuzbekova, I. (2019) Kak pererabatyvayut lampy i drugie rtutnye othody. *RBK* (How lamps and other mercury-containing waste are recycled). URL: https://www.rbc.ru/trends/green/5d6662ef9a7947434de36f14 (date of access: 19.04.2020)

Zholobova, E. (2019) «RosRAO» o sisteme utilizacii opasnyh othodov v Rossii — glavnoe. *7h7 Kirovskaya oblast*. URL: https://7x7-journal.ru/articles/2019/12/05/rosrao-kak-v-rossii-budet-rabotat-sistema-utilizacii-opasnyh-othodov-glavnoe (date of access: 26.03.2020)

18 The Impact of Hazardous Waste and Its Management in China

Zhiyang Xu, Qian Li and Yunlong Zhou

CONTENTS

18.1 INTRODUCTION

Currently, hazardous waste is one of the main global ecological hazards and the primary cause of environmental disasters. As hazardous waste is made up of corrosive, toxic, flammable, reactive or infectious solid or liquid waste (Trivedi et al., 2015), improper management, storage or transportation can result in human health hazards and environmental air and soil pollution (Higgins, 2017). As most countries have strict hazardous waste treatment laws and regulations (Callao et al., 2019), efficient hazardous waste management is vital to protect human and ecological health and mitigate environmental disaster risk.

The growth in both the global economy and the global population has led to a commensurate increase in medical and industrial hazardous waste, both of which are significant potential hazards to human and environmental health. It has been estimated that in the past few decades, around 90% of the hazardous waste has been generated from only 30 highly developed or industrialized Organization for Economic Cooperation and Development (OECD) countries.

Unfortunately, because of differences of development stage, economic and technical skill, there is a significant hazardous waste management gap between developed and developing countries (Tang et al., 2015). Further, there has been a persistent trend by developed nations to dump hazardous waste for processing in developing countries in which the economic benefits from accepting this waste often take precedence over the environmental health concerns. Therefore, while the management responsible for the transferred hazardous waste focus on the capital input and raw material benefits, they often lack the proper facilities or technology to properly treat or manage the hazardous waste, which has resulted in an increase in associated environmental disasters.

As the largest developing country and the world's second largest economy, China produces and uses significant hazardous waste (Mian et al., 2017). For example, there are nearly 3,000 hazardous chemicals produced and used in China, including 148 which are highly toxic (Wang, 2018). China produces around 8 billion tons of solid waste per year, of which about 40 million to 60 million tons are hazardous waste. Therefore, the hazardous waste management issues are complex in China as effective, efficient and safe hazardous waste treatment systems are necessary to protect human and ecological health and prevent environmental accidents.

This paper summarizes the current situation, types and hazards of hazardous waste in China, then analyses the various kinds of hazardous waste disposal method and the project operation model in China; finally, the problems that exist in the hazardous waste management in China are analysed, and countermeasures are given. It is of great significance to the improvement of the environment, the health of residents and the sustainable development of the economy.

18.2 OVERVIEW OF HAZARDOUS WASTE IN CHINA

Hazardous waste is a kind of special waste, which has the characteristics of flammability, explosion, strong corrosiveness and toxicity (Duan et al., 2008). It brings great harm and influence to life and the social environment (Li et al., 2002). According to the relevant laws and regulations issued in China, hazardous waste refers to waste with hazardous characteristics identified by the hazardous waste identification standards and methods listed in the national hazardous waste list. Therefore, special attention should be paid to the treatment of hazardous waste to avoid its harm to life and environment.

There are many kinds of hazardous waste and many classification methods. The hazardous waste can be divided into six categories: Inorganic waste, waste oil, organic waste, corrosive organic waste, large-quantity and low-harm waste and other wastes by using the classification system combining catalogue method and hazardous waste identification criteria; according to the different physical forms, it can be divided into solid, liquid, gas, sludge, mud floccule, barrel hazardous waste, etc.; according to the chemical elements contained in hazardous waste elements, it can be divided into clean hazardous waste, waste producing gaseous pollutants, waste containing heavy metals and waste containing alkali metals; according to the hazardous characteristics of hazardous wastes, hazardous wastes can be divided into flammable hazardous wastes, corrosive hazardous wastes, reactive hazardous wastes, leaching toxic hazardous wastes, acute toxic hazardous wastes and toxic hazardous wastes. In China, hazardous solid waste is generally divided into 47 categories, with a total of more than 600 kinds (Li et al., 2012). Among them, industrial and medical wastes are very common. From the perspective of production sources, this kind of waste can be divided into two types. One is industrial waste, the other is social waste. In terms of nature, this kind of waste can be divided into four major categories. The first is inorganic waste, the second is oil waste, the third is organic waste, and the last is other harmful waste. In summary, there are different treatment methods for each kind of hazardous waste. Of the 47 kinds of hazardous waste, 57.7% of the total waste comes from five main types. The first is waste alkali, the second is waste acid, the third is inorganic oxide, the fourth is copper-containing waste, and the fifth is inorganic cyanide. In recent years, with the rapid development of China's industrialization level, varieties of types of waste have been produced. At the end of 2018, China's annual output of hazardous waste was about 35 million tons, and there are various types of hazardous waste (Gu et al., 2014). Generally speaking, there are two kinds of hazardous waste, liquid and solid. Most of this hazardous waste is toxic, corrosive and infectious, which is very harmful to the human body. Compared with other conventional pollutants, hazardous waste has a great impact on economic development. In order to solve the problem of hazardous waste, in 2004, China set up relevant laws and regulations, and the law enforcement has been strengthened year by year, promoting the management and treatment efficiency of hazardous waste in China.

18.3 IMPACT OF HAZARDOUS WASTE

With the rapid development of China's national economy, the accompanying problems are increasingly prominent, and the growth trend of hazardous waste is worrying (Orloff and Falk, 2003). Many waste hazards are long-term and may lie dormant for a long time. In the event of leakage of hazardous waste, the air will be polluted by uncontrolled burning, wind-blowing, sublimation and weathering. The polluted atmosphere, on the one hand, can lead to indirect poisoning of people and animals, and serious combustion and explosion accidents. On the other hand, it can also pollute the soil by means of precipitation, surface and underground water. Over time, beyond the self-cleaning ability of the environment, there will be long-term, unpredictable consequences (Sun and Wu, 2007). The hazards caused by hazardous waste are as follows:

A. The random placement, storage and unscientific treatment of hazardous wastes directly cause environmental pollution. First, hazardous wastes can pollute the air, and even corrode steel and buildings. During the process of accumulation, the small particles and dust inside the hazardous waste, which react with the substances in the atmosphere to produce toxic and harmful substances and malodorous gases, evaporate into the atmosphere. Moreover, hazardous wastes without proper incineration can produce a highly toxic substance – dioxins. Second, leachate can be produced due to the random placement of hazardous waste, which can pollute groundwater and directly endanger human health. Third, the random placement of hazardous wastes not only occupies a large amount of land resources, but also damages the surrounding plant environment and breaks the microbial ecological balance in land resources. As a result, the microbial degradation ability of toxic and harmful substances in the soil is weakened and the purification ability of soil is reduced

B. Hazardous wastes are harmful to human health through the accumulation in the food chain. Direct human contact with hazardous wastes including inhalation, skin absorption, eye contact and so on brings about poisoning, worsening disease, genetic mutations and other hazards. In addition, it can also cause burning, explosion and other events. For human health, hazardous wastes are permanent, latent and even irreversible.

C. Long-term development of the economy can be damaged because of the harm to the human living environment and human health. At the present stage, the problems of water pollution and soil pollution have seriously restricted the sustainable development of China. Nowadays, the frequent problems of soil pollution, water pollution and air pollution cause immeasurable economic losses to the country every year

18.4 TREATMENT AND DISPOSAL TECHNOLOGY OF HAZARDOUS WASTE IN CHINA

The principles of hazardous wastes treatment are reduction, resource, harmlessness (Lin et al., 2020). In 1983, the American National Academy of Sciences proposed a three-stage model for hazardous wastes treatment. The first stage is source treatment, that is, reducing the output of hazardous waste from the source. The second stage is the disposal of hazardous waste. At last, the third stage is the end disposal of waste. Compared with foreign developed countries, hazardous waste disposal technology in China started late and developed slowly, so it is trying to catch up with the pace of foreign countries. At present, the technologies used in hazardous waste disposal projects in China mainly include comprehensive utilization (resource-based), physical, chemical, biological methods, solidification, incineration and safe landfill, as well as the emerging collaborative disposal of cement kilns, plasma melting, etc. The current pattern of hazardous waste disposal in China is mainly divided into large-scale centralized disposal centres built by the state (mainly for comprehensive utilization, landfill and incineration), relatively small-scale centralized disposal facilities built by local governments (mainly for landfill and incineration), and there are also state-supported and encouraged enterprises independently building disposal facilities for their own use (some socialized services can be provided).

18.4.1 Comprehensive Utilization

In theory, comprehensive utilization technology can be divided into energy substitution technology, physical and chemical treatment technology, material separation and recovery technology, material recovery and land reduction technology. At present, the domestic comprehensive utilization of hazardous waste is mainly focused on the metal, waste organic solvents, waste acid alkali waste, waste oil, the recycling of electronic waste, mainly using the physical and chemical treatment technology, and material separation and recovery technology, typically with cleaning, drying, crushing, separation, neutralization, flocculation and sedimentation, oxidation-reduction, crystallization, sintering, pyrolysis, distillation, adsorption, biological treatment, magnetic separation/electric separation, etc. Taking waste solvent and waste mineral oil as examples, the recovery and utilization of waste solvent mainly include adsorption, absorption, condensation, separation and purification. The recycling of waste mineral oil can be divided into two types: Purification (sedimentation, centrifugation, filtration and flocculation, etc.) and refining (mainly distillation). The advantages of the comprehensive utilization of hazardous wastes lie in the recycling of resources; the disadvantages are that the price of the resource-based products fluctuates greatly, the profit rate is not high and the domestic comprehensive utilization technology is relatively rough, and the regeneration process may bring secondary pollution. According to the *National Statistical Yearbook*, the proportion of industrial hazardous waste recycling disposal in 2016 was about 50% (Xu et al., 2019).

18.4.2 Pretreatment

The ultimate purpose of hazardous waste disposal is to make it harmless. However, in the process of disposal, some hazardous wastes may be corrosive, reactive, toxic or have other hazardous characteristics due to their unstable nature and state. Therefore, pretreatment of hazardous waste is necessary to make it harmless. The usual pretreatment means include changing its physical and chemical properties, reducing toxic effects, etc. According to the current commonly used pretreatment technology, it mainly includes physical, chemical, biological, curing and other technologies (Li, 2003).

18.4.2.1 Physical Treatment

Physical treatment is changing the structure of hazardous waste through concentration or phase change, so that it can be easily transported, stored, utilized or further disposed of (Gayathr and Kamala, 2005). Physical treatment methods mainly include adsorption, compaction, crushing, separation, extraction, solid-liquid separation, dialysis, cooling crystallization, reverse osmosis, air blowing off, etc. After the physical treatment of hazardous waste, the volume of hazardous waste can be greatly reduced, so that the solid and liquid hazardous waste can be easily transported, stored and reused, which is generally suitable for the treatment of sludge, industrial residue and other wastes with high moisture content.

18.4.2.2 Chemical Treatment

Chemical treatment refers to the treatment of hazardous wastes by chemical methods to change the chemical properties of the hazardous substances, or change them into forms suitable for further disposal (Guan, 2002). Its purpose is to change the chemical properties of the treated substance, reduce its harmfulness and change it into a form suitable for further treatment and disposal. It is the main pretreatment method before hazardous waste enters the final disposal stage, and is generally used to treat inorganic waste, such as waste acid, alkali, heavy metal, cyanide, emulsified oil, etc. For solid waste (or waste residue), common chemical treatment methods include heat treatment (pyrolysis, incineration), stabilization, etc.; while for liquid waste (or waste liquid), common chemical treatment methods include acid-base neutralization, oxidation-reduction, chemical precipitation, ion exchange, chemical adsorption, incineration, etc. The popular disposal technologies include rapid carbonation, plasma gasification and supercritical water oxidation.

The rapid carbonation technology was first proposed by Seifritz in 1990. Its main working principle is to expose the waste to an environment with a high concentration of CO_2, so that its reaction speed is faster and faster. This method was originally used in mineral carbonation. Most of the industrial wastes can react with CO_2, such as steel slag, waste lime, carbide slag, waste building materials, etc. About 80% of the heavy metals will be consumed after the carbonation reaction due to the high content of heavy metals in these substances. There are about 0.4–0.5 million tons of incineration fly ash produced in China's industry every year, which contains a certain amount of trace heavy metals. Some scholars have analysed the parameters such as the gas-solid ratio, carbonation time and pH value of carbonation reaction, and found that the sodium chloride and potassium chloride in the fly ash will test the experimental device, resulting in loose fly ash structure and further precipitation of heavy metals such as zinc and copper, so it is necessary to wash the fly ash. The fly ash after carbonation reaction can be used as building materials with good application effect. In addition, the carbon dioxide discharged from the cement kiln can continue to be used as the raw material for reaction, and there is a large amount of calcium carbonate in the fly ash, which can play the role of carbon dioxide storage.

Plasma gasification technology is a new natural and harmless hazardous waste treatment technology. Its working principle is to convert hazardous waste into a CO and H_2 gas mixture in a high-temperature anoxic environment. At present, the technology is mainly used in the treatment of agricultural straw waste and urban waste. Europe and other regions have used this technology in the treatment of petroleum fertilizer, construction waste and medical waste. This hazardous waste can be decomposed, evaporated and oxidized in a high-temperature environment because of its flammability, but the requirements for equipment are high. Compared with other treatment technologies, this technology does not require much water content in the treated materials. It can not only effectively transform sewage sludge into useful resources and reduce the solid volume, but also decompose and repair the toxic substances in the inert slag.

Supercritical water oxidation technology has been widely used in Europe, America, Japan and other developed countries, mainly in the treatment of organic wastewater, plastic degradation and biological sludge. Austin established a commercial treatment plant as early as 1995, which is mainly used for the treatment of long-chain organics and amines. The process of this method is relatively simple, but it will be affected by many factors. In the process of supercritical water oxidation, if the organic matter in the medium water reaches 2%, the oxidation process can achieve complete self-heating, which can effectively save energy compared with other methods.

18.4.2.3 Biological Treatment

Biological treatment is using the metabolism of microorganisms, animals and plants to degrade the organic substances contained in hazardous wastes, which is mainly limited to the treatment of organic waste liquid and wastewater. Compared with other pretreatment methods, biological treatment can not only degrade hazardous waste, but also realize resource recovery to a certain extent, which is mainly reflected in biogas production, fertilizer production and other aspects. The disadvantages of biological treatment are long treatment cycles, poor stability and the risk of pollutant diffusion to the environmental soil and water bodies in the process of treatment. It is necessary to record and control the process in detail.

18.4.2.4 Curing/Stabilizing Treatment

Solidification/stabilization is a process in which hazardous waste is fixed or encapsulated in a dense inert solid substrate by physical and chemical methods to achieve stabilization and minimize the migration rate of pollutants in the waste. The solidification/stabilization of hazardous waste is generally regarded as a necessary pretreatment step before safe landfill disposal. Solidification/stabilization technology is mainly used to treat wastes that are not suitable for incineration or inorganic treatment, such as radioactive wastes, concentrate, sludge containing heavy metals, incineration fly ash and slag, etc. Common treatment methods include cement solidification, asphalt solidification, glass solidification, plastic solidification, organic polymer solidification, etc.

In the 1980s, the relatively stable solidification technology was applied in solid waste disposal by some old industrial countries. At first, it was used in the treatment of radioactive substances such as chromium slag and electroplating sludge. After good application effect, it was gradually popularized and used. At present, this curing technology is generally used in the stabilization of various heavy metals. There are many kinds of solidification, such as cement solidification, glass solidification and medicament solidification. It is proved by experiments that this kind of solidification treatment can meet the safety landfill standard in the disposal of incineration residues. The defect of this method is that it cannot effectively treat sludge with low oil content, mainly because the sludge has low oil content, high water content and large particles. The improved solidification treatment can effectively solve this problem and provide feasible measures for the treatment of large-scale oily sludge in oil fields.

18.4.3 Sanitary Landfill

Among the current disposal technologies, landfill is the most widely used disposal method. The common method is to reduce the volume of hazardous waste through physical disposal technology, then pave it into a thin layer, and compact it and finally cover it with a soil layer of a certain thickness. The landfill method mainly includes site selection, landfill design, landfill construction, environmental protection and other aspects (Slack et al., 2005). The landfill method consists of two parts: Sanitary landfill and safe landfill. For ordinary garbage and urban garbage in daily life, sanitary landfill is the main treatment method; for other special hazardous wastes, safe landfill technology is needed. Among them, safe land burying requires higher treatment requirements and stricter treatment technology. For example, a more standardized treatment site is needed, the leaching solution of hazardous waste needs a special collection system and centralized treatment, the groundwater level is strictly controlled, and the gas produced by hazardous waste can only be discharged after centralized treatment. At present, it is generally believed that the safe landfill technology is the final disposal method of hazardous waste at home and abroad. In order to control, reduce and eliminate the hazards of hazardous wastes after pretreatment (such as dehydration, neutralization, composting, solidification and stabilization), safe landfill is the final disposal process of hazardous wastes.

The core of the hazardous waste landfill is the anti-seepage system. The main function of the landfill is to block the hydraulic connection between the waste and the external environment, that is, to prevent the leachate produced by the waste from polluting the groundwater and other surrounding water bodies. The common anti-seepage system adopts a double artificial lining layer. From the bottom to the top, the layers are the foundation layer, groundwater drainage layer, compacted clay lining layer, high-density polyethylene film (HDPE), protective layer on the film, secondary drainage layer of leachate, high-density polyethylene film (HDPE), protective layer on the film, primary collection and drainage layer of leachate, geotextile and hazardous waste.

Generally speaking, the main advantages of the landfill method include: The treatment method is universal, and the cost is less, and at the same time, it is not limited by the types of waste, it is suitable for the disposal of large-scale waste, and the land after landfill can be used to build some public facilities such as activity sites. There are also some shortcomings in the safe landfill method, mainly because the leaching solution from hazardous wastes may leak and pollute the soil and groundwater. In addition, the landfill method occupies a lot of valuable land resources.

18.4.4 Heat Treatment

The heat treatment of hazardous waste is a comprehensive process of high-temperature decomposition and deep oxidation, which can completely oxidize and decompose combustible hazardous waste, so as to greatly reduce volume and completely remove toxicity (Delgado, 2007).

18.4.4.1 Rotary Kiln Incineration

The structure of 'rotary kiln + second combustion chamber' is generally adopted for the incineration of hazardous wastes. The operation temperature of the rotary kiln is 800–1,000°C, the retention time of waste in the kiln is 40–75 min, and when it reaches the outlet, it is completely transformed into high-temperature flue gas and ash. The high-temperature flue gas enters into the secondary combustion chamber and is further fully combusted under the action of excess air. The residence time is more than 2 s to ensure the full decomposition of trace organic matter and dioxins at high temperature. The ash is discharged from the rotary kiln and cooled before entering the ash bin. Generally, the heat recovery and purification process of 'waste heat boiler + quench + deacidification of slaked lime + activated carbon injection + bag filter + flue gas washing' is adopted for rotary kiln incineration. After the flue gas leaves the second combustion chamber, it enters the waste heat boiler, the temperature is reduced from 1,100–1,200°C to about 550°C, and then it enters the quench tower. In the quench tower, the evaporation of atomized water drops makes the flue gas temperature drop to 200°C in an instant (generally about 0.6 s). Because the residence time of flue gas between 200–500°C is less than 1 s, thus, the re-synthesis of dioxins is prevented. The deacidification of slaked lime is mainly to remove the acid gas, and the activated carbon injection is mainly to adsorb the toxic and harmful components.

The rotary kiln incineration technology of hazardous wastes is mature. Compared with other technologies, the rotary kiln is simple in operation, stable in operation, convenient in control and good in continuity. However, in the actual operation process, there are still many problems, such as poor adaptability of materials, serious slagging, frequent start-up and stop and excessive discharge.

The secondary treatment of fly ash is the main problem that limits the use of rotary kiln incineration technology. The fly ash produced by rotary kiln incineration accounts for 3–5% of the hazardous waste disposal. According to the relevant national standards, it is still hazardous waste, and is generally treated by safe landfill. However, due to less and less available land resources and many restrictions on the site selection of hazardous waste landfill, it is more and more difficult to dispose of the incineration fly ash in a safe way. At present, there are also a variety of incineration fly ash reduction, harmless and resource-based disposal technologies, such as solidification technology, heavy metal chelating stabilization treatment technology, melting solidification treatment technology and resource-based treatment technology for building materials or ceramics, among which solidification technology and heavy metal chelating stabilization treatment technology are most widely used, but these two technologies are still used for incineration fly ash after treatment. Landfill treatment is required, and there is a risk of heavy metal leaching when environmental factors change. Therefore, the rotary kiln incineration technology of hazardous wastes is essentially a reduction treatment, and the harmless treatment is not thorough enough. The scale of rotary kiln incineration disposal technology is generally small, usually 15–40 t/d. The high investment cost, long construction period (two to three years) and long ramp-up period (one to two years) also limit its application.

18.4.4.2 Collaborative Disposal of Cement Kiln

The collaborative disposal of hazardous wastes in cement kilns refers to the process of putting hazardous wastes that meet the requirements of the kiln after pretreatment into the cement kiln or cement mill to realize the harmless disposal of hazardous wastes for clinker or cement production. Developed countries, such as Europe, America and Japan, have been using cement kilns to dispose of solid waste in coordination for more than 30 years, with mature technology. The collaborative disposal of solid wastes by cement kilns in China started relatively late. Since 2015, the government has vigorously supported the collaborative disposal of solid wastes by cement kilns. *The guidelines for the review of the business license for collaborative disposal of hazardous wastes by cement kilns* issued in May 2017 standardized the approval of the license for the collaborative disposal of solid wastes by cement kilns and promoted the development of collaborative disposal.

The specific operation steps are as follows: The hazardous waste enters the cement rotary kiln through the dosing point after the pretreatment system, the temperature in the kiln is about 1,450°C, the gas stays in the kiln for more than 4 s, the material stays for more than 30 min, the gas-solid two-phase mixture is uniform, and the hazardous waste material is completely incinerated. The cement kiln is in alkaline atmosphere, which can absorb the acid gas produced in the waste incineration process.

The advantages of cement kiln collaborative disposal technology are as follows: (i) Complete harmless treatment. The hazardous wastes are disposed of by the cement kiln in coordination, the ash and slag directly enter the cement clinker, and the heavy metal elements are solidified in the mineral lattice of the cement clinker, which can realize the thorough harmless treatment of the hazardous wastes and avoid secondary treatment. The cement clinker enters the grate cooler after being discharged from the kiln. Within a few minutes, the clinker is suddenly cooled from 1,300–1,400°C to below 100°C, and the temperature of the waste gas is about 200°C, crossing the generation range of dioxin (200–500°C), avoiding the re-synthesis of dioxin in the waste gas. (ii) Large amounts of hazardous waste can be disposed of. The cement kiln for the collaborative disposal of hazardous wastes is a new dry process cement kiln with a designed clinker production scale of not less than 2,000 tons/day. The proportion of hazardous wastes with different calorific value and composition cannot be higher than 22% of clinker production capacity. If done at 10%, the quantity of hazardous wastes that can be disposed of by a 2,000 t/d cement kiln production line is 200 t/d. Therefore, the collaborative treatment of cement kiln is suitable for the treatment of large amounts of hazardous wastes with stable performance. (iii) The construction period is short, and the capacity is released quickly. The EIA period of the collaborative disposal system of a hazardous waste cement kiln is about 6 months, the reconstruction period is about 8–10 months, the ramp-up period of production capacity is short, and the release rate of production capacity is fast.

The disadvantages of cement kiln collaborative disposal technology are as follows: (i) There are strict requirements for hazardous waste materials. It is forbidden to dispose of radioactive waste, explosive and reactive waste, unwrapped electronic waste, mercury-containing thermometers, sphygmomanometers, fluorescent tubes and switches, chromium slag and unknown waste with unknown characteristics. In the actual operation, the collaborative disposal of the hazardous wastes with high content of sulphur and chlorine and the hazardous wastes with strong acidity need to be strictly tested and trial fired in the kiln, and the proportion of dosing should be strictly controlled in the operation process. (ii) Potential safety hazards: Various trace or excess elements contained in the hazardous wastes and their recombinants have an impact on the long-term safety performance of the cement products, or cause corrosion and harm to the cement production system equipment, resulting in safety problems.

18.4.4.3 Plasma High-Temperature Melting Treatment

The technology of plasma treatment of hazardous waste is to use high-temperature hot plasma (temperature in the furnace is 1,500–1,700°C) to decompose and destroy hazardous waste rapidly, in which organic matter is decomposed into combustible small molecular matter, and inorganic matter is melted at high temperature to form vitreous-like residue. Besides, the plasma melting furnace is divided into a plasma arc melting furnace and a plasma torch melting furnace. It was first used for the treatment of biochemical weapons and highly toxic hazardous wastes. It is recommended to be used for the treatment of hazardous wastes with high toxicity and stable chemical properties, which can exist in the environment for a long time. It is especially suitable for the disposal of fly ash after waste incineration, smashed electronic waste, liquid or gaseous toxic hazardous waste, etc. It has poor adaptability to the complex shape, not easy to break into small particles of waste.

The specific operation steps are as follows: The hazardous waste materials are pretreated and enter the plasma high-temperature melting furnace through screw feeding, and the furnace is in an anoxic environment with a central temperature of 1,450–1,600°C. The organics in hazardous wastes are decomposed into small molecule combustible gases (CO, H_2, CH_4, etc.) in the high-temperature melting furnace, and enter the secondary combustion chamber from the top of the furnace. The inorganic substances are melted at high temperature and converted into a completely inert, non-toxic and very low permeability glass material. The combustion temperature in the secondary combustion chamber is controlled above 1,100°C, the residence time of flue gas is more than 2 s, and harmful components are fully decomposed and eliminated in the secondary combustion chamber. The flue gas with complete combustion enters the waste heat boiler and purification system after leaving the second combustion chamber.

The advantages of plasma treatment technology lie in safety, high efficiency, no secondary pollution and wide applicability, which can achieve the maximum reduction and harmlessness. The disadvantage is that the plasma melting furnace uses electric power as the energy source and has high economic cost; the technology is still in the initial stage, and the process control requires a high degree of automation, so the large-scale promotion needs a solid engineering foundation. China is organizing the formulation of the national standard '*technical requirements for vitrification products of solid waste*', which will clearly define the vitreous substances produced by plasma melting and further promote the resource utilization of the molten vitreous of hazardous waste.

The disadvantages of plasma high-temperature melt disposal technology are as follows: (i) High operation cost. The hazardous waste disposal technology of plasma high-temperature melting uses electric power as the energy source, with a thermal efficiency of about 70% and high energy consumption; in the operation process, the cathode and anode of the plasma torch need to be replaced regularly, so the use cost of accessories is high, resulting in high operation cost. (ii) Plasma high-temperature melting technology is a new type of technology for the domestic hazardous waste disposal industry, with little experience to learn from.

18.4.4.4 Other Technologies

In addition to the above three mainstream heat treatment technologies, the heat treatment technologies for hazardous wastes also include pyrolysis incineration technology, electrolytic oxidation technology, wet air oxidation technology, etc., which are gradually being eliminated and replaced (Yasuda and Tanaka, 2006).

18.5 OPERATION MODE OF HAZARDOUS WASTE DISPOSAL PROJECTS IN CHINA

In the operation mode of hazardous waste disposal projects, BOT (Build Operate Transfer), PPP and BOO (Build Own Operate) are mainly used (Chinese Research Academy of Environmental Sciences, 2004). The analysis of each mode and its main advantages and disadvantages are as follows.

18.5.1 BOT Mode

In short, BOT mode is the form of private capital participation in construction. As a new special investment method of public service, it is a new market financing method for attracting capital and advanced technology, which mainly includes three links of construction, operation and transfer, with priority recourse characteristics (Chinese Research Academy of Environmental Sciences, 2004). In this mode, the government makes the framework regulation, which does not directly participate in the project investment, and the enterprise operates independently. In the meantime, enterprises are responsible for daily operation, which is conducive to giving full play to their special advantages, and the government can better fulfil its supervision function. After the expiry of the concession period, the government can recycle the hazardous waste treatment facilities. However, in terms of disadvantages, the government has weak control over the project, and the approval process of hazardous waste treatment is cumbersome. If you are not familiar with the enterprise process, it can affect the early work efficiency. Moreover, the franchise period of this model is long, so there may be a greater risk of hidden danger in the actual operation of the government and enterprises. After expiration, there are many uncertainties in the delivery of hazardous waste facilities.

18.5.2 PPP Mode

PPP mode is the combination of the government and enterprises, which can meet the public needs of the government and the needs of project management (Chinese Research Academy of Environmental Sciences, 2004). Under this mode, the government's participation in investment can avoid policy and technical risks to a certain extent, and improve the efficiency of engineering construction. The government has certain control over the actual project construction and can make full use of the professional technology of the enterprise. However, PPP as a mode of cooperation between government and enterprises, especially for some enterprises unfamiliar with business, will greatly extend the period of adaptation.

18.5.3 BOO Mode

This mode is mainly the independent operation of enterprises, which can effectively give play to the advantages of enterprise management and strengthen the supervision function of the government (Chinese Research Academy of Environmental Sciences, 2004). The enterprise has the ownership of hazardous waste equipment and does not need to deliver, so there is no limitation on the term of the concession. The government's ability to control the project is weak, while the approval process of the hazardous disposal project is complex and cumbersome. For enterprises unfamiliar with business operations, the efficiency of self-construction in the early stage is relatively low.

18.5.4 Comparison and Selection of Hazardous Waste Operation Modes

It can be seen from various project operation modes that the BOT mode is more widely used in sewage and domestic waste treatment processes. For the operation of hazardous waste projects, the production and collection rate of hazardous waste is not stable, and the utilization rate of production capacity cannot be guaranteed, resulting in the government's inability to make a clear guarantee for the franchise, so the BOT mode is generally not used for the operation of hazardous waste projects. The BOO mode refers to the independent investment and construction of enterprises, the project development rights granted by the government and the right of project charge during the operation of enterprises. The government mainly plays the role of supervision and regulation. Enterprises with the ownership of hazardous waste equipment can give full play to their own technology and management role, which is more suitable for the market development law. In the BOO model, investors need to set up a special subsidiary to operate. For example, Veolia Environmental Services China has taken this approach. The PPP mode is widely used in the construction of public works, and it is also involved in the operation of hazardous waste projects. The form of equity participation mainly includes funds and land use rights. Among them, capital participation is simpler and does not need to be converted, but it will increase the financial burden of the government. Shares participating in the use of land use rights can not only reduce financial expenditure, but also improve the utilization rate of resources, so as to solve the problems such as land project indicators, but also negotiate with enterprises in terms of discount. No matter which method is adopted, the government will have certain project management rights, which is conducive to later control and risk aversion. Combined with the operation characteristics of hazardous waste projects, such as long cycles, large investment, collection and transportation, pricing and other factors, the government needs to take certain control power and decision-making power, and ensure the normal operation of the project through project supervision. At present, the hazardous waste disposal industry is in the initial stage of development, with a good development scenario, but there are also many adverse factors, so the PPP model is more appropriate.

18.6 HAZARDOUS WASTE MANAGEMENT IN CHINA

18.6.1 Importance of Hazardous Waste Management

Through the analysis of the hazards of hazardous waste, we can see that it will have a great impact on people's normal life, health and social environment (Li and Wang, 2005). If it is not handled properly, it will cause huge harm to the whole human society. With the rapid development of China's industrialization process, the amount of hazardous waste produced in China has increased dramatically (Duan et al., 2005). According to statistics, more than 100 million tons of hazardous waste are produced every year, which seriously affects people's normal life. Hazardous waste has the characteristics of strong corrosiveness, toxicity, flammability and explosion, so the treatment of hazardous waste has become a very difficult problem. Therefore, hazardous waste treatment has become a very important environmental protection task at this stage. Compared with developed countries, the development of the hazardous waste treatment industry in China is relatively late, and there are many areas in which to learn from developed countries.

18.6.2 Foreign Hazardous Waste Management

An increasing number of studies have started to pay attention to hazardous waste disposal and management, due to the heated discussions among the general public in both developed countries and developing countries (Hanifzadeh et al., 2017). Both developed and developing countries are concerned about using sustainable and suitable methods to deal with hazardous waste challenges (Shekdar, 2009). As the second largest economic power with fast industrialization and urbanization, China also faces problems of hazardous waste management and disposal (Li et al., 2015). Hazardous waste management practice in the developed world is more standardized and mature (Mian et al., 2017). Since most developed countries have established comprehensive waste recycling systems, waste generation can be seen as a significant part of developed countries' waste disposal systems (Sun et al., 2016).

The US and the EU governments established waste recycling systems through market and policy approaches to achieve effective waste recycling. OECD countries, especially the United States and European countries, have high waste generation rates as the rate in OECD countries was 1.430 kg/person/day, the rate in the United States was 2.080 kg/person/day, and the rate in the EU was 1.510 kg/person/day. Developing countries have much lower waste generation rates than developed countries since the average waste generation rate of 23 developing countries was 0.759 kg/person/day according to the 2002 United Nations Environment Programme data. It is believed that the average income level is positively related to the waste generation rate and the performance of waste management in western countries (Arbulú et al., 2015).

A similar situation can also be seen in developed countries in East Asia. Compared with the western countries, East Asian developed countries lack land and natural resources but have greater population density. Therefore, they try to develop waste generation to increase energy supply and to reduce possible land pollution brought by waste landfill (Othman et al., 2013). Besides that, Japan, South Korea, Singapore, Hong Kong and Taiwan are using various methods such as the '3R' strategy (reduce, reuse and recycle), the '3P' strategy (public, private and public, government) and volume-based waste classification systems to achieve waste minimization (Memon, 2010).

Hazardous waste management and disposal problems are more obvious in developing countries. Compared with developed countries, developing countries face more institutional and financial challenges in addressing hazardous waste problems. In many developing countries, agricultural waste disposal has also been a traditional problem. Backwards agricultural production methods without environmental consideration not only produce solid waste such as straw and cow manure directly, but also the waste disposal method (normally burning) can cause secondary pollution (air pollution).

Besides traditional agricultural solid waste disposal problems, many Southeast Asian developing countries also do not have specific policies to manage hazardous waste effectively in urban areas. Hazardous waste management policies in other South and Southeast Asian developing countries also do not have much practical effect. Malaysia has shaped a private-based hazardous waste management system that allows individuals and households to collect and recycle waste based on human labour; Indonesia also established quasi-private and community-based hazardous waste management systems by involving the non-government sectors. Malaysia and Indonesia's hazardous waste disposal policies encourage individual and community participation in hazardous waste disposal. However, neither the Malaysian nor Indonesian government has mandatory rules to punish individuals or small enterprises if the disposal result does not meet relevant environmental requirements. The negative result of private-based hazardous waste disposal policies in Malaysia and Indonesia is that only some types of hazardous waste with high economic value or that are easily collected by human labour are recycled while other types of hazardous waste are not. Without effective government involvement, hazardous waste disposal does not achieve its environmental purpose.

As a large developing country with a huge population, India made policies and attempted to establish mechanical waste processing programmes in big cities in the 1980s (Rathi, 2006). Since then, those programmes and policies failed due to information asymmetry and the incompatibility of facilities since the government did not consider that the structures of Indian cities were not suitable for waste recycling vehicles and large waste disposal factories and machines. Indian officials also could not find experienced workers to operate and maintain waste disposal machines. India still has serious hazardous waste challenges and secondary pollution. Various approaches and policies have been made to increase India's hazardous waste disposal abilities (Srivastava et al., 2015).

Compared with developed countries, developing countries face greater challenges in the effective use of incineration, since their hazardous waste contents have lower heating value and most of the countries do not have clean incineration technology (Nizami et al., 2017). Besides institutional and policy reasons, poor environmental education is another factor that affects hazardous waste disposal. Farmers, small waste recycling business owners and even some officials did not get sufficient education and training in hazardous waste disposal. This makes hazardous waste disposal in developing countries more challenging.

Although hazardous waste disposal does not work well in some developing countries, many countries such as China have established their own waste recycling system (Fei et al., 2016). Some hazardous waste is recycled as resources. As mentioned, recycling waste in some developing countries is a commercial activity carried out by poor individuals or families as small businesses (Moh and Manaf, 2014). This recycling method can deal with some specific types of wastes effectively. Meanwhile, some other wastes, especially those which do not have much value for recyclers, may still lead to environmental problems (Pappu et al., 2007).

Mexico and China are two developing countries which share similar characteristics with developed countries in hazardous waste disposal. This is because the social and economic developmental levels of the two countries are closer to developed countries. Fast economic and social development in the two countries offers more resources and a better institutional environment to improve hazardous waste disposal (Mathews and Tan, 2016).

As mentioned above, hazardous waste disposal development in China has made some progress in the past 30 years. Meanwhile, China still faces various challenges in establishing an effective and sustainable hazardous waste disposal system (Chen et al., 2010). China faces both technological and institutional problems for hazardous waste disposal. Compared with developed countries, the contents of hazardous waste in China are more similar to other developing countries. The moisture content of hazardous waste in China is much higher than that in developed countries while the heating value is much lower (Zhao et al., 2014). China is still not able to recycle some types of hazardous waste, which can be recycled in the US or Japan. Although China faces technical bottlenecks such as using low heating value waste effectively and using machines to classify different types of hazardous waste, it has established waste power stations and several mechanized recycling plants to increase the effectiveness of hazardous waste disposal. At the start of the 21st century, China could dispose of 54.2% of its total municipal solid waste and 50.4% of its total industrial solid waste. About 44% of hazardous waste was also successfully recycled (China Environmental Yearbook, 2011). According to *China Statistical Yearbook 2016*, the municipal waste harmless treatment rate has reached 94.1% and the comprehensive utilization rate of industrial solid waste was over 70% in 2015.

In the 21st century, China has adopted a series of policies, plans and laws to regulate and control hazardous waste disposal. The government encouraged the development of the circular economy and green industries. It also has taken action to shut down polluting enterprises and push the upgrading of current equipment and industrial systems (Zhao et al., 2014). However, those ambitious plans did not have a practical basis since the Chinese government has little experience or capital to achieve environmental goals within a short time period. Hazardous waste disposal in China still has a long way to go.

18.6.3 Problems in Hazardous Waste Management

18.6.3.1 Weak Concept of Hazardous Waste Management

For the treatment of hazardous waste, first of all, we need to distinguish between the concepts of government and businesses. If we lack the support of the concept, the environmental management and treatment of hazardous waste will not be in place. At present, the management departments of the country pay little attention to the hazardous waste treatment work and lack of supervision. Due to the lack of management personnel and the wide variety of hazardous wastes, regulators in many regions only supervise large factories, but ignore small and scattered factories. Because of the government's weak supervision of small factories and enterprises, many small enterprises' hazardous wastes cannot be treated in time and effectively.

18.6.3.2 Poor Treatment of Hazardous Waste

Today, with the rapid development of industrialization, treatment facilities and systems of hazardous waste have been established in all cities in China. However, after investigation and statistics, there are still many problems in the treatment system of hazardous waste in China: On the one hand, due to the imperfection and poor connectivity of the treatment system, large amounts of the information transmission and interoperability are poor, so most regions are unable to move waste out of the urban area for treatment or face problems with untimely treatment and the phenomenon of hazardous waste accumulating in the urban area. On the other hand, due to the low technical level and management level of hazardous waste treatment in China, the level of centralized treatment and recycling of the existing treatment system is relatively low, and the waste treatment system in many small and medium-sized areas cannot completely clean the hazardous waste. In addition, at present, in some remote areas in China, because of the lack of economy and lack of effective ideas, there is not even a treatment system for hazardous waste, or hazardous waste is directly treated as general waste.

18.6.3.3 Unbalanced Region and Unreasonable Structure

Different regions in China give different levels of attention and supervision to waste management, and invest in different technology and equipment, which leads to the problem of big differences in processing capacity. The common phenomenon is that in some areas, the treatment capacity of hazardous waste is surplus, and the utilization rate of the resources invested is low; in some areas, the treatment capacity of hazardous waste is insufficient, which easily causes the problem of waste accumulation or inadequate treatment; or in some areas, in order to avoid the investigation and punishment of the regulatory authorities, the illegal transfer of hazardous waste increases the harm of the transfer process of hazardous waste, seriously affecting the local environment, causing pollution and affecting the healthy life of local people.

18.6.4 Countermeasures for Environmental Management of Hazardous Waste

Through the analysis of the current situation and problems of hazardous waste in China, we can see that there are still major problems in the treatment of hazardous waste in China. If not treated in time, it will seriously harm people's lives and bring about greater environmental pollution. Therefore, through comprehensive investigation and in-depth study, this paper puts forward some corresponding countermeasures.

18.6.4.1 Improve the Management System of Hazardous Waste

For the management of hazardous waste, the first thing is to improve its management system and laws and regulations. On the one hand, it is necessary to clarify the regulations on the management of hazardous wastes at this stage, strengthen the management work of the Environmental Protection Department, improve the management system and formulate detailed rules. For different types of hazardous wastes, detailed management regulations should be formulated. For example, some valuable recyclable wastes can be reused through green treatment technology. On the other hand, in the process of management, we will gradually do a good job in the revision of laws and regulations. In the implementation and management process of the measures for the management of hazardous waste business licenses, we will continue to make adjustments according to the actual situation, and constantly establish and improve the legal system. At the same time, the management department needs to organize regular learning and on-the-spot investigation to comprehensively improve the management's awareness of hazardous waste disposal, and lay a theoretical and practical foundation for better revision of relevant management methods.

18.6.4.2 Increase Technology Research and Development, and Improve Hazardous Waste Treatment Capacity

For the treatment capacity of hazardous waste, there are still big problems in the previous analysis. Relevant departments need to strengthen management and supervision, increase the number of disposal enterprises of hazardous wastes in various regions and have a deep understanding of the treatment level and situation of hazardous wastes in various enterprises. On the one hand, enterprises need to constantly improve their own technical strength and the introduction of advanced equipment, increase the strength of technological research and development, combine scientific and technological innovation, increase research and development funds, drive technical talents, constantly improve the technical level of waste treatment and improve the recycling technology of waste. For some valuable metal waste, it is necessary to recycle as much as possible through advanced technology use, and then reduce the pressure of environmental management at the grassroots level. On the other hand, in order to improve the level of waste treatment, hardware and equipment are the foundation. Enterprises need to continuously introduce high-quality equipment according to the needs to improve the efficiency of hazardous waste treatment.

18.6.4.3 Strengthen the Management of Hazardous Waste Collection and Transportation to Reduce Environmental Risks

Industrial hazardous waste collection requires special containers and classified collection according to the composition. The collection and transportation unit of hazardous waste shall have corresponding facilities and licenses, have a certain number of specific professional skills and obtain personnel engaged in the collection and transportation of hazardous waste. The classified collection of hazardous waste must be combined with comprehensive utilization and treatment and disposal, and mixed collection is not allowed. Classified collection can be carried out by setting different types of collection containers. The hazardous waste generated by the family shall be combined with the classified collection of the garbage, and the recycling rate and resource utilization of household hazardous waste, such as waste batteries, waste fluorescent tubes, pesticide containers, etc., shall be improved through the classified collection, and the recycling network of the household hazardous waste shall be gradually established and improved. Develop a safe and efficient transport system for hazardous wastes, encourage the development of various forms of closed vehicles, eliminate open transport vehicles for hazardous wastes and reduce secondary pollution and environmental risks in the transport process.

18.6.4.4 Construction of Landfill Facilities for Hazardous Waste to Realize Safe Disposal of Hazardous Waste

Safe landfill is the final disposal method of hazardous waste. Safe landfill disposal is applicable to the hazardous wastes whose useful components and energy cannot be recycled, including the residues and fly ash in the incineration process. The business unit of the safe landfill site shall have a certain number of personnel with professional skills and obtain postal permit to engage in the landfill operation of hazardous wastes, and shall obtain the corresponding business license and engage in the treatment and disposal of hazardous wastes within the scope specified in the license. The planning, site selection, construction and operation management of safe landfills shall be carried out in strict accordance with the requirements of relevant national standards. The safe landfill must receive the hazardous waste according to the regulations of the business license. Before the landfill, the hazardous waste must be pretreated by means of oxidation-reduction, neutralization or solidification/stabilization, so as to make it reach the admission control standard of the safe landfill. The double liner artificial anti-seepage method must be adopted at the bottom of the safe landfill. The leachate from the hazardous waste landfill and the waste water in the production process shall be treated and meet the relevant discharge standards. The bottom of the safe landfill site shall be provided with drainage facilities according to the terrain to guide the bottom of the site to the waste water regulating pool. The leachate can be discharged into the surface water after proper treatment. The safe landfill operation shall be carried out in strict accordance with the operation procedures, with good compaction and coverage, clean water and sewage diversion, and the production of leachate shall be reduced. After the completion of the landfill site, the landfill site will be closed to cover the farmland and restore the ecological environment; the leachate shall continue to be guided and treated; the groundwater, surface water and atmosphere shall be regularly monitored.

18.6.4.5 Develop Incineration Disposal of Hazardous Waste to Reduce and Recycle Hazardous Waste

The purpose of incineration disposal of hazardous waste is to reduce and make harmless hazardous waste and recycle the residual heat. Except for explosive waste, all waste can be incinerated. Incineration is suitable for the hazardous waste which cannot be recycled and has certain calorific value. The construction, operation and pollution control of incineration facilities shall comply with relevant national standards. The design and construction unit of hazardous waste shall hold a design certificate and construction certificate of the corresponding category and level. The business organization of hazardous waste incineration facilities shall have a certain number of personnel with professional skills and must obtain a work license for hazardous waste incineration operation. The incineration plant for hazardous waste shall be engaged in the incineration of hazardous waste within the scope specified in the permit, and shall abide by special and strict regulations. It is encouraged to transform and use rotary kilns for cement production to burn hazardous wastes. Incinerators must have a tail gas purification system, alarm system and emergency disposal device. The flue gas emission should be regarded as a key index of hazardous waste incineration disposal facilities, and advanced technical means should be adopted for strict treatment, so as to make it stably meet the requirements of pollutant control standards before discharge. Residues from incineration and fly ash from flue gas treatment shall be safely landfilled as hazardous wastes. Special facilities should be built for the incineration of hospital clinical wastes and some hazardous wastes with high toxicity or persistent organic compounds.

18.6.4.6 Effective Control of Special Hazardous Waste and Reduction of Environmental Pollution

A hospital's clinical waste shall be disposed of by special medical waste incineration facilities, which shall meet the requirements of the environmental health standard for medical waste incineration and the pollutant control standard for hazardous waste incineration. It is not allowed to use other methods of disposal. Each city must build centralized disposal facilities, collect and dispose of hospital clinical waste in the city and the area where the city is located and prohibit the regeneration and recycling of hospital clinical waste.

Waste lead-acid batteries for motor vehicles must be recycled, and other methods are not allowed to be used for disposal. Hazardous waste management must be included in the collection and transportation. The state plans to build and encourage the development of waste lead-acid battery recycling enterprises with an annual treatment capacity of more than 20,000 tons. We will eliminate small-scale lead enterprises and implement the business license system in the whole process of collection, transportation and recycling of waste lead-acid batteries. It is strictly forbidden for unlicensed traders to collect and transport waste lead-acid batteries. It is encouraged to adopt a clean wet recycling lead production process.

Waste acid and waste alkali shall be comprehensively utilized and treated in the vicinity (referring to the source of production). The waste acid and waste alkali need to be neutralized before landfilling.

Organic solvent waste and waste mineral oil should be recycled first, and the residue should be incinerated.

The waste emulsion needs to be separated first. The dross is incinerated, and the rest is treated in wastewater treatment facilities.

New chemical waste shall not be disposed of before its composition and hazardous characteristics have been identified. It shall be temporarily stored and disposed of after its characteristics have been identified.

18.6.4.7 Strengthen the Supervision of Hazardous Waste

For the treatment of hazardous waste, it is not enough to rely on the self-consciousness of enterprises. Relevant management departments should strengthen supervision and formulate detailed and strict management systems. First of all, from the macro level, managers should recognize their responsibilities, shoulder their responsibilities, strengthen management, strictly abide by the rules and regulations in the process of supervision, not play favouritism or commit fraud and achieve the goal according to law. Second, from the micro aspect, managers should strengthen learning and field research, understand the situation of enterprises under their jurisdiction and master the types and quantities of major hazardous wastes of local enterprises, so as to increase the management and control of enterprises. At the same time, the punishment should be strengthened. If it is found that a declaration is not made truthfully, it should be dealt with seriously and reported to the public. In addition, the management department should improve the management level and efficiency, and can use advanced network technology to master and supervise the situation of hazardous waste of local enterprises through information means.

18.7 CONCLUSION

With the rapid growth of China's population and the acceleration of industrialization, the output of hazardous waste is increasing day by day. It brings serious harm to the ecological environment, human health and the sustainable development of our country. According to different methods, hazardous waste can be classified into different categories and forms. The harm of hazardous waste is mainly manifested in the irreversible damage to the air quality, buildings, water environment and soil quality in the environment on which human beings rely for survival; it will cause great harm to human survival and health through direct contact and food chain accumulation; after environmental pollution, it will bring serious constraints to the sustainable development of China's economy. At present, large and medium-sized enterprises in China mainly deal with hazardous wastes through intermediate and final disposal, generally through physicochemical methods, resource-based methods, incineration, landfill and other ways. However, at present, the disposal technology is relatively traditional and has many disadvantages. Therefore, we should pay attention to the technological innovation of each disposal measure, overcome the defects and disadvantages of the existing technology and realize the double harvest of economy and environmental protection.

Since the quantity of hazardous waste continues to increase, new requirements are put forward for the current management system. Based on this, it is necessary to do a good job in the construction of management institutions, establish a sound supervision and management system, strengthen the training and education of personnel and improve the management level and comprehensive quality of personnel. At the same time, law enforcement should be strengthened and waste supervision and management centres should be established in cities to realize the full-time development of hazardous waste management. For the classification of all kinds of hazardous waste, a complete identification index system should be established to improve personnel identification efficiency and improve the energy efficiency of hazardous waste management. In addition, the Environmental Protection Department should improve law enforcement for the illegal operation of enterprises, further strengthen the law enforcement, and enforce monetary measures as well. And according to the current generation and development of hazardous waste, the corresponding treatment technology must be optimized for the future improvement of China's urban environmental security to provide a guarantee.

REFERENCES

Arbulú I., Lozano J., Rey-Maquieira J. (2015) Tourism and solid waste generation in Europe: A panel data assessment of the environmental Kuznets curve. *Waste Manag.* 46, 628–636.

Callao C., Martinez-Nuñez M., Latorre M.P. (2019) European countries: Does common legislation guarantee better hazardous waste performance for European Union member states? *Waste Manag.* 84, 147–157.

Chen X., Geng Y., Fujita T. (2010) An overview of municipal solid waste management in China. *Waste Manag.* 30, 716–724.

Chinese Environmental Yearbook Compiling Committee. (2011) *China Environmental Yearbook 2011.* China Environmental Science Press, Beijing, China.

Chinese Research Academy of Environmental Sciences. (2004) *Construction Layout for Hazardous Waste Treatment and Disposal in Beijing.* Beijing Solid Waste Management Center, Beijing, pp. 33–37 (in Chinese). (Chinese Research Academy of Environmental Sciences, 2004)

Delgado O.B. (2007) Comparative analysis of hazardous household waste in two Mexican regions. *Waste Manage.* 27(6), 792–801.

Duan H., Huang Q., Wang Q., Zhou B., Li J. (2008) Hazardous waste generation and management in China: A review. *J. Hazard. Mat.* 158(2).

Duan H., Wang Q., Huang Q. (2005) The research of China hazardous waste list. *Sichuan Environ.* 24(3), 94–97 (in Chinese).

Fei F., Qu L., Wen Z., Xue Y., Zhang H. (2016) How to integrate the informal recycling system into municipal solid waste management in developing countries: Based on a China's case in Suzhou urban area. *Resour. Conserv. Recycl.* 110, 74–86.

Gayathr V., Kamala P. (2005) Biomedical solid waste management in an Indian hospital: A case study. *Waste Manage.* 25, 592–599.

Gu B., Zhu W., Wang H., et al. (2014) Household hazardous waste quantification, characterization and management in China's cities: A case study of Suzhou. *Waste Manag.* 34(11), 2414–2423.

Guan X. (2002) The status of hazardous waste pollution control and its countermeasures. *Coal Mine Environ. Prot.* 16, 12–14 (in Chinese).

Hanifzadeh M., Nabati Z., Longka P., Malakul P., Apul D., Kim D.S. (2017) Life cycle assessment of superheated steam drying technology as a novel cow manure management method. *J. Environ. Manag.* 199, 83–90.

Higgins T.E. (2017) *Hazardous Waste Minimization Handbook.* Boca Raton, FL: CRC Press.

Li J. (2003) *The Management and Treatment and Disposal Techniques of Hazardous Waste.* Chemical Industry Press, Beijing, pp. 10–15 (in Chinese).

Li J., Bai Q., Nie Y. (2002) Future solutions for the treatment and disposal of hazardous wastes in China. *Environ. Manag.* 29(5).

Li S., Wang H. (2005) Current situation of treatment and management countermeasures for medical waste. *Chin. J. Hosp. Admin.* 25, 73–74 (in Chinese).

Li Y., Li J., Chen S., Diao W. (2012) Establishing indices for groundwater contamination risk assessment in the vicinity of hazardous waste landfills in China. *Environ. Pollut.* 165.

Li Y., Zhao X., Li Y., Li X. (2015) Waste incineration industry and development policies in China. *Waste Manag.* 46, 234–241.

Lin X., Li M., Chen Z., et al. (2020) Long-term monitoring of PCDD/Fs in soils in the vicinity of a hazardous waste incinerator in China: Temporal variations and environmental impacts. *Sci. Tot. Environ.* 713.

Mathews J.A., Tan H. (2016) Lessons from China: The country consumes the most resources in the world and produces the most waste—But it also has the most advanced solutions. *Nature* 531, 440–443.

Memon M.A. (2010) Integrated solid waste management based on the 3R approach. *J. Mater. Cycles Waste Manag.* 12, 30–40.

Mian M.M., Zeng X., Nasry A.N.B., et al. (2017) Municipal solid waste management in China: A comparative analysis. *J. Mater. Cycles Waste Manag.* 19(3), 1127–1135.

Moh Y.C., Manaf L.A. (2014) Overview of household solid waste recycling policy status and challenges in Malaysia. *Resour. Conserv. Recycl.* 82, 50–61.

Nizami A.S., Rehan M., Waqas M., et al. (2017) Waste biorefineries: Enabling circular economies in developing countries. *Bioresour. Technol.* 241, 1101–1117.

Orloff K., Falk H. (2003) An international perspective on hazardous waste practices. *Int. J. Hyg. Environ. Health.* 206, 291–302.

Othman S.N., Noor Z.Z., Abba A.H., Yusuf R.O., Hassan M.A.A. (2013) Review on life cycle assessment of integrated solid waste management in some Asian countries. *J. Clean. Prod.* 41, 251–262.

Pappu A., Saxena M., Asolekar S.R. (2007) Solid wastes generation in India and their recycling potential in building materials. *Build. Environ.* 42, 2311–2320.

Rathi S. (2006) Alternative approaches for better municipal solid waste management in Mumbai, India. *Waste Manag.* 26, 1192–1200.

Shekdar A.V. (2009) Sustainable solid waste management: An integrated approach for Asian countries. *Waste Manag.* 29, 1438–1448.

Slack R.J., Gronow J.R., Voulvoulis N. (2005) Household hazardous waste in municipal landfills: Contaminants in leachate. *Sci. Total Environ.* 337(1–3), 119–137.

Srivastava V., Ismail S.A., Singh P., Singh R.P. (2015) Urban solid waste management in the developing world with emphasis on India: Challenges and opportunities. *Rev. Environ. Sci. Bio/Technol.* 14, 317–337.

Sun L., Wu Sh. (2007) Analysis and thought about nationwide general investigation on hazardous waste disposal facilities. *Nonferrous Met. Eng. Res.* 28(3), 7–17 (in Chinese).

Sun X., Li J., Zhao X., Zhu B., Zhang G. (2016) A review on the management of municipal solid waste fly ash in American. *Procedia Environ. Sci.* 31, 535–540.

Tang Z., Huang Q., Yang Y. (2015) China: Overhaul rules for hazardous chemicals. *Nature* 525(7570), 455.

Trivedi A., Singh A., Chauhan A. (2015) Analysis of key factors for waste management in humanitarian response: An interpretive structural modeling approach. *Int. J. Disaster Risk Reduc.* 14, 527–535.

Wang B. (2018) The future of hazardous chemical safety in China: Opportunities, problems, challenges and tasks. *Sci. Total Environ.* 643(12), 1–11.

Xu L., Deng Y., Mancl K. (2019) Environmental disaster risk reduction-oriented centralized treatment of hazardous wastes: A novel approach for production-distribution decision optimization in China. *Int. J. Disas. Risk Red.* 40(63), 5000.

Yasuda K., Tanaka M. (2006) Report on hazardous household waste generation in Japan. *Waste Manage. Res.* 24(4). 397–401.

Zhao P., Shen Y., Ge S., Chen Z., Yoshikawa K. (2014) Clean solid biofuel production from high moisture content waste biomass employing hydrothermal treatment. *Appl. Ener.* 131, 345–367.

Zhao X., Li H., Wu L., Qi Y. (2014) Implementation of energy-saving policies in China: How local governments assisted industrial enterprises in achieving energy-saving targets. *Ener. Pol.* 66, 170–184.

19 Management of Hazardous Waste in Brazil: Reality, Perspectives and Case Studies

Luíza Santana Franca, Ingrid R.F.S. Alves, Isabelli Dias Bassin and João Paulo Bassin

CONTENTS

LIST OF SYMBOLS AND ACRONYMS

Abilumi Brazilian Association of Manufacturers and/or Importers of Lighting Products
Abilux Brazilian Association of the Lighting Industry
ABREE Green Eletron and the Brazilian Association for the Recycling of Electronics and Appliances
ANEEL Brazilian Electricity Regulatory Agency
CNI National Confederation of Industries
CO_2 carbon dioxide
CREA Municipal Environment Foundation, Regional Council of Engineering and Agronomy
CTF Federal Technical Register
EE electronic equipment
FEAM State Environment Foundation
GDP gross domestic product
inpEV National Institute of Empty Packaging Processing
MHDI Municipal Human Development Index
MMA Ministry of Environment
MRRJ Metropolitan Region of Rio de Janeiro
MSW municipal solid waste
NSWP National Solid Waste Policy
Planares National Plan for Solid Waste
Reciclus Management of Reverse Logistics of Lighting Products
RDF refuse-derived fuel
SNIR National Solid Waste Information System
WDP waste delivery points
WtE waste-to-energy
WWTP wastewater treatment plant

19.1 BACKGROUND AND INSIGHTS INTO BRAZILIAN HAZARDOUS WASTE MANAGEMENT

The generation of solid waste is directly associated with the purchasing power and forms of consumption of the population, which will demand from the productive sector the supply of goods for direct and indirect use, in order to satisfy its needs. Therefore, solid waste will be generated from the first stages of the consumer goods production chain, starting with the extraction and processing of raw materials, through the stages of processing and manufacturing primary and secondary products, until reaching the final consumer through transport and distribution operations. Therefore, there is generation of waste at each stage of this chain and its qualitative and quantitative characteristics are associated with the type of product manufactured and consumed.

According to the classification determined by the Brazilian standard NBR 10.004 (ABNT, 2004), hazardous waste, classified as solid waste class I, is considered to be that which may, depending on its physical, chemical and biological properties, pose a risk to public health and the environment, besides causing mortality and disease incidence. These residues have the characteristics of being flammable, corrosive, reactive, toxic and pathogenic and, therefore, they must be handled and managed in the safest possible way, minimizing the risks of accidents.

Flammable hazardous wastes are those that have a low flash point in the liquid phase and are capable of producing fire by friction, moisture absorption or spontaneous chemical changes. On the other hand, corrosive residues exhibit a very low or very high pH. Reactive residues are unstable, react with water to form explosive mixtures and generate gases with a high potential for contamination. Toxic substances are characterized as such through leaching tests and present toxicity assessed through concentration, nature, bioaccumulation potential, persistence and harmful teratogenic, mutagenic, carcinogenic or ecotoxicological effects. Finally, pathogenic residues contain or may contain pathogenic organisms, viral proteins, recombinant DNA or RNA, genetically modified organisms, plasmids, chloroplasts, mitochondria or toxins capable of producing diseases in humans.

According to the National Solid Waste Policy (NSWP) (BRASIL, 2010), which is the legal framework for the sector in Brazil, hazardous solid waste must be handled and managed separately from non-hazardous solid waste. If these two types of waste are brought into contact with each other, the hazardousness characteristics of one type of waste will be transferred to the waste that does not contain it. This occurs because hazardous waste has a very high potential for contamination, and as such, it is capable of contaminating the solid waste with less hazardous potential.

Industrial activities vary according to their fields of activity, such as metallurgical, chemical, petrochemical, paper and cellulose, food and mining, among others, and each of these types is responsible for the generation of hazardous waste throughout its production chain. In addition, the generation of this type of waste can also occur in the urban environment through the disposal of products after consumption that contain substances with a great potential for contamination and environmental impact, such as batteries, light bulbs, electrical and electronic materials, paints and oils residues, among others.

The preliminary diagnosis of the generation of industrial solid waste in Brazil (IPEA, 2011) observed the absence of a National Inventory of Industrial Waste and the lack of standardization in state inventories, which are often outdated. Not all state inventories present data on the generation of all industrial types indicated in the Brazilian CONAMA Resolution No. 313/02. It is worth mentioning that, in order to compile this data, it is necessary for industries to register their activities and their generation of solid waste in the Federal Technical Register (CTF), which is operated by the National Solid Waste Information System (SINIR), linked to the Ministry of Environment (MMA). It is through the data recorded in this information system that federal public managers can compile and disseminate information about the generation and management of hazardous solid waste in the country.

The National Inventory of Industrial Solid Waste does not yet exist, but the National Plan for Solid Waste (Planares), which presents the generation of solid waste in Brazilian states and across the country, was recently published and is currently under public consultation.

According to Planares (MMA, 2020), the five main categories of industries in Brazil that generate the most solid waste, in terms of volume, are chemistry (27.17%), food products and beverages (24.92%), metallurgy (17.26%), rubber (8.42%) and mechanics (6.08%). It is worth mentioning that, in terms of mass generated, the chemical industry has the largest share in the generation of industrial solid waste. There is a total of 308,835 m^3 of hazardous solid waste generated in all industrial units, representing 79.95% of the total volume of hazardous waste generated in 2016.

According to NSWP, the industries generate large volumes of specific types of solid waste and, therefore, they must be responsible for their management, starting with the processes of conditioning and internal storage of waste within the factories. Thereafter, industries must hire specific companies to collect and transport the waste to the treatment units, according to each type of waste generated. It is essential that the collection of hazardous solid wastes aggregate only those wastes that have compatible physicochemical and biological characteristics and that can also be subjected to the same treatment process.

The treatment step aims to reduce the potential for contamination of hazardous solid waste considering its characteristics, aiming to reduce its potential for flammability, corrosivity and reactivity, among others, as indicated by NBR 10.004 (BRASIL, 2004). On the other hand, following the principles of the circular economy, the treatment process may also take advantage of these characteristics so that this type of waste becomes an input for a new process. For example, if waste has potential for combustion, instead of being neutralized, the treatment process can make use of this property to generate energy.

As presented by Planares (MMA, 2020), in Brazil, in the year 2016, 31.57% of the volume of industrial solid waste was declared and subjected to physicochemical treatment processes, 28.95% to unspecified biological treatment processes and 6.81% was used as fuel, but not in direct incineration, or other means of generating energy. Finally, operations to recover specific components, among which are solvent reuse/regeneration, acid or base regeneration, recovery of components used to reduce pollution and recovery of catalyst components, represented a small share (1.23%) of the total volume generated.

After receiving the environmentally appropriate and economically viable treatment, hazardous solid waste must be sent to its final disposal. According to Planares (MMA, 2020), the forms of final disposal adopted by the industrial sector for these residues are ordered distribution in the soil, in depth or at the surface, treatment in soil, deep injection, surface confinement, specially designed landfills and permanent storage. The disposal in the soil accounts for 1.55% of the total volume generated, considering the data provided throughout 2016. It is observed, therefore, that in this year, approximately 8,000 m^3 of industrial waste, plus 104.5 million tons and 199,000 units of industrial waste generated in the country, were sent for final disposal in soil. It is worth mentioning that in the preliminary diagnosis of IPEA (2011), the final disposal of solid industrial waste can take place in industrial landfills (class I), which require elaborate design and execution, due to the types of materials received, and must have at most 70% humidity, be stable, poorly soluble and non-volatile. Acids, strong bases, flammable materials, explosives and radioactive waste cannot be disposed of in class I landfills unless special pre-treatment techniques are employed.

In addition to solid industrial waste, some waste generated in urban environments can also be considered hazardous waste. According to article 33 of the NSWP, manufacturers, importers, distributors and traders of the following products are required to structure and implement reverse logistics systems for some types of hazardous waste, upon return of the products after use by the consumer, independently of the urban cleaning and solid waste collection public service:

- Agrochemicals (pesticides), their residues and packaging
- Batteries
- Lubricating oils, their residues and packaging
- Fluorescent, sodium and mercury vapor and mixed light lamps
- Electronics products and their components

From the creation of sectoral agreements between the public authorities and the companies responsible for their manufacture and distribution to the consumer, it was possible that these solid residues could receive a differentiated collection to be sent to specific treatment systems so that these products could return to their respective production chains.

For the case of agrochemical (pesticides), their residues and packaging, the 'Campo Limpo' System is the Brazilian reverse logistics programme that brings together the industry that manufactures these products, the distribution channel, farmers and the government, each with their own role and responsibilities defined by legislation, to promote reverse logistics and give this packaging an environmentally appropriate destination. The National Institute of Empty Packaging Processing (inpEV) is responsible for the entire operation and planning of this system, with more than 400 fixed units, being stations or receiving centres (INPEV, 2019). According to ABRELPE (2019), of the total packaging collected, 94% is reinserted in the production process through recycling, transforming into new packaging and lids for pesticides or other materials for the civil construction and the automotive and energy industries, among other sectors. The material that cannot be recycled (6%) is sent to incineration. At the end of 2019, 45,563 tons of material were collected (MMA, 2020).

As for the batteries, their manufacturers, importers, distributors and traders make available to consumers places to receive the waste batteries. Consumers must take them to the nearest waste delivery points (WDP), which store the received batteries, and, upon reaching a certain amount, these residues are sent to the collection and sorting system. The management entity responsible for the operationalization of this system is Green Eletron, which registered, in 2019, 155 tons collected and properly disposed of and 1,648 WDPs installed in 560 municipalities (MMA, 2020).

Regarding waste lubricating oils and their packaging, 'Jogue Limpo' is the system managing body responsible for the reverse logistics of the used plastic packaging and used or contaminated lubricating oil (Jogue Limpo Institute, 2020). In 2018, the programme received 4,774 tons of plastic packaging, equivalent to 95.48 million containers, and 572,114 litres of used or contaminated lubricating oil. Of the total plastic packaging collected, 98% had an environmentally appropriate destination, of which 4,568 were recycled. Regarding the total oil collected, 557,597 litres were destined for re-refining.

In addition, in 2014, another sectoral agreement was signed for the implementation of the reverse logistics system for fluorescent lamps of sodium and mercury vapor and mixed light lamps. The operationalization of the system is under the responsibility of the Brazilian Association for the Management of Reverse Logistics of Lighting Products (Reciclus), a non-profit organization founded in November 2015 by companies linked to the Brazilian Association of the Lighting Industry (Abilux) and the Brazilian Association of Manufacturers and/or Importers of Lighting Products (Abilumi). Reciclus is present in 33 cities in 25 states in the country (except Acre and Tocantins). According to the 2017 Annual Activity and Results Report, the system ended that year with 304 collection points installed throughout Brazil; 2,963,646 lamps were used in an environmentally friendly manner, 65% of which were compact and 35% tube lamps (ABRELPE, 2019).

Finally, for waste electrical and electronic products and their components, the management entities responsible for the operation of the reverse logistics system are Green Eletron and the Brazilian Association for the Recycling of Electronics and Appliances (ABREE). The results for the year 2019 indicated the receipt of 332 tons to be properly allocated, through the installation of 228 WDPs installed in 70 municipalities served (MMA, 2020).

19.2 HAZARDOUS WASTE TREATMENT

The most used processes for the treatment of hazardous waste in Brazil are incineration and co-processing. Both are considered by the National Confederation of Industries (CNI) as promising technologies not only as ways of treating waste, but also as processes responsible for generating by-products, as fuels, following the principles of the circular economy (CNI, 2019). CNI also describes the gasification and pyrolysis processes as quite promising for use in Brazil, but still little used.

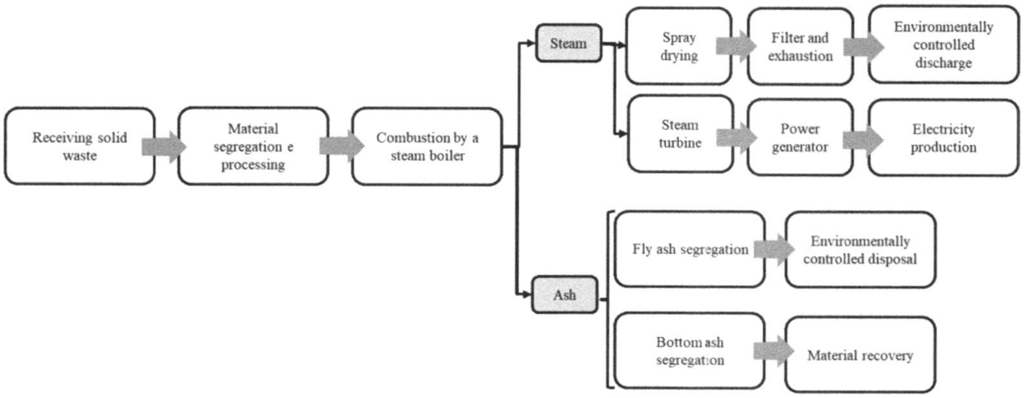

FIGURE 19.1 Incineration process scheme. (Source: Adapted from CNI, 2019.)

These technologies are also called waste-to-energy (WtE) and their development is motivated by the creation of Interministerial Ordinance No. 274, on 30 April 2019 (BRASIL, 2019).

Incineration is a process that involves the combustion of the waste at high temperatures (such as 850°C) to generate thermal and electrical energy. Incineration plants are also known as WtE plants. They have a boiler that captures and converts the heat generated into electricity and steam; they also have a robust air pollution control system that treats the flue gases before releasing them into the atmosphere through a chimney, in order to comply with environmental regulations. In addition to the gases emitted, the process also generates a material composed of bottom ash, of higher density, and fly ash, of low molecular weight. In order to illustrate the incineration process, Figure 19.1 shows a scheme of the treatment steps. A plant usually processes between 50,000 and 300,000 tons of combustible materials per year (CNI, 2019). The types of waste that can be used as fuels in this case are:

- Municipal solid waste (MSW)
- Industrial and commercial waste
- Refuse-derived fuel (RDF), usually produced from urban residues, after previous mechanical-biological treatment

In this context, WtE technologies have become an important bridge between energy policy and the circular economy in Brazil, two themes that should be prioritized on the country's agenda. The supply of local and sustainable energy from waste that would otherwise go to landfills contributes to the local climate and air quality goals.

Co-processing in clinker furnace is considered and classified as a waste treatment process similar to incineration. This is mainly due to the final temperatures of the clinker manufacturing process and the length of the furnace, which implies residence times and temperatures higher than those of conventional incinerators. This process is called co-processing because the waste can be used as a partial or total substitute for the fossil fuel used in these ovens, with petroleum-derived coke being normally used. For this to happen, it must supply the same amount of thermal energy to the process. The maximum rate at which the waste is fed to the oven is established through mass balances and taking into account its physicochemical characterization, based on tests previously carried out, and comparing with the performance of the process when adopting only the original fuel.

Cement production consists of obtaining minerals, via limestone, clay, silica and aluminium mining, which go through a crushing and grinding process to which corrective agents (sand, iron

TABLE 19.1
Requirements for Co-Processing Facilities

	Combustion chamber	Kiln inlet	Calciner	Main burner
Size	<1,500 mm	<800 mm	<100 mm	<15–30 mm
Time	Requires minutes for burnout		Requires 5–8 s of retention time	Easy and fast to ignite
More specification				Must be free of impurities which effect the clinker burning process

Source: Adapted from TUB (2020).

and aluminium ore, high-grade limestone, etc.) can be added until a very fine powder, known as 'flour', is obtained. After homogenization, the flour passes through the oven, where it undergoes a controlled thermal process (minimum temperature required 1450°C) to obtain a by-product called clinker. The ashes resulting from the combustion process are also reincorporated in the process. Clinker, together with plaster and other additives, is finally ground, giving rise to different types of cement.

Co-processing activity is regulated in Brazil through CONAMA Resolution 264/99, and it prohibits the use of the following types of waste: Household raw waste, health services waste, explosives, radioactive waste, organochlorine waste, pesticides and similar waste. Not all types of waste can be used in a clinker oven, often due to environmental restrictions or due to the clinker manufacturing process itself. Hazardous waste that has a high calorific value, such as oily lees, paint lees, wastewater treatment plant (WWTP) sludge, contaminated plastics and papers and scrap tires, among others, can be used as fuel for this process. Some specific fuel requirements to guarantee the efficiency of the process are enumerated in Table 19.1.

Finally, the pyrolysis and gasification processes allow the use of solid waste as a fuel for the production of 'synthesis gas' or syngas, which has applications in the generation of electric energy or can be used as a raw material for the production of methane, biofuels, hydrogen or other chemical derivatives. The gasification and pyrolysis plants perform the thermal treatment of the fuels allowing sufficient oxygen to enter for complete combustion. They are generally of smaller capacities compared to incineration plants, consuming between 25,000 and 150,000 tons/year of waste, but some reach 350,000 tons/year (CNI, 2019). The types of waste used in this process can be:

- MSW
- Industrial and commercial waste
- RDF
- Other waste such as wood and other types of biomass

In Brazil, RDF can typically be used in cement furnaces and in the chemical and food processing industries, which need to generate heat/steam in their chemical processes. RDF production plants can be installed under the same conditions as other types of waste recovery plants, located in industrial areas, or together with waste incineration plants.

19.3 CASE STUDIES ON HAZARDOUS WASTE MANAGEMENT IN BRAZIL

Some case studies related to the management of hazardous waste in the Brazilian context are described below.

Ottoni et al. (2020) proposed a scenario analysis to support decision-making in the management of electronic waste (e-waste) in the metropolitan region of Rio de Janeiro (MRRJ). This study analysed the generation of such waste, the location of recycling companies in this segment and the MRRJ collection routes. The results indicate that the city of Rio de Janeiro is the largest producer of e-waste from the MRRJ, producing approximately 127 tons/day, corresponding to 251 collection points, which represents a high discrepancy in relation to other municipalities in the region. The conclusions pointed out that indicators, such as the gross domestic product (GDP) and the Municipal Human Development Index (MHDI), could be used as an alternative of more precise analysis for a sustainable urban grid project.

The study carried out by Ottoni et al. (2020) also identified 24 electronic waste recycling industries that meet the current demand of MRRJ, most of them concentrated in its central and southern portions, which coincide with the municipalities with the highest e-waste generation rates. Some hotspots, however, are isolated from the recycling industries, as noted at the extremes of MRRJ. Urban e-waste mining has been indicated as an interesting alternative for secondary mining and recovery of raw materials, especially in a country with an emerging economy like Brazil, which consumes huge amounts of electronic devices and requires solutions compatible with the requirements of environmental legislation.

Rodrigues et al. (2020) characterized the flow of domestic electronic waste in the city of São Paulo, Brazil, describing the stages of product acquisition, out-of-use storage and final destination at the end of their useful life through a cross-sectional study based on a population survey, using a statistically representative sample. Home interviews were conducted to collect information on characteristics, behaviours and amounts of primary data related to electronic equipment (EE) within the city, using a questionnaire survey to clarify the domestic flow of e-waste. This study revealed an average of 21 items, with a per capita average of 5.3 EE, in use. Interestingly, residents often tried to repair damaged EE; 50.6% of respondents have always tried to have their damaged EE repaired, while 30.6% have only tried it a few times. Although an average of 2.5 household EE was stored out of use, the average EE discarded was only 1.2 items. This is possibly related to the lack of specific legislation, collection systems and accessible points for delivery/discharge, as pointed out by the study.

Disposal for donation in sorting cooperatives predominates in comparison to other destination routes. Among EE sent as municipal waste, small items were particularly common. The study by Rodrigues et al. (2020) revealed clear trends related to the domestic flow, and it is hoped that the conclusions of this study can support decision-making for public policies related to electronic waste in Brazil and also in other countries where reverse logistics systems are in the initial phase of implementation.

Leopoldino et al. (2019) analysed the disposal of fluorescent lamps from user industries, located in the metropolitan area of Belo Horizonte, in the state of Minas Gerais, Brazil. The study was carried out with a quantitative and qualitative approach, with a semi-structured interview applied to representatives of public environmental agencies and a random sample of 370 companies in the industrial sector that consume fluorescent lamp products. The results showed that the common waste was the predominant option of disposal (33%) of the lamps, although recycling has obtained a significant percentage (29%). The companies that recycle do not obtain economic benefits, since they assume the costs of hiring a third-party company for treating the rejected lamps. In addition, there are many difficulties in implementing reverse logistics, and the absence of public policies that encourage correct disposal was cited by almost all companies (95%). The latter was also confirmed by respondents from environmental public bodies, since the public authority has no programmes or open and clear communication regarding waste disposal. Therefore, there is a lack of government support and initiative by companies to comply with the legislation.

Silva et al. (2019) conducted exploratory, descriptive and qualitative research with analysis in a case study at the main hazardous waste collection company in Palmas – Tocantins, Brazil. To this end, the forms of storage, collection, transport and final disposal of this waste were analysed. Studies have indicated that about half of the waste collected corresponds to health care waste

(48.4%), usually hospitals, clinics and laboratories, followed by mechanical workshops (34.4%) with materials contaminated with oils and greases. Printers, pest control and industries that produce packaging contaminated with oil and chemicals amounted to 10.7%. The remaining residues (6.5%) correspond to those produced by gas stations and car washes (oily lees, fluorescent lamps, batteries and sand, packaging, tow contaminated with oil and grease), agricultural houses and construction stores (packaging contaminated with chemical products) and supermarkets (products beyond the expiration date).

Companies in the health service sector stand out as the biggest customers of the main hazardous waste collection company in the municipality, probably due to the greater rigor in the treatment of health service waste, whose pathogenic contaminants can offer high risk to human health and the environment. After analysing the hazardous solid waste scenario, it was observed that the municipality of Palmas has shown a trend towards environmental regularization of the hazardous waste-generating companies due to the active inspection of competent bodies (Municipal Environment Foundation, Regional Council of Engineering and Agronomy of Tocantins (CREA) and Municipal Health Surveillance) (Silva et al., 2019).

Deschamps et al. (2012) evaluated the management of solid waste generated by the pharmaceutical industry in the state of Minas Gerais, Brazil, based on data from the State Environment Foundation (FEAM). The survey showed that the pharmaceutical industry in Minas Gerais is composed of 114 companies and employs approximately 10,800 people. It focuses almost exclusively on the domestic market, and its main products are generic medicines and herbal medicines. Most of the companies are concentrated in the Metropolitan Region of Belo Horizonte (50%), about 20% are located in the south of Minas Gerais, and the rest are distributed in the rest of the state. The production of waste is quite diverse, with about 100 different types of waste produced. Among the ten most produced, there are three types classified as hazardous waste, which correspond to 73.6% of the total waste generated. The main form of disposal of hazardous waste has been incineration, and non-hazardous waste is sent for recycling as well as to landfills. However, complaints to the environmental agency FEAM have indicated that a significant number of companies do not properly manage their waste, and the disposal of hazardous waste is usually carried out in landfills and even in dumps.

Although most pharmaceutical companies in the state of Minas Gerais use conventional sewage treatment, studies carried out using mass spectrometry with electrospray ionization (ESI-MS) showed that the treatment routes adopted by two selected pharmaceutical industries were not sufficiently effective, as antibiotic residues and degradation products were detected in the final effluent. For there to be an adequate management of hazardous waste in the state of Minas Gerais, it is urgent that more research be carried out in the treatment of solid waste from the pharmaceutical industry in the state.

19.4 FUTURE PERSPECTIVES

The federal government authorized financial institutions to create special credit lines to finance waste reuse, recycling and recovery activities, including the development and implementation of new technologies in the waste management sector in Brazil. In addition, the Brazilian Electricity Regulatory Agency (ANEEL), in order to encourage the construction of WtE plants, aims to offer exemption from electric energy transmission and distribution fees for companies that invest in these technologies. This exemption will be available for plants with an installed capacity of up to 30 MW that use solid waste as fuel in a proportion of insertion in the system of more than 50% (CNI, 2019).

One of the options for the future development of this sector in Brazil is the installation of WtE plants in municipalities or consortia of municipalities, at the regional or national level. The choice of technology depends on factors such as the population served and the generation of waste, which will define the capacity of the system, the characteristics of solid waste, the level of development of the region, the national technology production market, the technical training of public and private entities responsible for the operation of the system and, finally, the political and strategic will to invest in a certain type of technology. It is worth mentioning that the correct segregation of hazardous solid waste from non-hazardous waste is essential so that their physicochemical and biological characteristics are not changed. If inappropriate and undesirable materials are added to the mass of hazardous waste, it is possible that the efficiency of the treatment system may be compromised and, with that, the quality of the final by-products.

According to the energy recovery guide for decision makers, prepared by CNI (2019), RDF production technology offers advantages in cases where it can be incinerated with other fuels in existing industrial plants equipped with control of atmospheric emissions, in accordance with the emission standards of volatile metals, dioxins and particulates. RDF is promising, with a net calorific value between 15 and 25 kJ/kg, which can be used as a partial or total substitute for fuel in high-energy industrial processes. There are many advantages associated with the adoption of this strategy: Contribution to the reduction of CO_2 emissions by power plants, savings in the use of fossil fuels in the face of their substitution by RDF and possibility of treatment with a great reduction in the mass of hazardous solid waste. The RDF is not typically produced to be used in WtE plants; however, due to the eventual low quality or due to the lack of other consumers, it can be used by the waste energy recovery plants. It is even possible that plants dedicated to gasification or even pyrolysis may be able to recover RDF.

For the Brazilian context, the direct combustion of residues, as received at the plant, is feasible and can be widely applied in scenarios with a capacity from 500 tons/day (over 900,000 inhabitants), being less expensive to implement than the combination of an RDF production plant and a WtE plant for RDF. However, it is necessary to take into account that a high moisture content of the waste, which may be due to inappropriate materials present in this mass of waste, can reduce its calorific value incineration, in addition to the need for drying pre-treatment. For cities with a population below 100,000 inhabitants, the CNI (2019) identifies that WtE technologies may not be cost-effective solutions for waste treatment, but should not be excluded.

Finally, other barriers to the development of these technologies in the country can be listed here (CNI, 2019):

- Taxes for the services associated with the solid waste treatment and final destination in Brazil are very low, which can increase the return time for investment in technology
- High taxes on energy commercialization and waste management services, which approach 20% of the total gross revenue
- WtE technologies do not generate enough revenue to meet operating costs
- Little knowledge and awareness of the population about technological options for energy recovery
- The cost of credit coupled with a high interest rate can hinder the attractiveness to investors

REFERENCES

ABNT NBR 10004, de 31 de maio de 2004. (2004). Norma brasileira: Resíduos Sólidos – Classificação. Brasil.

ABRELPE – Associação Brasileira de Empresas de Limpeza Pública e Resíduos Especiais. (2019). Panorama dos Resíduos Sólidos no Brasil 2018/2019. São Paulo.

BRASIL. LEI N° 12.305/2010. (2010). Institui a Política Nacional de Resíduos Sólidos. Presidência da República do Brasil.

BRASIL. Portaria Interministerial N° 274. (2019). Disciplina a recuperação energética dos resíduos sólidos urbanos. Presidência da República do Brasil.

CNI – Confederação Nacional da Indústria. (2019). Recuperação Energética de Resíduos Sólidos - Um Guia para Tomadores de Decisão. Brasília, BR.

CONAMA N° 264, de 26 de agosto de 1999. (1999). Licenciamento de fornos rotativos de produção de clínquer para atividades de co-processamento de resíduos. Presidência da República do Brasil.

CONAMA Nº 313, de 29 de outubro de 2002. (2002). Dispõe sobre o Inventário Nacional de Resíduos Sólidos Industriais. Presidência da República do Brasil.

Deschamps, E., Vasconcelos, O., Lange, L., Donnici, C. L., Silva, M. C., & Sales, J. A. (2012). Management of effluents and waste from pharmaceutical industry in Minas Gerais, Brazil. *Brazilian Journal of Pharmaceutical Sciences*, *48*. doi:10.1590/S1984-82502012000400017.

INPEV - Instituto Nacional de Processamento de Embalagens Vazias. (2019). Relatório de Sustentabilidade 2019. Available at: https://inpev.org.br/noticias-publicacoes/relatorio-sustentabilidade/index. Accessed: 24/08/2020.

IPEA – Instituto de Pesquisa Econômica Aplicada. (2011). Caderno de Diagnóstico: Resíduos Sólidos Industriais. Brasil.

Jogue Limpo Institute. (2020). Cartilha institucional. Available at: https://www.joguelimpo.org.br/institucional/promocional.php. Accessed: 24/08/2020.

Leopoldino, A. C. L., Mendoça, F. M., Siqueira, P. H. L., & Borba, E. L. (2019). The disposal of fluorescent lamps of industries of the metropolitan region of Belo Horizonte – MG. *Journal of Cleaner Production*, *233*, 1486–1493.

MMA – Ministério do Meio Ambiente. (2020). Plano Nacional de Resíduos Sólidos, em processo de consulta pública. Presidência da República do Brasil.

Ottoni, M., Dias, P., & Xavier, L. H. (2020). A circular approach to the e-waste valorization through urban mining in Rio de Janeiro, Brazil. *Journal of Cleaner Production*, *261*, 120990.

Rodrigues, A. C., Boscov, M. E. G., & Gunther, W. M. R. (2020). Domestic flow of e-waste in São Paulo, Brazil: Characterization to support public policies. *Waste Management*, *102*, 474–485.

Silva, R. B. R., Nunes, R. S. B., Silva, K. B. R., Guarda, E. A. G., & Serra, J. C. V. (2019). Estudo de Caso: Levantamento das Empresas Geradoras de Resíduos Perigosos em Palmas. TO Congresso Técnico Científico da Engenharia e da Agronomia, Brasil.

Index

I

J

K

L

M

T

U

V

W

Z